Pitman Research Notes in Mathematics Series

Submission of proposals for consideration

Suggestions for publication, in the form of outlines and representative samples, are invited by the Editorial Board for assessment. Intending authors should approach one of the main editors or another member of the Editorial Board, citing the relevant AMS subject classifications. Alternatively, outlines may be sent directly to the publisher's offices. Refereeing is by members of the board and other mathematical authorities in the topic concerned, throughout the world.

Preparation of accepted manuscripts

On acceptance of a proposal, the publisher will supply full instructions for the preparation of manuscripts in a form suitable for direct photo-lithographic reproduction. Specially printed grid sheets are provided and a contribution is offered by the publisher towards the cost of typing. Word processor output, subject to the publisher's approval, is also acceptable.

Illustrations should be prepared by the authors, ready for direct reproduction without further improvement. The use of hand-drawn symbols should be avoided wherever possible, in order to maintain maximum clarity of the text.

The publisher will be pleased to give any guidance necessary during the preparation of a typescript, and will be happy to answer any queries.

Important note

In order to avoid later retyping, intending authors are strongly urged not to begin final preparation of a typescript before receiving the publisher's guidelines and special paper. In this way it is hoped to preserve the uniform appearance of the series.

Longman Scientific & Technical
Longman House
Burnt Mill
Harlow, Essex, UK
(tel (0279) 26721)

Titles in this series

D F Griffiths and G A Watson (Editors)

University of Dundee

Numerical analysis 1989

Proceedings of the 13th Dundee
Conference, June 1989

Longman
Scientific &
Technical

Copublished in the United States with
John Wiley & Sons, Inc., New York

Longman Scientific & Technical,
Longman Group UK Limited,
Longman House, Burnt Mill, Harlow
Essex CM20 2JE, England
and Associated Companies throughout the world.

Copublished in the United States with
John Wiley & Sons, Inc., 605 Third Avenue, New York, NY 10158

© Longman Group UK Limited 1990

First published 1990

AMS Subject Classification: 65–06

ISSN 0269-3674

British Library Cataloguing in Publication Data
Numerical analysis 1989.
1. Numerical analysis
I. Griffiths, David F. (David Francis), *1945–* II. Watson,
G.A. (G Alistair)
519.4

ISBN 0-582-05923-2

Library of Congress Cataloging-in-Publication Data
Dundee Conference on Numerical Analysis (13th: 1989: University of
   Dundee)
    Numerical analysis 1989: proceedings of the 13th Dundee
Conference, June 1989 / D.F. Griffiths and G.A. Watson, editors.
      p. cm.-- (Pitman research notes in mathematics, ISSN
0269-3674: 228)
    ISBN 0-470-21586-0 (Wiley)
    1. Numerical analysis--Congresses. I. Griffiths, D.F. (David
Francis) II. Watson, G.A. III. Title. IV. Series: Pitman
research notes in mathematics series: 228.
    QA297.D85   1989
519.4--dc20   89-13163
                                            CIP

Printed and bound in Great Britain
by Biddles Ltd, Guildford and King's Lynn

Contents

Preface

The 13th Dundee Biennial Conference on Numerical Analysis was held at the University of Dundee during the 4 days June 27-30, 1989. The meeting was attended by around 200 people from over 20 countries, with over half the participants coming from outside the UK. The technical program consisted of 16 invited talks and 98 submitted talks, the latter being presented in up to 4 parallel sessions. This was by far the largest number of talks offered at any of the Dundee meetings, and it particularly gratifying to us as organizers to see the continued support given by the numerical analysis community. This proceedings volume contains full versions of the invited papers; the titles of all contributed talks given at the meeting, together with the names and addresses of the presenters, are also listed here.

We would like to take this opportunity of thanking all the speakers, including the after dinner speaker at the conference dinner, Professor J C Mason, all chairmen and participants for their contributions. We are also grateful for help received from members of the Department of Mathematics and Computer Science before and during the Conference, in particular the secretarial assistance of Miss S Scott. The conference is also indebted to the University of Dundee for making available various University facilities throughout the week, and for the provision of a Reception for the participants in University House.

Finally, we would like to thank the publishers of these Proceedings, Longman Scientific and Technical, for their co-operation during the pre-publication process.

D F Griffiths

G A Watson
November 1989

Invited Speakers

M. J. Baines: Mathematics Department, University of Reading, Whiteknights, P.O. Box No 220, Reading RG6 2AX, UK.

J. H. Bramble: Department of Mathematics, Cornell University, Ithaca, New York 14853, USA.

K. Burrage: Department of Mathematics & Statistics, University of Auckland, Auckland, New Zealand.

T. F. Coleman: Department of Computer Science, Cornell University, Ithaca, New York 14853, USA.

P. J. Deuflhard: Konrad-Zuse-Zentrum für Informationstechnik, Heilbronner Strasse 10, D-1000 Berlin 31, F.R. Germany.

R. Fletcher: Department of Mathematics and Computer Science, University of Dundee, Dundee DD1 4HN, Scotland, UK.

B. Fornberg: Exxon Research and Engineering Co., Clinton Township, Route 22 East, Annandale, New Jersey 08801, USA.

I. G. Graham: School of Mathematics, University of Bath, Claverton Down, Bath BA2 7AY, UK.

N. J. Higham: Department of Mathematics, University of Manchester, Manchester M13 9PL, UK.

C. Johnson: Chalmers University of Technology, 41296 Göteborg, Sweden.

T. Lyche: Institute of Informatics, University of Oslo, P O Box 1080 Blindern, 0316 Oslo 3, Norway.

J. C. Mason: Computational Mathematics Group, Royal Military College of Science, Shrivenham, Swindon SN6 8LA, UK.

M. J. D. Powell: Department of Applied Mathematics and Theoretical Physics, University of Cambridge, Silver Street, Cambridge, CB3 9EW, UK.

T. F. Russell: Department of Mathematics, University of Colorado at Denver, 1200 Larimer Street, Campus Box 170, Denver, Colorado 80204-5300, USA.

D. C. Sorensen: Department of Mathematical Sciences, Rice University, P.O. Box 1829, Houston, Texas 77251-1829, USA.

A. Spence: School of Mathematics, University of Bath, Claverton Down, Bath BA2 7AY, UK.

G. A. Watson: Department of Mathematics and Computer Science, University of Dundee, Dundee DD1 4HN, Scotland, UK.

M.J. BAINES
Moving Finite Elements and the Legendre Transformation

1. Introduction

In 1981 Miller [22],[23] introduced the Moving Finite Element (MFE) method as follows. For the scalar partial differential equation in one dimension

$$u_t - \mathscr{L}u = 0 \ , \tag{1.1}$$

where $\mathscr{L}$ is an operator involving space derivatives but not time, seek a finite element approximation

$$u \sim U = \Sigma \ U_j \alpha_j$$

where U_j are time dependent coefficients and α_j are linear basis functions also dependent on time through moving nodes (see Fig. (1.1)). Then it can be shown that

$$u_t = \Sigma \ \dot{U}_j \alpha_j + \Sigma \ \dot{X}_j (-U_x) \alpha_j \tag{1.2}$$

where X_j is the coordinate of node j and the dot means differentiation with respect to time on the moving grid. Let R be the residual of equation (1.1). Then minimisation of the L_2 norm of R over $\dot{U}_j$ and $\dot{X}_j$ leads to the normal equations

$$\langle \ R, \ \alpha_i \ \rangle = 0 \qquad \langle \ R, (-U_x) \alpha_i \ \rangle = 0 \tag{1.3}$$

for the determination of $\dot{U}_j$ and $\dot{X}_j$, using the usual inner product notation. These ODE's are then solved in time by finite differences to evolve the solution. The method is an extension of the method of lines (MOL) to a method of trajectories.

The residual minimisation central to the method allows the inclusion of weights and/or penalty terms. The method was introduced with the aim of resolving steep fronts tracking across a region by moving the nodes automatically to where the action is.

Several authors [22]–[25],[7],[16],[17],[27],[28],[34],[4],[11], [15],[19], have used the method in both one and two dimensions. Most use a stiff solver for the time–stepping and penalty functions with tunable parameters [20],[5]. Good results have been obtained for a range of time dependent problems.

The aims of this paper are, first, to bring out the relationship

1

between the MFE method, characteristics and the Legendre Transformation, secondly to discuss the claim that nodes move to where the action is, and thirdly, to discuss time-stepping strategies.

We recall the Lagrangian framework and approach to the derivation of MFE given by Mueller and Carey in ref. [28]. Define a coordinate transformation (assumed non-singular) between x, t and new independent variables ξ, τ by

$$x = \hat{x}(\xi, \tau) \; , \quad t = \tau \; ; \quad \hat{u}(\xi, \tau) = u(x, t) \tag{1.4}$$

for which the partial derivatives satisfy

$$\frac{\partial u}{\partial t} = \frac{\partial \hat{u}}{\partial \tau} + \frac{\partial \hat{u}}{\partial \xi} \frac{\partial \xi}{\partial t} = \frac{\partial \hat{u}}{\partial \tau} - \frac{\partial u}{\partial x} \frac{\partial \hat{x}}{\partial \tau} \; , \quad \frac{\partial \hat{u}}{\partial \xi} = \frac{\partial \hat{x}}{\partial \xi} \frac{\partial u}{\partial x}$$

Then (1.1) becomes, in a Lagrangian frame,

$$\frac{\partial \hat{u}}{\partial \tau} - \frac{\partial u}{\partial x} \frac{\partial \hat{x}}{\partial \tau} = \mathcal{L}u \tag{1.5}$$

which, using the notation[1] $\dot{u} = \dfrac{\partial \hat{u}}{\partial \tau}$, $\dot{x} = \dfrac{\partial \hat{x}}{\partial \tau}$, $u_x = \dfrac{\partial u}{\partial x}$, may be written as

$$\dot{u} - u_x \dot{x} - \mathcal{L}u = 0 \; . \tag{1.6}$$

To determine both $\dot{u}$ and $\dot{x}$ from (1.6) requires a further equation.

If $\hat{u}$ and $\hat{x}$ are restricted to functions $\hat{U}$ and $\hat{X}$, belonging to sets of admissible trial functions, (1.6) becomes a residual,

$$\frac{\partial \hat{U}}{\partial \tau} - \frac{\partial U}{\partial x} \frac{\partial \hat{X}}{\partial \tau} - \mathcal{L}\hat{U} = R \left[\frac{\partial \hat{U}}{\partial \tau}, \frac{\partial \hat{X}}{\partial \tau} \right] \tag{1.7}$$

say, no longer zero in general. There are however special situations where R will be zero and these are considered in detail in the next section. It turns out that there is enough generality in these cases to provide a useful framework for comparison with more general situations.

When $R \neq 0$ the problem of determining the time derivatives of $\hat{U}$ and $\hat{X}$ may be cast as a least squares variational problem by minimising

---

[1]Here and in what follows u_x is a notation for $\hat{u}_\xi / \hat{x}_\xi$ (see (1.3)).

the L_2 norm of R over $\frac{\partial \hat{U}}{\partial \tau}$ and $\frac{\partial \hat{X}}{\partial \tau}$. This gives the weak forms

$$\langle R, \psi \rangle = 0 \qquad \langle R, u_x \chi \rangle = 0 \qquad (1.8)$$

for all admissible test functions $\psi (= \delta \dot{u})$ and $\chi (= \delta \dot{x})$,

where $\qquad \langle R,S \rangle = \int_{\xi_1}^{\xi_2} R S \, w \, d\xi \qquad (1.9)$

(w being a weight function)

and $\qquad ||R||_2^2 = \langle R,R \rangle \qquad (1.10)$

is the L_2 norm of R with weight w defined over a suitable range (ξ_1, ξ_2) of ξ . Thus (1.8) gives two equations to determine $\dot{\hat{U}}$ and $\dot{\hat{X}}$.

Introducing finite element basis functions $\hat{\alpha}_j(\xi)$ in the ξ space (see fig 1.1) we can write $\hat{U}$ and $\hat{X}$ in the forms

$$\hat{U} = \sum_j \hat{U}_j(\tau) \hat{\alpha}_j(\xi) \qquad \hat{X} = \sum_j \hat{X}_j(\tau) \hat{\alpha}_j(\xi) \qquad (1.11)$$

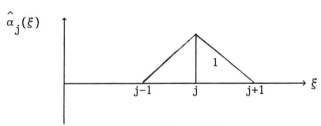

Fig. 1.1

where the $\hat{U}_j(\tau)$ are time-dependent coefficients, and the τ time derivatives are

$$\dot{\hat{U}} = \sum_j \dot{\hat{U}}_j \hat{\alpha}_j \qquad \dot{\hat{X}} = \sum_j \dot{\hat{X}}_j \hat{\alpha}_j \qquad (1.12)$$

where $\qquad \dot{\hat{U}} = \frac{\partial \hat{U}}{\partial \tau} \, , \quad \dot{\hat{X}} = \frac{\partial \hat{X}}{\partial \tau} \, , \quad \dot{\hat{U}}_j = \frac{\partial \hat{U}_j}{\partial \tau} \, , \quad \dot{\hat{X}}_j = \frac{\partial \hat{X}_j}{\partial \tau} \qquad (1.13)$

Then with $\psi = \chi = \hat{\alpha}_i$ the weak forms (1.8) become

$$\left. \begin{array}{l} \langle R, \hat{\alpha}_i \rangle = \langle \dot{U} - U_x \dot{X} - \mathscr{L}U, \hat{\alpha}_i \rangle = 0 \\[2mm] \langle R, U_x \hat{\alpha}_i \rangle = \langle \dot{U} - U_x \dot{X} - \mathscr{L}U, U_x \hat{\alpha}_i \rangle = 0 \end{array} \right\} \quad \forall i \qquad (1.14)$$

3

which can be seen to be equivalent to minimising the L_2 norm

$$||\dot{U} - U_x\dot{X} - \mathcal{L}U||_2 \tag{1.15}$$

over $\dot{U}_i$ and $\dot{X}_i$, i.e. Miller's method [22],[23]. Constraints, for example a lower bound on the Jacobian of the transformation or upper bounds on the relative nodal velocities, may be introduced through the use of penalty functions (see [22],[23],[28],[17],[20]).

2. Semidiscrete Exact Solutions in One Dimension

Writing (1.7) in the form

$$\dot{U} - U_x\dot{X} - \mathcal{L}U = R \tag{2.1}$$

we first investigate those forms of $\mathcal{L}U$ for which there exist $\dot{U},\dot{X}$ belonging to the space S_h of piecewise linear continuous functions, for which R vanishes identically [1].

Since U_x is piecewise constant $(= U_\xi/X_\xi)$ the Lagrangian derivative $\dot{U} - U_x\dot{X}$ lies in the space of piecewise linear discontinuous functions, D_h say. Thus, from (2.1), if $\mathcal{L}U \in D_h$ we can find $\dot{U},\dot{X} \in S_h$ such that R vanishes. We demonstrate how this may be done for the equation

$$u_t + H(x,u,u_x) \equiv \dot{u} - u_x\dot{x} + H(x,u,u_x) = 0 \tag{2.2}$$

with $H(x,u,ux)$ <u>linear</u> in x and/or u (but not bilinear), in which case (2.1) with $R = 0$ becomes

$$\dot{U} - U_x\dot{X} + H(X,U,U_x) = 0 \tag{2.3}$$

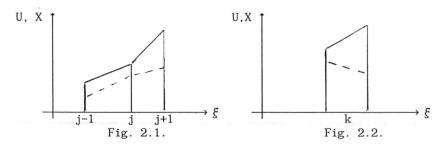

U, X

j-1   j   j+1
Fig. 2.1.

U,X

k
Fig. 2.2.

In (2.3) the first term $\in S_h$ and the other terms $\in D_h$. To extract the nodal values $\dot{X}_j,\dot{U}_j$ consider the jumps in each term across a node j (see Fig. 2.1.). Using the square bracket notation $[\,]_j$ for such jumps we have from (2.3)

$$0 - [U_x]_j \dot{X}_j + [H]_j = 0 \qquad (2.4)$$

since $\dot{U}$ and $\dot{X}$ are continuous at each node. Provided that $[U_x]_j \neq 0$ we find that

$$\dot{X}_j = [H]_j/[U_x]_j \qquad (2.5)$$

Dividing (2.3) by U_x and considering the jumps in each term of the resulting equation across node j gives

$$[U_x]_j^{-1} \dot{U}_j - 0 + [U_x^{-1} H]_j = 0 \qquad (2.6)$$

from which

$$\dot{U}_j = -[U_x^{-1} H]_j/[U_x^{-1}] \qquad (2.7)$$

Now, considering jumps in U_x, U and X across an element k (see Fig. 2.2), using the bracket notation $\{ \}_k$ for such jumps, we have using (2.3)

$$(\dot{U}_x)_k = \frac{\partial}{\partial \tau}\left[\{U\}_k/\{X\}_k\right] = \left[\{\dot{U}\}_k - U_x\{\dot{X}\}_k\right]/\{X\}_k \qquad (2.8)$$

$$= - \{H\}_k/\{X\}_k \qquad (2.9)$$

since $H(U_x)$ is constant across an element.

Taking the limit of (2.5) as $[U_x]_j \to 0$ with X_j fixed, and the limit of (2.9) as $\{X\}_k \to 0$ with $(U_x)_k$ fixed, leads to the equations

$$\dot{x} = \frac{\partial H}{\partial u_x} \qquad\qquad \dot{u}_x = -\frac{\partial H}{\partial x} \ . \qquad (2.10)$$

Similarly, the limit of equation (2.7) as $[U_x]_j \to 0$ with X_j fixed is

$$\dot{u} = \frac{-\partial}{\partial u_x^{-1}} (u_x^{-1} H) = - H + u_x \frac{\partial H}{\partial u_x} \ . \qquad (2.11)$$

The solutions of (2.5),(2.7),(2.9) therefore approximate the properties of the solution of (2.2), by characteristics [10],[13], although the domain of dependence is spread either side of the characteristic.

In classical PDE theory [10], if u is absent from $H(x,u,u_x)$ in (2.2), equations (2.10) form a Hamiltonian system, decoupled from (2.11), for the independent calculation of x and u_x: the function u may ultimately be calculated from (2.11). Similarly, in the discrete case (2.3), with $\dot{U}, \dot{X} \in S_h$, $\dot{U}_x \in D_h$, the functions $\dot{X}$ and $\dot{U}_x$ can be

5

calculated independently and the function $\dot{U}$ may ultimately be constructed from (2.8).

3. Local Projections

If $\mathcal{L} U \in D_h$ the above results are not valid since (2.3) can no longer be satisfied with $\dot{U}, \dot{X} \in S_h$. However we may project $\mathcal{L} U$ into D_h and then proceed as before. Since D_h is the space of piecewise linear discontinuous functions the projection may be carried out in each element separately.

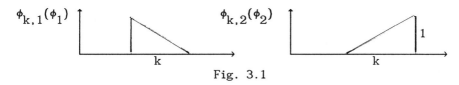

$$\text{Fig. 3.1}$$

For example, in the L_2 case the normal equations are

$$< W_1\phi_1 + W_2\phi_2 - \mathcal{L}U , \phi > = 0 \tag{3.1}$$

where ϕ is one of two possible linear basis functions in a single element (Fig. 3.1) and the W's are coefficients to be determined. By combining the ϕ's to form $\hat{\alpha}_i$ and $- U_x\hat{\alpha}_i$ (c.f. Figures (1.1),(3.1)), we see that (3.1) is equivalent to minimising (1.15) over $\dot{U}_i$ and $\dot{X}_i$, i.e. Miller's method [22],[23]. However we emphasize that (3.1) is a local elementwise projection involving only the solution of a 2×2 system (see refs. [2],[3]).

Consider now the equation

$$u_t + H(x,u,u_x) \equiv \dot{u} - u_x\dot{x} + H(x,u,u_x) = 0 \tag{3.2}$$

where H is a general function. Let the result of a set of local projections of $H(X,U(X),U_x)$ into D_h be $\tilde{H}(X,U_x)$ where $\tilde{H}$ is linear in X in each element. (Note that $\tilde{H}$ will not contain the function U within the element in question although it will contain values of U (and X) at the end points of the element). Then, by the argument in section 2, for the projected function $\tilde{H}$

$$\dot{X}_j = [\tilde{H}]_j/[U_x]_j \qquad (\dot{U}_x)_k = - \{\tilde{H}\}_k/\{X\}_k \tag{3.3}$$

6

$$\dot{U}_j = -[U_x^{-1} \tilde{H}]_j / [U_x^{-1}]_j \tag{3.4}$$

We recall that if u is absent from (3.2), equations (3.3) form a decoupled Hamiltonian-type system.

For example, for the conservation law

$$u_t + f(u)_x = 0 \tag{3.5}$$

we take the local projection of $f(u)_x$ which in the L_2 case is given by, say

$$f(U)_x = - W_1\phi_1 - W_2 \phi_2 \tag{3.6}$$

where ϕ_1, ϕ_2 are local elementwise basis functions (see fig. 3.1) and W_1, W_2 are coefficients satisfying

$$C \begin{bmatrix} W_1 \\ W_2 \end{bmatrix} = - \int_1^2 f(U)_x \begin{bmatrix} \phi_1 \\ \phi_2 \end{bmatrix} dx \tag{3.7}$$

over each element, where $C = \{C_{k\ell}\}$; $C_{k\ell} = \langle \phi_k, \phi_\ell \rangle$, from which, in particular

$$\dot{U}_x = - \{f(U)\}/\{X\} = \{W\}/\{X\} = \frac{-6}{\{X\}^2}[f(U)_2 + f(U)_1 - \frac{2}{\{X\}} \int_1^2 f(U) \, dx] . \tag{3.8}$$

Now consider the set of local elementwise projections of $H(X,U(X), U_x)$ into T_h , the set of piecewise constant functions on the mesh. Denote this weaker projection by $\bar{H}(U_x)$. Then by the argument in section 2, for the weaker projected function $\bar{H}$

$$\dot{X}_j = [\bar{H}]_j/[U_x]_j \qquad (\dot{U}_x)_k = 0 \tag{3.9}$$

$$\dot{U}_j = - [U_x^{-1} \bar{H}]_j/[U_x^{-1}]_j . \tag{3.10}$$

For this weaker projection, therefore, the slopes U_x do not alter with τ .

In the above example, take (3.6) with $W_1 = W_2 = W$. From (3.9),

$$W = - \{f(U)\}/\{X\} \qquad \dot{X}_j = - \frac{[\{f(U)\}/\{X\}]_j}{[U_x]_j} \tag{3.11}$$

The use of penalty functions in the projection will generally destroy its local character. Other constraints, such as demanding that the nodes be fixed, will do the same.

4. Legendre Transformations and the Envelope Construction

We now introduce the Legendre transformation which plays a central role in describing the structure and its approximation.

The relationship (first of (2.10))

$$\dot{x} = \frac{\partial H}{\partial u_x} \qquad (4.1)$$

from the theory of characteristics for the PDE (2.3) generates a Legendre transformation [30] between u_x and $\dot{x}$ which has the inversion

$$u_x = \frac{dG}{d\dot{x}} \qquad (4.2)$$

where $G(\dot{x})$ is the Legendre dual function of $H(u_x)$ with the values of

$$u_x \dot{x} - H . \qquad (4.3)$$

From (2.2) we identify G as $\dot{u}(\dot{x})$. In dynamics G and H are the Lagrangian and Hamiltonian functions.

The Legendre transformation has an envelope construction which will be useful in characterising approximations ([30],[31]). We describe this construction in relation to equation (4.5). For each point P of initial data we may evaluate u_x and $H(u_x)$ and plot them as a point L in u_x, H space (Fig. 4.1). Moreover, for each value of u_x, H , (2.2) is the equation of a line, denoted by ℓ , in $\dot{x}, \dot{u}$ space (Fig. 4.2). As P varies along the curve of inital data, the point L in Fig.4.1 traces out a curve and the line ℓ in Fig.4.2 traces out a pencil of lines (envelope). Fig.4.1 is the point-line dual of Fig.4.2 (ref. [30]). The construction exhibits the Legendre Transformation between the dual functions $H(u_x)$ and $\dot{u}(\dot{x})$ geometrically.

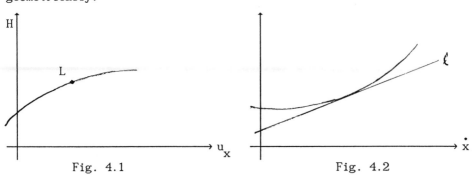

Fig. 4.1   Fig. 4.2

In the discrete case discussed in section 2, $U_x = M$ is piecewise constant and therefore takes only discrete values, one for each cell, as does $H(M)$ (see Fig. 4.3). The data is therefore represented in m, H space as a series of points L_k, each of which determines a line ℓ_k in the dual $\dot{x}, \dot{u}$ space (fig. 4.4). (If the discrete points in u_x, H space are simply sampled values taken from the continuous data curve $H(u_x)$, then they are the vertices of a chordal polygon connecting points on the data curve.)

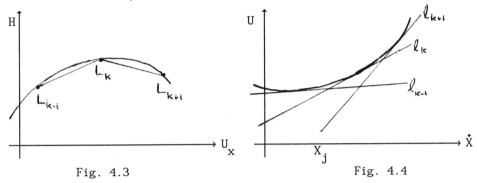

Fig. 4.3   Fig. 4.4

From equation (2.5) the velocity $\dot{X}_j$ of node j is given by the slope of the line joining two of the points, L_k and L_{k-1} say, on the $H(m)$ curve. By the duality of the Legendre transformation it follows that $\dot{X}_j$ is at the point of intersection of the two corresponding tangents ℓ_k and ℓ_{k-1} to the envelope in $\dot{x}, \dot{u}$ space (see Figs 4.1, 4.2). Moreover $\dot{U}_j$ must also be at this intersection since it satisfies (2.3).

Since $\dot{X}$ and $\dot{U}$ are linear in ξ (and therefore in each other) within each element, they are given by intermediate points along each tangent, corresponding in the dual diagram (Fig. 4.3) to sets of lines through each discrete point on the $H(m)$ curve.

For more general H the projection of section 3 is needed to produce $\tilde{H}$ in the right space to enforce the properties

$$\dot{X}, \dot{U} \in S_h, \quad M \in T_h.$$

5. Second Order Equations in 1-D

For piecewise linear approximation the second order operator u_{xx}

exists only in the sense of delta functions but L_2 projection into the space D_h is possible. For example if we write

$$\mathcal{L}\tilde{U} = W_{k,1}\,\phi_{k,1} + W_{k,2}\,\phi_{k,2} \tag{5.1}$$

(c.f. (3.7)) for the L_2 projection of $\mathcal{L}u$ into D_h , then using Miller's mollification procedure [23] or the Hermite cubic recovery method [19] in (1.20), the $W_{k,1}$, $W_{k,2}$ of (5.1) are

$$\left.\begin{aligned}
W_{k,1} &= \frac{1}{h_k}(-M_{k+1} + 3M_k - 2M_{k-1})\\[2mm]
W_{k,2} &= \frac{1}{h_k}(2M_{k+1} - 3M_k + M_{k-1})
\end{aligned}\right\} \tag{5.2}$$

where M_k denotes U_x in element k and $h_k = \{X\}_k = X_k - X_{k-1}$ is the length of the element.

This gives from (3.4)

$$-(M_k - M_{k-1})\,\dot{X}_j = \frac{1}{h_k}(-M_{k+1}+3M_k- 2M_{k-1}) - \frac{1}{h_{k-1}}(2M_k- 3M_{k-1}+ M_{k-2}) \tag{5.3}$$

and

$$h_k^2\dot{M}_k = 3(M_{k+1}- 2M_k + M_{k-1}) . \tag{5.4}$$

The latter is evidently the difference form of a linear heat equation, although we should note that h_k varies with τ . The former is a difference form of

$$-h_j m_j'\dot{X}_j = m_{j+\frac{1}{2}}'- m_{j-\frac{1}{2}}' \tag{5.5}$$

where $m' = \dfrac{dm}{dx}$ and $h_j = \frac{1}{2}(h_k+ h_{k-1})$, taking upwind forms of the derivatives on the right hand side.

Clearly from (5.3) there is no finite solution for $\dot{X}_j$ when $M_k = M_{k-1}$. Moreover in this case $\dot{X}_{j+1}$ satisfies

$$-(M_{k+1}- M_k)\,\dot{X}_{j+1} = \frac{1}{h_{k+1}}(-M_{k+2}+3M_{k+1}- 2M_k) - \frac{2}{h_k}(2M_{k+1}- M_k) \tag{5.6}$$

so that, if $M_k \neq M_{k+1}$

$$\dot{X}_{j+1} = \frac{2}{h_k} - \frac{1}{h_{k+1}}\frac{(-M_{k+2}+3M_{k+1}-2M_k)}{M_{k+1}-M_k} . \tag{5.7}$$

Adding a multiple of $M_k - M_{k-1}$ gives

$$\dot{X}_{j+1} = \frac{2}{h_k} - \frac{1}{h_{k+1}} \frac{(-M_{k+2}+3M_{k+1}-3M_k+M_{k-1})}{M_{k+1}-M_k} . \qquad (5.8)$$

$$= \frac{2}{h_k} - h_{k+1} \frac{m'''(\theta)}{m'(\phi)} \qquad (5.9)$$

by the Mean Value Theorem, where $\theta, \phi \in (X_k, X_{k+1})$. It follows that if

$$h_k h_{k+1} < 2 \min_{\phi} |m'(\phi)|/\max_{\theta} |m'''(\theta)| \qquad (5.10)$$

then $\dot{X}_{j+1} > 0$ and node $j+1$ moves away from the singular point j. By a similar argument node $j-1$ also moves away from node j, which therefore has the character of an anti-cluster singular point.

It is interesting to note that the leading term in the Taylor expansion of the right hand side of (5.3) is $- 2h_j m''$ whereas that of (5.5) is $h_j m''$, the difference arising from the subtraction of the approximations to m' on the right hand side of (5.3). The implication is that if m' were better approximated (perhaps by higher order finite elements) then the sign of $\dot{X}_j$ would be reversed and a singular node j for which $M_k = M_{k-1}$ (near zero curvature) would become a cluster point, completely at variance with how nodes might be expected to move.

This is not so surprising since, in order to approximate $\dot{u}_x = 0$ (see (2.10)) (which is evidently unnatural in a diffusion process), nodes would have to move immediately to points of maximum slope where the curvature is zero. In practice, linear finite element approximation fortuitously reverses this tendency, leaving however a question mark over higher order approximations.

A consequence of the anti-cluster property of the singular points, as pointed out by A.Wathen [36], is that nodes do not pass through zero curvature points and are confined to regions between then.

Now suppose generally that one of the slopes, say M_k, becomes very large. Then from (5.3), we have approximately

$$
\begin{aligned}
-M_k \dot{X}_j &= \frac{3M_k}{h_k} - \frac{2M_k}{h_{k-1}} \\
M_k \dot{X}_{j+1} &= \frac{2M_k}{h_{k+1}} - \frac{3M_k}{h_k}
\end{aligned} \Bigg\} \tag{5.11}
$$

Moreover, if the large slope is associated with small h_k (when nodes become very close but function values stay apart) we have the approximations

$$
\dot{X}_j = -\frac{3}{h_k} \qquad\qquad \dot{X}_{j+1} = \frac{3}{h_k} \tag{5.12}
$$

indicating from the signs a natural springing effect, first noticed by P. Jimack [18].

6. Higher Dimensions

The argument in this section is given for two dimensions only but generalises to any number of dimensions.

The PDE corresponding to (2.2) in two dimensions, using an extension of the argument in section 1, is

$$
u_t + H(u_x, u_y) \equiv \dot{u} - u_x \dot{x} - u_y \dot{y} + H(u_x, u_y) = 0 \tag{6.1}
$$

and, corresponding to (2.1), (2.3) we have

$$
R = \dot{U} - U_x \dot{X} - U_y \dot{Y} + H(U_x, U_y) \tag{6.2}
$$

where $\dot{U}, \dot{X}, \dot{Y} \in S_h$, $H(U_x, U_y) \in T_h$, these spaces being respectively the space of piecewise linear continuous functions and piecewise linear constant functions on a triangular partition of the (ξ, η) plane (where η is the second reference variable).

In one dimension it was possible to find $\dot{U}, \dot{X}$ such that $R \equiv 0$ but in higher dimensions this is no longer true. The reason is that (6.2) belongs to a particular subset E_h of D_h , the space of piecewise linear discontinuous functions on the triangulation. The subset E_h , which depends on U_x, U_y , does not necessarily contain $H(U_x, U_y)$, which $\in T_h$. Moreover, if H is $H(x, y, u_x, u_y)$ where H is linear in x and y , then $H(X, Y, U_x, U_y) \in D_h$ but again $H \in E_h$ in general, so that we cannot find $\dot{U}, \dot{X}, \dot{Y}$ such that $R \equiv 0$.

The difficulty can be resolved by a projection of H into E_h , reminiscent of the projections $H \to \tilde{H}$ and $H \to \bar{H}$ discussed earlier. The projection into E_h is however a minor projection in comparison with the earlier ones, which were from infinite spaces to finite ones. (see ref. [3]).

Thus in the case of a general H we would first project $H \to \tilde{H}$ locally in each element, obtaining $\tilde{H} \in D_h$. Then a second projection of $\tilde{H}$ from D_h to E_h would be carried out, enabling $\dot{U}, \dot{X}, \dot{Y}$ to be chosen such that $R \equiv 0$.

To make the necessity for this second projection clearer consider a node j and let the suffix k range over its surrounding elements. After the projection $H \to \tilde{H}(X, Y, U_x, U_y)$ in which $\tilde{H}$ is a linear function of X and Y, for R to be zero we require

$$\dot{U}_j - (U_x)_k \dot{X}_j - (U_y)_k \dot{Y}_j + \tilde{H}(X_j, Y_j, (U_x)_k, (U_y)_k) = 0 \qquad (6.3)$$

for all k. In a standard triangulation k runs from 1 to 6, but there are only three unknowns $\dot{U}_j, \dot{X}_j$ and $\dot{Y}_j$. Hence the need for a further projection. If we use L_2 projection with area weighting we have the least squares equations

$$\left.\begin{array}{l}
(\Sigma A_k)\dot{U}_j - (\Sigma M_k A_k)\dot{X}_j - (\Sigma N_k A_k)\dot{Y}_j + \Sigma A_k \tilde{H}_k = 0 \\[2mm]
-(\Sigma M_k A_k)\dot{U}_j - (\Sigma M_k^2 A_k)\dot{X}_j + (\Sigma M_k N_k A_k)\dot{Y}_j - \Sigma M_k A_k \tilde{H}_k = 0 \\[2mm]
-(\Sigma N_k A_k)\dot{U}_j + (\Sigma M_k N_k A_k)\dot{X}_j + (\Sigma N_k^2 A_k)\dot{Y}_j - \Sigma N_k A_k \tilde{H}_k = 0
\end{array}\right\} \qquad (6.4)$$

where $\dot{U}_j, \dot{X}_j, \dot{Y}_j$ are now averaged values, A_k is the area of an adjacent triangular element and we have used the notation M_k, N_k for $(U_x)_k$, $(U_y)_k$. The two projections $H \to \tilde{H}$ and the finite L_2 projection above are combined into a single L_2 projection into the space E_h.

A notable exception to the necessity for projection is the equation

$$u_t + auu_x + buu_y \equiv \dot{u} - u_x \dot{x} - u_y \dot{y} + auu_x + buu_y = 0 \qquad (6.5)$$

where a, b are constants, for which

$$R = \dot{U} - U_x \dot{X} - U_y \dot{Y} + aUU_x + bUU_y \qquad (6.6)$$

vanishes identically with

$$\dot{U}_j = 0 \ , \qquad \dot{X}_j = aU_j, \qquad\qquad \dot{Y} = bU_j \ . \qquad\qquad (6.7)$$

On the other hand the same example with au, bu replaced by nonlinear
functions a(u), b(u) certainly requires projection.

Geometrically the effect of the projection required to make R = 0 in
(6.2) is to convert the trace on the (ξ,η) plane of the intersections
of a set of planes into a triangulation. The projection is needed to
complete the cycle of "triangulation – projection into D_h – projection
into E_h – new triangulation" sequence. It therefore modifies the
tangent plane construction, the two-dimensional equivalent of Fig.(4.4).

7. Time-Stepping

As with many time dependent finite element algorithms time-stepping in
the standard MFE method is carried out by finite differences. Having
generated the coupled system of nonlinear ODE's (1.16) (or an extended
version of the same form) most authors then apply an implicit stiff
solver to the system on the grounds that it is expected to be stiff due
to the abundance of degrees of freedom. Only one approach [3],[19] uses
an explicit solver (the simplest Euler one step method) with however
sufficient success as to cast doubt on the need for an expensive stiff
solver. (In view of the relationship of the ODE's with Hamilton's
equations it is clear that time-stepping for symplectic systems may also
have an important part to play [14].)

Within the present framework we can in one dimension use the
equations (2.5),(2.9) to do the time-stepping. Consider (2.2) with
$H(X,U_x)$ of the form

$$H(X,M) = H_o(M) + X H_1(M) \qquad\qquad (7.1)$$

in each element k . For this function the time derivative of U_x is
given by

$$\dot{M}_k = - H_1(M_k) \qquad\qquad (7.2)$$

which may be time-stepped independently for each k using any standard
ODE solver for a single equation.

For equations of the form (3.2) we first perform a projection into

14

the space D_h giving

$$\tilde{H}_o(M) + X \tilde{H}_1(M) \tag{7.3}$$

in each element k . Equation (7.2) above still holds (with a tilda on H_1) , but it should be realised that the projected functions depend on values of U,X at end nodes of the element k and these must be frozen in the integration if iteration is to be avoided.

For the second order operator u_{xx} we have seen that the corresponding equation for M is

$$\dot{M}_k = \frac{3}{h_k^2} \delta^2 M_k \tag{7.4}$$

(see (5.4)). As remarked earlier, (7.11) resembles a semi-discrete approximation of the linear heat equation for m (see refs. [3],[11]) . The ideal time-stepping for this equation is the one step fully implicit algorithm, since this maintains the inherent maximum principle contained in the equation (even with h_k^2 variable).

8. Approximation of the Legendre Transformation

In some applications of numerical methods it is necessary to be able to construct an approximate Legendre transformation (see e.g. ref.[9]). The basic construction involves numerical approximation of (4.1)-(4.3) or its two-dimensional generalisation. We have already described a method of carrying out this transformation approximately when $\dot{U},\dot{X}, \epsilon S_h$ and $H,U_x \epsilon T_h$. This transformation is unsymmetrical in the sense that the two sets of functions belong to different spaces, of piecewise linear continuous and piecewise constant functions, respectively, but this seems not unnatural in view of the point/line duality of the transformation (see [30],[8]).

We illustrate the approximate Legendre transformation in a neutral notation [30],[6]. Let X,Y be functions of single variables x,y , respectively, and let the exact Legendre transformation be described by the equations

$$x = \frac{dY}{dy} \qquad\qquad y = \frac{dX}{dx} \qquad\qquad X + Y - xy = 0 \tag{8.1}$$

Suppose now that $X, x \epsilon S_h$ and $Y, y \epsilon T_h$ and that in terms of piecewise linear basis functions $\hat{\alpha}_j$

$$x = \sum_j x_j \, \hat{\alpha}_j \qquad\qquad X = \sum_j X_j \, \hat{\alpha}_j \qquad\qquad (8.2)$$

(c.f. (1.11)). Then in one dimension we have from (2.5) and (2.7) that the approximate transformation is given by

$$x_j = [Y]_j/[y]_j \qquad\qquad X_j = - [y^{-1}Y]_j/[y^{-1}]_j \qquad (8.3)$$

together with the third of (8.1). This corresponds to the tangent construction described in section 4.

In higher dimensions let X, Y be functions of $\underline{x}, \underline{y}$, respectively, where $\underline{x} = \{x_i\}$, $\underline{y} = \{y_i\}$, ($i = 1, 2, \ldots d$), d is the number of dimensions, and let the exact transformation be given by

$$x_i = \frac{\partial Y}{\partial y_i} \qquad\qquad y_i = \frac{\partial X}{\partial x_i} \qquad\qquad X + Y - \underline{x}.\underline{y} = 0 \qquad (8.4)$$

Using the corresponding spaces S_h, T_h and basis functions $\hat{\alpha}_j$ we cannot now satisfy the third of (8.4) everywhere, except in an averaged sense (see section 6). Choosing an L_2 average with area (A_k) weighting in adjacent elements k gives equations corresponding to (6.4), namely

$$\begin{bmatrix} \Sigma A_k & - \Sigma A_k \underline{y}_k^T \\ -\Sigma A_k \underline{y}_k & \Sigma A_k \underline{y}_k \underline{y}_k^T \end{bmatrix} \begin{bmatrix} \tilde{X}_j \\ \tilde{\underline{x}}_j \end{bmatrix} = \begin{bmatrix} -\Sigma A_k Y_k \\ \Sigma A_k \underline{y}_k Y_k \end{bmatrix} \qquad (8.5)$$

$\tilde{X}_j$, $\tilde{\underline{x}}_j$ being the averaged values of X_j, $\underline{x}_j$. Equation (8.5) and gives the solution in S_h closest in the L_2 norm to the tangent construction of section 4.

In [9] the approximations to the dual variables are a little different, being polygonal intersections of planes (in two dimensions) in one of the variables and point values in the other. However, the only difference between the procedure described there and that described here is in the projection used above, which converts the polygonal approximation into a triangulation. Although there is an additional degree of approximation involved, computations carried out on the piecewise linear representation on triangles are much easier.

9 Conclusion
In this paper we have sought first of all to bring out the

16

correspondence between the Moving Finite Element method and the method of characteristics for first order PDE's by stressing the similarities as strongly as possible. This has meant relying on a model equation,

$$u_t + \tilde{H}(x,u,u_x) = 0 \tag{9.1}$$

with $\tilde{H}$ linear in x and/or u, in one dimension and showing that the MFE equations closely resemble the characteristic equations in this instance. The connection is traced to the commonality of a Legendre transformation between the data and the characteristics/nodal trajectories when the approximating spaces are S_h and T_h. For more general equations a finite element type projection may be used to carry H into the $\tilde{H}$ of (9.1).

The benefits of this approach are that the discretised forms of the characteristic equations are separated from the projection details, that the goals of the MFE procedure are clarified, and that the distinct aspects of the procedure may be analysed separately for the purposes of predicting nodal speeds, time-stepping and, it is anticipated, error analysis. This approach is thought to be more fruitful than a combined approach in which these various aspects are studied together.

For example, an important feature of the MFE method has always been how to cope with the so-called "parallelism" singularity which occurs when the matrix A of (1.18) becomes singular as a result of the matrix M losing rank. This corresponds to collinear nodes or some degree of local 'flatness' of the solution U. It has motivated both the introduction of pragmatic penalty functions and other procedures which recognise the impossibility of solving the MFE equations in this event. However, once we know that the goal of the method is the calculation of $\dot{x}$ given by the first of (2.16), the remedy is straightforward, as follows.

If $[U_x]_j$ is very small the first of (3.4) has a very small denominator, but according to (2.16) the right hand side should be approximating the derivative $\partial H/\partial u_x$. Thus we expect $[\tilde{H}]_j$ to also be very small. The ratio of two very small numbers is hard to compute and usually leads to considerable error. Moreover $[\tilde{H}]_j$ will not necessarily tend to zero as $[U_x]_j$ tends to zero in practice because of the approximate nature of $\tilde{H}$ depending on the approximation of x and u . Therefore, for $(U_x)_k - (U_x)_{k-1}$ less than a certain tolerance, we should, in order to avoid the parallelism "singularity", evaluate $\dot{X}_j$ as $\partial H/\partial u_x$ with the values of X,U taken at the point j . A similar argument applies to the second of (3.3). Again the recommendation is that, for $[X]_k$ smaller than a certain tolerance, we should evaluate $(\dot{U}_x)_k$ as $- \partial H/\partial u_x$ with x,u,u_x taken as their values at, say, $\frac{1}{2}(x_{j+1} + x_j)$. This resolves the parallelism singularity. Similar arguments hold in two-dimensions when the system of (6.4) becomes singular.

An alternative numerical approach can also be suggested for the calculation of $\dot{X}_j$. By taking a weak form of the first of equations (2.10) directly, namely,

$$\langle \hat{\alpha}_i, \dot{x} - \frac{\partial H}{\partial u_x} \rangle = 0 , \tag{9.2}$$

we may replace the second of equations (1.14) and avoid the singularity altogether. Even more directly the strong form of the first of (2.10) may be used.

The same problem of "parallelism" arises in the treatment of the second order operator u_{xx} but this time it is not solved so easily. In line with the approach of this report we observe that the parallelism singularity is a feature only of the mapping of elementwise information on to the nodal velocities. It is not present in the local projection of section 3, nor does it appear in the time-stepping. It arises in the calculation of the $\dot{U}, \dot{X}$ variables from U_x and H .

To avoid the singularity we may add a penalty P to the minimisation (1.15). For example, if $P = \dot{x}^2$ this leads in the case of a local projection [3],[26] to the discrete equation (c.f. [32])

$$\begin{bmatrix} h_k + h_{k-1} & -M_k h_k - M_{k-1} h_{k-1} \\ -M_k h_k - M_{k-1} h_{k-1} & M_k^2 h_k + M_{k-1}^2 h_{k-1}^2 + \epsilon (h_k + h_{k-1}) \end{bmatrix} \begin{bmatrix} \dot{U}_j \\ \dot{X}_j \end{bmatrix}$$

$$= \begin{bmatrix} h_k W_{k,1} + h_{k-1} W_{k-1,2} \\ -M_k h_k W_{k,1} - M_k h_{k-1} W_{k-1,2} \end{bmatrix} \tag{9.3}$$

Note that when $\epsilon = 0$ the solution of (9.3) for $\dot{X}_j$ is consistent with (5.3). If $\epsilon \neq 0$ however, we find that

$$\dot{X}_j = - \frac{(W_{k,1} - W_{k-1,2})(M_k - M_{k-1})h_k h_{k-1}}{(M_k - M_{k-1})^2 h_k h_{k-1} + \epsilon(h_k + h_{k-1})^2} \tag{9.4}$$

$$= - \frac{(W_{k,1} - W_{k-1,2})}{M_k - M_{k-1} + \epsilon(h_k + h_{k-1})(\frac{1}{h_k} + \frac{1}{h_{k-1}})/(M_k - M_{k-1})} \tag{9.5}$$

which approximates

$$\dot{x} = - \frac{m''}{m' + \dfrac{\epsilon}{m'}} . \tag{9.6}$$

(c.f. (5.5)). The factor ϵ has the dimensions of M^2 and, from (9.5), we see that the speed $\dot{X}_j \to 0$ under any of the three conditions (i) M_k, $M_{k-1} \to \infty$, (ii) $M_k - M_{k-1} \to 0$, (iii) one of h_k, $h_{k-1} \to 0$. Asymptotically the directions of the node velocities are unaltered but they are prevented from becoming infinite near the points of zero curvature. The approximate form (9.5) retains the properties of nodal speed direction and nodal springing described in section 5.

The ultimate accuracy of the method is left to the appropriate Galerkin equation for $\dot{U}_j$. In the case of the local projection this is obtainable from (9.3) as

$$(h_k + h_{k-1})\dot{U}_j = W_{k,1} h_k + W_{k-1,2} h_{k-1} + [M_k h_k + M_{k-1} h_{k-1}]\dot{X}_j \tag{9.8}$$

which directly approximates (2.11).

Moving on to other aspects of the paper, we have observed that the mapping from the data to the nodal speeds in the case of equation (9.1) is an example of a Legendre transformation. Where additional projections are incorporated, as for example in the case of more general functions H and in the case of higher dimensions to map the speeds into S_h , the mapping is an approximate Legendre transformation. In the latter case we extracted the procedure to present a general approximate Legendre transformation between dual functions in spaces S_h and T_h (section 8).

The accuracy of semi-discrete Galerkin methods is well-known. It is the time-stepping which destroys this accuracy on fixed grids but in the present method we expect to better maintain in the accuracy by the integration along characteristic trajectories. Also on the subject of accuracy it is anticipated that the structure presented here will stimulate the development of error analysis for this method. It has already been shown [21],[33] that, for nodes moved along characteristics, much higher accuracy is achieved than for fixed nodes in convection and convection-diffusion equations.

References

[1] M.J. BAINES, On approximate solutions of time dependent partial differential equations by the moving finite element method, Numerical Analysis Report 1/85 (1985), Department of Mathematics, University of Reading, U.K.

[2] M.J. BAINES, Locally adaptive moving finite elements, Numerical Methods for Fluid Dynamics II (K.W. MORTON AND M.J. BAINES (eds.)), Oxford University Press, 1986, pp.1-14

[3] M.J. BAINES AND A.J. WATHEN, Moving finite elements for evolutionary problems I, theory, Journal of Computational Physics 79(1988), pp.245-269.

[4] M.J. BAINES AND A.J. WATHEN, Moving finite element modelling of compressible flow, Applied Numerical Mathematics, 2(1986), pp. 495-514.

[5] M.J. BAINES, Recent developments in moving finite element methods, Proceedings of the VIth MAFELAP Conference, Brunel (J. WHITEMAN (ed.)), Academic Press, 1987, pp. 387-396.

[6] M.J. BAINES, Numerical Approximation of the Legendre transformation, Moving Finite Elements and Approximate Legendre Transformations. Numerical Analysis Report 15/88 (1988), and 5/89 (1989) Department of Mathematics, University of Reading, U.K.

[7] N. CARLSON AND K. MILLER, The gradient weighted moving finite element method in 2-D, Center for Pure and Applied Mathematics Report PAM-347 (1986), University of California, Berkeley, USA.

[8] S. CHYNOWETH & M.J. SEWELL, Mesh Duality and Legendre Duality, Department of Mathematics, University of Reading, U.K. (1989).

[9] S. CHYNOWETH AND M.J. BAINES, Legendre Transformation solutions to Semi-Geostrophic Frontogenesis, Proceedings of 7th International Conference on Finite Elements in Fluids, University of Alabama Press, 1989. (see also S. CHYNOWETH, Ph.D. thesis, Dept. of Maths., University of Reading, U.K. (1987)).

[10] R. COURANT AND D. HILBERT, Methods of Mathematics Physics, Volume II, Interscience (1962), Wiley Classics Library (1989).

[11] J.D.P. DONNELLY, A version of moving finite elements for 1-D parabolic equations, Oxford University Computing Laboratory Numerical Analysis Report 19/86 (1986), UK.

[12] M.G. EDWARDS, The mobile element method for systems of conservation laws, Proceedings of the VIIth GAMM Conference (M. DEVILLE (ed.)), Vieweg (1988), pp. 72-79.

[13] M.G. EDWARDS, Ph.D. thesis, Department of Mathematics, University of Reading, U.K. (1987).

[14] K. FENG, Difference Schemes for Hamiltonian Formalism and Sympletic Geometry, J. Comput. Maths., 4, 279-289 (1986).

[15] R.M. FURZELAND, J.G. VERWER AND P.A. ZEGELING, A numerical study of three moving grid methods for one-dimensional partial differential equations which are based on the method of lines, Report NM-R8806 (1988), Centre for Mathematics and Computer Science, Amsterdam.

[16] R. GELINAS, S.K. DOSS AND K. MILLER, The moving finite element method: applications to general partial differential equations with multiple large gradients, Journal of Computational Physics 40(1981), pp. 202-249.

[17] A.H. HRYMARK, G.J. MCRAE AND A.W. ESTERBERG, An implementation of a moving finite element method, Journal of Computational Physics 163 (1986), pp. 168-190.

[18] P. JIMACK, Private Communication, School of Mathematics, University of Bristol, U.K.

[19] I.W. JOHNSON, A.J. WATHEN AND M.J. BAINES, Moving finite element methods for evolutionary problems II: applications, Journal of Computational Physics 79(1988), 270-290.

[20] D. JUAREZ-ROMERO, R.H.W. SARGENT AND W.P. JONES, The behaviour of the moving finite element method in 1-D problems: error analysis and control, Department of Chemical Engineering and Chemical Technology, Imperial College, London (1987).

[21] B. LUCIER, A Moving Mesh Numerical Method for Hyperbolic Conservation Laws, Math. Comp., 46, 59-69 (1986).

[22] K. MILLER AND R.N. MILLER, Moving finite elements part I, SIAM Journal of Numerical Analysis 18(1981), pp. 1019-1032.

[23] K. MILLER, Moving finite elements part II, SIAM Journal of Numerical Analysis 18 (1981), pp. 1033-1057.

[24] K. MILLER, Alternate modes to control the nodes in the moving finite element method, Adaptive Computational Methods for Partial Differential Equations (I. BABUSKA, J. CHANDRA and J. FLAHERTY (eds.)), University of Maryland, SIAM (1983), pp. 165-182

[25] K. MILLER, Recent results on finite element methods with moving nodes, Accuracy Estimates and Adaptive Refinement in Finite Element Calculations, Proceedings of the ARFEC Conference, Lisbon, (I. BABUSKA, O.C. ZIENKIEWICZ (eds.)), Wiley, (1986), pp. 325-338.

[26] K. MILLER, On the Mass Matrix Spectrum Bounds of Wathen and the Local Moving Finite Elements of Baines, Report PAM-430, Center for Pure & Applied Mathematics, University of California, Berkeley, U.S.A. (1988).

[27] M.C. MOSHER, A variable node finite element method, Journal of Computational Physics 57 (1985), pp. 157-187.

[28] A.C. MUELLER AND G.F. CAREY, Continuously deforming finite elements for transport problems, International Journal for Numerical Methods in Engineering 21 (1985), pp. 2099-2126.

[29] O. OLEINIK, Discontinuous solutions of nonlinear differential equations, Usp. Mat. Nauk, 12, 3-73 (1957).

[30] M.J. SEWELL, Legendre transformations and extremum principles, In Mechanics of Solids (H.G. HOPKINS AND M.J. SEWELL (eds.)), Pergamon Press, Oxford (1982), pp. 563-605.

[31] M.J. SEWELL, Maximum & Minimum Principles. CUP (1988).

[32] P.K. SWEBY, Numerical Analysis Report 13/87, Department of Mathematics, University of Reading, U.K. (1987).

[33] Y. TOURIGNY AND E. SULI, Private Communication, Oxford University Computing Laboratory, U.K. (1989).

[34] A.J. WATHEN AND M.J. BAINES, On the structure of the moving finite element equations, IMA Journal of Numerical Analysis (1985), 161-182.

[35] A.J. WATHEN, M.J. BAINES AND K.W. MORTON, Moving Finite Element Methods for the Solution of Evolutionary Equations in One and Two Dimensions, Proceeding of the Vth MAFELAP Conference, Brunel (J.R. WHITEMAN (ed.)), Academic Press, (1984), pp. 421-430.

[36] A. WATHEN, Private Communication, School of Mathematics, University of Bristol, U.K.

M J Baines
Department of Mathematics
University of Reading
P O Box 220, Reading, RG6 2AX, UK.

J.H. BRAMBLE, J.E. PASCIAK and J. XU

Parallel Multilevel Preconditioners

1. Introduction

In this paper, we shall report on some techniques for the development of preconditioners for the discrete systems which arise in the approximation of solutions to elliptic boundary value problems. These techniques are analysed in [**11**]. Here we shall only state the resulting theorems; complete proofs can be found in [**11**].

It has been demonstrated that preconditioned iteration techniques often lead to the most computationally effective algorithms for the solution of the large algebraic systems corresponding to boundary value problems in two and three dimensional Euclidean space (cf. [**3**] and the included references). The use of preconditioned iteration will become even more important on computers with parallel architecture. This paper discusses an approach for developing completely parallel multilevel preconditioners. In order to illustrate the resulting algorithms, we shall describe the simplest application of the technique to a model elliptic problem. Let Ω be a polygonal domain in R^2 and consider the problem of approximating the solution u of

$$Lu = f \text{ in } \Omega,$$
$$u = 0 \text{ on } \partial\Omega, \tag{1.1}$$

where

$$Lu = -\sum_{i,j=1}^{2} \frac{\partial}{\partial x_i} a_{ij} \frac{\partial u}{\partial x_j} + au.$$

We assume that the matrix $\{a_{ij}(x)\}$ is symmetric and uniformly positive definite and $a(x) \geq 0$ in Ω.

We first define a sequence of multilevel finite element spaces in the usual way. Since Ω is polygonal, we can define a 'coarse' triangulation $\tau_1 = \cup_l \tau_1^l$ where τ_1^l represents an individual triangle and τ_1 denotes the triangulation. Successively finer triangulations $\{\tau_k, k = 2, \ldots, J\}$ are defined by breaking each triangle of a coarser triangulation into four triangles by connecting the midpoints of the edges. The space $\mathcal{M}_k$ is defined to be

This manuscript has been authored under contract number DE-AC02-76CH00016 with the U.S. Department of Energy. Accordingly, the U.S. Government retains a non-exclusive, royalty-free license to publish or reproduce the published form of this contribution, or allow others to do so, for U.S. Government purposes. This work was also supported in part under the National Science Foundation Grant No. DMS84-05352 and by the U.S. Army Research Office through the Mathematical Science Institute, Cornell University.

Typeset by $\mathcal{A}_{\mathcal{M}}S$-TEX

23

the continuous functions defined on Ω which are piecewise linear with respect to τ_k and vanish on $\partial\Omega$. We shall be interested in developing a preconditioner for the solution of the Galerkin equations on the J'th subspace, i.e. $U \in \mathcal{M}_J$ satisfying

$$A(U, \phi) = (f, \phi), \qquad \text{for all } \phi \in \mathcal{M}_J. \tag{1.2}$$

Here $A(\cdot, \cdot)$ denotes the generalized Dirichlet integral defined by

$$A(u, v) = \sum_{i,j=1}^{2} \int_{\Omega} a_{ij} \frac{\partial u}{\partial x_i} \frac{\partial v}{\partial x_j} \, dx + \int_{\Omega} a u v \, dx \tag{1.3}$$

and $(\cdot, \cdot)$ denotes the L^2 inner product on Ω.

Let $\{\phi_k^l\}$ denote the usual nodal basis for the subspace $\mathcal{M}_k$, i.e. the l'th basis function is one on the l'th node of k'th triangulation and vanishes on all others. The preconditioner $\mathcal{B}$ is defined by

$$\mathcal{B}v = \sum_{k=1}^{J} \sum_{l} (v, \phi_k^l) \phi_k^l. \tag{1.4}$$

The above preconditioner is simply a double sum, the terms of which can be computed concurrently. This results in an inherently parallel algorithm.

As is well known, the rate of convergence of an iterative method can be estimated in terms of the condition number of the preconditioned system. In [11], a theory for the estimation of the condition number for this type of multilevel preconditioner in terms of a number of *a priori* assumptions is given. In the above example, this theory can be used to show that the relevant condition number is at worst $O(J^2)$. Moreover, these results hold for problems in two, three and higher dimensions as well as problems with only locally quasi-uniform mesh approximation.

We note that many alternative preconditioning techniques have been proposed for such discrete systems. For example, domain decomposition preconditioners have been developed ([5],[6],[7],[8],[14], and the included references). These domain decomposition preconditioners are inherently parallel however become somewhat complex in three dimensional applications. Alternatively, multigrid [4],[9], [15],[18] and hierarchical multigrid [2],[21] techniques give rise to different multilevel preconditioners. The standard multigrid algorithms do not allow for completely parallel computations since the computations on a given level use results from the previous levels. Theoretical results for the usual multigrid algorithms are available, in general, for problems in any number of spatial dimensions but only for quasi-uniform mesh approximation. Good results hold for the hierarchical basis method in two dimensions with refined meshes but degenerate when applied to three dimensional problems. Finally, preconditioners based on approximate LU factorization are often proposed however a comprehensive theory is yet to be developed [12],[13], [19].

The outline of the remainder of the paper is as follows. A general abstract theory for the development of parallel multilevel preconditioners is discussed in Section 2. In Section 3, this theory is applied to second order elliptic boundary value problems and the serial and parallel complexity of the resulting algorithms is discussed. We apply

24

the abstract theory to a second order problem with a locally refined mesh in Section 4. Finally, the results of numerical experiments illustrating the theory of the earlier sections are given in Section 5.

2. General theory.

In this section, we discuss a general theory for the construction of parallel multilevel preconditioners. This theory is presented in an abstract setting to most clearly illustrate the relevant analytic assumptions. The development of this class of preconditioners is based on a certain orthogonal decomposition of the approximation space. The parallel multilevel preconditioners are then abstractly defined in terms of this decomposition by the replacement of orthogonal projections by more computationally efficient operators. Applications to second order elliptic boundary value problems are given in Sections 3 and 4.

We start with the basic abstract framework. We assume that we are given a nested sequence of finite dimensional spaces,

$$\mathcal{M}_1 \subset \mathcal{M}_2 \subset \ldots \subset \mathcal{M}_J \equiv \mathcal{M}, \quad J \geq 2. \tag{2.1}$$

The space $\mathcal{M}$ and hence all of its subspaces are equipped with two inner products $(\cdot, \cdot)$ and $A(\cdot, \cdot)$. The first part of this section will consider properties of a certain orthogonal decomposition of $\mathcal{M}$ with respect to the inner product $(\cdot, \cdot)$ and the sequence of spaces (2.1). We shall investigate the spectral properties of these spaces with respect to the form $A(\cdot, \cdot)$ since, ultimately, we are interested in computing the solution to the Galerkin equations: Given $f \in \mathcal{M}$, find $U \in \mathcal{M}$ satisfying

$$A(U, v) = (f, v) \qquad \text{for all } v \in \mathcal{M}. \tag{2.2}$$

We shall use the following notation in the development. For each $k = 1, \ldots, J$, we introduce the following operators:

(1) The projection $P_k : \mathcal{M} \longrightarrow \mathcal{M}_k$ is defined for $u \in \mathcal{M}$ by

$$A(P_k u, v) = A(u, v), \qquad \text{for all } v \in \mathcal{M}_k.$$

(2) The projection $Q_k : \mathcal{M} \longrightarrow \mathcal{M}_k$ is defined for $u \in \mathcal{M}$ by

$$(Q_k u, v) = (u, v), \qquad \text{for all } v \in \mathcal{M}_k.$$

(3) The operator $A_k : \mathcal{M}_k \longrightarrow \mathcal{M}_k$ is defined for $u \in \mathcal{M}_k$ by

$$(A_k u, v) = A(u, v), \qquad \text{for all } v \in \mathcal{M}_k.$$

We shall also denote $A = A_J$ and define

$$\mathcal{O}_k = \{\phi | \phi = (Q_k - Q_{k-1})\psi, \ \psi \in \mathcal{M}\},$$

25

where $Q_0 = 0$. We shall discuss the spectral properties of A with respect to the decomposition

$$\mathcal{M} = \mathcal{O}_1 + \cdots + \mathcal{O}_J. \tag{2.3}$$

It follows from the above definitions that

$$Q_k A = A_k P_k$$
$$Q_k Q_l = Q_l Q_k = Q_l \quad \text{for } l \leq k. \tag{2.4}$$

From the second equation of (2.4), it follows that

$$(Q_k - Q_{k-1})(Q_l - Q_{l-1}) = 0$$

if $k \neq l$ and hence the decomposition (2.3) is orthogonal, i.e. $(u, v) = 0$ whenever $u \in \mathcal{O}_l$, $v \in \mathcal{O}_k$ with $l \neq k$.

We consider first the operator

$$B = \sum_{k=1}^{J} \lambda_k^{-1}(Q_k - Q_{k-1}), \tag{2.5}$$

where λ_k denotes the spectral radius of A_k. Clearly, B is symmetric and positive definite and

$$A(BAv, v) = \sum_{k=1}^{J} \lambda_k^{-1} \|(Q_k - Q_{k-1})Av\|^2, \tag{2.6}$$

where $\|\cdot\|^2 = (\cdot, \cdot)$. Note that B is block diagonal with respect to the decomposition (2.3) and each diagonal block is a multiple of the identity matrix.

The operator B may be thought of as an "approximate inverse" for A. Thus, we shall state theorems estimating the condition number $K(BA)$ of BA. We note that $K(BA) \leq c_1/c_0$ for any positive constants c_0, c_1 satisfying

$$c_0 A(v, v) \leq A(BAv, v) \leq c_1 A(v, v), \quad \text{for all } v \in \mathcal{M}. \tag{2.7}$$

REMARK 2.1: The form of the operator B can be motivated by the spectral decomposition of the operator A. Indeed, for a special example, namely, $\mathcal{M}_k$ the space spanned by the eigenvectors corresponding to the smallest k distinct eigenvalues of A, the operator B defined by (2.5) is in fact equal to A^{-1}.

It is straightforward to show that (cf. [11])

$$A(BAv, v) \leq JA(v, v), \quad \text{for all } v \in \mathcal{M}. \tag{2.8}$$

The lower estimate of (2.7) will require some additional hypotheses concerning the spaces $\mathcal{M}_k$. We first consider the following assumptions on the operators Q_k: For $k = 1, \ldots, J$, there exists a constant $C_1 > 0$ such that

$$\|(I - Q_{k-1})v\|^2 \leq C_1 \lambda_k^{-1} A(v, v), \quad \text{for all } v \in \mathcal{M}. \tag{A.1}$$

We have the following theorem and corollaries.

26

Theorem 1. *Assume that (A.1) holds. Then*

$$C_1^{-1}J^{-1}A(v,v) \le A(BAv,v) \le JA(v,v), \qquad \text{for all } v \in \mathcal{M}. \qquad (2.9)$$

Corollary 1. *For any real s,*

$$B^s = \sum_{k=1}^{J} \lambda_k^{-s}(Q_k - Q_{k-1}). \qquad (2.10)$$

Moreover, for any $s \in [0,1]$,

$$J^{-s}(A^s v, v) \le (B^{-s}v, v) \le (C_1 J)^s (A^s v, v), \qquad \text{for all } v \in \mathcal{M}. \qquad (2.11)$$

We have included Corollary 1 for the purpose of future applications which will not be described in this paper. In particular, it will be used for the development of preconditioners for certain boundary operators which arise in domain decomposition techniques for second order boundary value problems [**10**].

In the next corollary, we consider the case of the sum of two operators. Let $\hat{A}(\cdot, \cdot)$ be another symmetric positive definite form and let $\hat{A}$, $\{\hat{A}_k\}$ and $\{\hat{\lambda}_k\}$ be defined analogously in terms of $\hat{A}(\cdot, \cdot)$. Consider the operator $\bar{B} : \mathcal{M} \mapsto \mathcal{M}$ defined by

$$\bar{B} = \sum_{k=1}^{J} (\lambda_k + \hat{\lambda}_k)^{-1}(Q_k - Q_{k-1}).$$

Theorem 1 immediately implies the following corollary.

Corollary 2. *Assume that (A.1) holds for both A and $\hat{A}$. Then,*

$$J^{-1}((A + \hat{A})v, v) \le (\bar{B}^{-1}v, v) \le C_1 J((A + \hat{A})v, v), \qquad \text{for all } v \in \mathcal{M}.$$

The most natural application of the above corollary is to the discrete systems which arise in parabolic time stepping algorithms. At each time level, a function $U^n \in \mathcal{M}$ satisfying

$$(I + \tau A)U^n = F^n,$$

with known $F^n \in \mathcal{M}$ must be computed. Here τ is a positive number which is related to the time step size. We shall not consider further application of Corollary 2 in this paper.

We next apply the above results to analyze parallel multilevel preconditioners for A. An operator $\mathcal{B} : \mathcal{M} \mapsto \mathcal{M}$ is a good preconditioner for A if it satisfies:

(1) The action of $\mathcal{B}$ on vectors of $\mathcal{M}$ is economical to compute.
(2) The condition number $K(\mathcal{B}A)$ of the preconditioned system is not too large.

Item (1) above guarantees that the cost per iteration in a preconditioned scheme using B for solving (2.2) will not be unreasonable. Item (2) guarantees that the number of iterations in a preconditioned scheme will not be too large. Note that by Theorem 1, B satisfies (2). B may in fact satisfy (1) in many applications but generally it is desirable to avoid evaluating the action of Q_k. Hence we shall develop more computationally effective algorithms by modifying (2.5).

To get a computationally effective preconditioner, we write (2.5) in the form

$$B = \sum_{k=1}^{J-1} (\lambda_k^{-1} - \lambda_{k+1}^{-1}) Q_k + \lambda_J^{-1} I.$$

Notice that if $\{\lambda_k\}_{k=1}^J$ satisfies the growth condition $\lambda_{k+1} \geq \sigma \lambda_k$, for $\sigma > 1$ then the operator

$$\hat{B} = \sum_{k=1}^{J} \lambda_k^{-1} Q_k \tag{2.12}$$

satisfies

$$(1 - \sigma^{-1})(\hat{B}u, u) \leq (Bu, u) \leq (\hat{B}u, u), \qquad \text{for all } u \in \mathcal{M}.$$

We consider a slightly more general operator defined by replacing $\lambda_k^{-1} I$ in (2.12) with a symmetric positive definite operator $R_k : \mathcal{M}_k \mapsto \mathcal{M}_k$, i.e.

$$B = \sum_{k=1}^{J} R_k Q_k. \tag{2.13}$$

Clearly, B is symmetric and positive definite on $\mathcal{M}$. The cost of evaluating the action of the preconditioner B on a vector in $\mathcal{M}$ will be discussed in later sections but will obviously depend on an appropriate choice of R_k.

For our subsequent development, we shall need to make the following assumption concerning the operator R_k. We assume that

$$C_2 \frac{\|u\|^2}{\lambda_k} \leq (R_k u, u) \leq C_3 (A_k^{-1} u, u), \qquad \text{for all } u \in \mathcal{M}_k, \tag{A.2}$$

where C_2 and C_3 are positive constants not depending on J. Clearly the choice $R_k = \lambda_k^{-1} I$ corresponding to (2.12) satisfies (A.2).

The preconditioner (2.13) can be thought of as a parallel version of a V-cycle multigrid algorithm. The operator R_k plays the role of a smoothing procedure. The major difference between (2.13) and the V-cycle multigrid scheme is that the smoothing on every level of (2.13) is applied to the original fine grid residual. In contrast, the multigrid V-cycle applies the smoothing to the residual computed using the corrections from the previously visited grid. Obviously, the different terms in (2.13) can be computed in parallel while, in contrast, computations on a given grid level in a standard multigrid algorithm must wait for the results from previous levels. The connection between (2.13) and the multigrid V-cycle will be more fully discussed in Section 3. However, it is not surprising that assumptions which are equivalent to (A.2) have been made in the analysis of the usual serial multigrid algorithms [4],[9],[16],[17].

REMARK 2.2: A particularly interesting choice of R_k can be motivated as follows. As noted above, $R_k = \lambda_k^{-1} I$ satisfies (A.2). Let $\{\psi_k^l\}$ be an orthonormal basis for $\mathcal{M}_k$. Then

$$\lambda_k^{-1} u = \lambda_k^{-1} \sum_l (u, \psi_k^l) \psi_k^l, \qquad \text{for all } u \in \mathcal{M}_k. \tag{2.14}$$

In practice, an orthonormal basis for $\mathcal{M}_k$ is seldom available. However, for finite element applications with quasi-uniform grids, the right hand side of (2.14) with normalized nodal basis functions $\{\bar{\psi}_k^l\}$ defines an R_k satisfying (A.2) (see Section 3). Moreover, we note that for $u \in \mathcal{M}$,

$$R_k Q_k u = \lambda_k^{-1} \sum_l (u, \bar{\psi}_k^l) \bar{\psi}_k^l$$

and hence $R_k Q_k$ is computable without the solution of gram matrix systems. This will be discussed in more detail in Section 3.

With $\mathcal{B}$ defined in (2.13), we have the following corollary.

Corollary 3. *Under assumptions (A.1) and (A.2),*

$$C_1^{-1} C_2 J^{-1} A(v, v) \le A(\mathcal{B} A v, v) \le C_3 J A(v, v), \qquad \text{for all } v \in \mathcal{M}. \tag{2.15}$$

We next provide an alternative hypothesis for a lower estimate in (2.15). This is the so called "regularity and approximation" assumption often used in multigrid analysis (cf. [4],[15],[18]). We assume that for a fixed $\alpha \in (0, 1]$, there exists a positive constant C_4 not depending on $k = 1, \ldots, J$ satisfying

$$A((I - P_{k-1})v, v) \le (C_4 \lambda_k^{-1} \|A_k v\|^2)^\alpha A(v, v)^{1-\alpha}, \qquad \text{for all } v \in \mathcal{M}_k, \tag{A.3}$$

where $P_0 = 0$. In finite element applications, the above assumption is usually proved by using elliptic regularity for the continuous problem and the approximation properties of the space $\mathcal{M}_{k-1}$ [1],[4]. In such applications, assumption (A.3) may be stronger than (A.1), e.g. when $\alpha = 1$, (A.3) implies (A.1).

Theorem 2. *Assume that (A.2) and (A.3) hold. Then*

$$C_2 C_4^{-1} J^{1-1/\alpha} A(v, v) \le A(\mathcal{B} A v, v) \le C_3 J A(v, v), \qquad \text{for all } v \in \mathcal{M}.$$

Remark 2.3: Included in (A.1) and (A.3) is the implicit assumption that C_1 and C_4 are greater than or equal to $K(A_1)$. In finite element applications, $K(A_1)$ will not be large if the grid size of the coarsest grid is of unit size. However, if a good preconditioner R_j is available for any finer grid, i.e. R_j satisfies in addition,

$$(R_j^{-1} u, u) \le C_5(A_j u, u), \tag{2.16}$$

then it suffices to use

$$\mathcal{B} = \sum_{k=j}^J R_k Q_k.$$

In such applications, (A.1) or (A.3) need only be satisfied for $k > j$. Note that $R_j = A_j^{-1}$ will be computationally economical provided that the j'th grid size is relatively small. Many alternative choices are possible.

3. The quasi-uniform finiet element application.

In this section, we shall illustrate the application of the abstract theory and algorithms discussed in the previous section to a second order elliptic boundary value problem approximated using finite element functions on a quasi-uniform mesh. We note that the hypotheses of the previous section are satisfied. We also consider the computational complexity of the resulting algorithm in both serial and parallel computing applications. For brevity, we consider only the most basic finite element applications. Many other applications are possible including examples of elliptic problems in higher dimensions.

Let $\mathcal{M}_1 \subset \cdots \subset \mathcal{M}_J \equiv \mathcal{M}$ be the finite element spaces defined in the introduction subsequent to (1.1), $A(\cdot, \cdot)$ be the generalized Dirichlet form defined in (1.3) and $(\cdot, \cdot)$ be the L^2 inner product on Ω.

We will apply the results stated in Section 2 to Problem (1.2) with the above sequence of spaces. Let h_k denote the size of the k'th triangulation. It easily follows that there are constants c_0 and c_1, not depending on k and satisfying

$$c_0 h_k^{-2} \leq \lambda_k \leq c_1 h_k^{-2}. \tag{3.1}$$

Inequality (A.1) with $k > 2$ is well known. For $k = 1$, we have that

$$\|v\|^2 \leq \Lambda^{-1} A(v, v), \qquad \text{for all } v \in \mathcal{M}$$

where Λ is the smallest eigenvalue of A and is obviously bounded from away from zero (independently of J). We shall suppose in this application that $\mathcal{M}_1$ is such that h_1 is proportional to the diameter of Ω so that $C_1 \geq \lambda_1 / \Lambda$ which is not large.

We next consider the operator R_k motivated by Remark 2.2, i.e.

$$R_k v = \sum_l (v, \phi_k^l) \phi_k^l, \text{ for } v \in \mathcal{M}_k, \tag{3.2}$$

where the sum is taken over all nodes of τ_k. As observed in Remark 2.2, the action of $R_k Q_k$ can be computed without explicitly computing Q_k. Moreover, using R_k defined by (3.2) in (2.13) leads to the preconditioner of (1.4) It is shown in [11]that (A.2) holds for this R_k.

For this problem, (A.3) will always be satisfied for some $\alpha \in (0, 1]$, (cf. [1],[4]). The size of α depends on the elliptic regularity of Problem (1.1). Thus, in the case when Ω is a convex polygonal domain and the coefficients defining L are smooth, (A.3) holds with $\alpha = 1$ and we conclude from Theorem 2 that

$$K(\mathcal{B}A) \leq cJ.$$

In the case of a so called crack problem (with smooth coefficients), the largest interior angle is 2π and the regularity of (1.1) is such that (A.3) does not hold for $\alpha \geq 1/2$. Hence Corollary 3 yields the better estimate and shows that

$$K(\mathcal{B}A) \leq CJ^2.$$

Remark 3.1: It is possible to apply the theory of Section 2 to elliptic problems in three or more dimensions. Many examples are possible but we consider only the simplest. In three dimensions, we let the coarse mesh be a union of equally sized cubes. Finer meshes are obtained by breaking each cube of a coarser mesh into eight smaller cubes in the obvious way. The subspaces $\{\mathcal{M}_k\}$ are defined to be the functions on Ω which are continuous and piecewise trilinear with respect to the k'th mesh and vanish on $\partial\Omega$. The nodes of these spaces are the vertices of the cubes defining the mesh. We may take

$$\mathcal{B}u = \sum_{k=1}^{J} h_k^{-1} \sum_l (u, \phi_k^l)\phi_k^l, \tag{3.3}$$

where $\{\phi_k^l\}$ denotes the set of nodal basis functions. We emphasize here again that all the terms in (3.3) are independent and hence may be computed concurrently.

Remark 3.2: Assumption (A.1) is often easier to verify than (A.3). For example, we consider the two dimensional problem (1.1) when the coefficients of the operator L are discontinuous. If the jumps in the coefficients are only along the lines of the coarse mesh, then it is possible to prove that (A.1) holds with $C_1 \le CJ$ where the constant C depends on the local variation of the coefficients of L on the coarse grid triangles but not on the magnitude of the jumps across triangles [20]. This leads to a conditioning result of the form

$$K(\mathcal{B}A) \le CJ^3.$$

The dependence of constant C_4 (in (A.3)) on the size of the jumps is a much more difficult question since it requires the knowledge of the dependence of the elliptic regularity constants on such jumps.

In the remainder of this section, we consider computational issues involved in implementing the above algorithm in serial and parallel computing architectures. However, before proceeding, we make the following observation. Even though we have defined $\mathcal{B}$ as an operator on $\mathcal{M}$, in a preconditioned iterative scheme we are only required to compute $\mathcal{B}v$ given the data $W_J^l = (v, \phi_J^l)$. This is because when $v = A_J\theta$, we always compute $\{(A_J\theta, \phi_J^l) = A(\theta, \phi_J^l)\}$ and hence avoid the solution of the gram matrix problem required for the computation of $A_J\theta$.

We first consider the serial version of the algorithm. Let $v \in \mathcal{M}$ be given and define $W_k^l = (v, \phi_k^l)$. Let W_k denote the vector with entries $(W_k)_l = W_k^l$. We need to compute the action of $\mathcal{B}v$ given W_J. We define W_{k-1} from W_k in a recursive manner. Note that each basis function in $\mathcal{M}_{k-1}$ can be written as a local linear combination of basis functions for $\mathcal{M}_k$. Thus, each value of W_{k-1}^l can be written as a local linear combination of values of W_k. Moreover, the work involved in computing W_{k-1} from W_k is proportional to the number of unknowns in $\mathcal{M}_{k-1}$. Consequently, the work involved in computing the vectors $\{W_k\}$, $k = 1, \dots, J$ bounded by a constant times the number of unknowns in $\mathcal{M}$. Once the vectors $\{W_k\}$ are known, we are left to compute the representation of $\mathcal{B}v$ in the basis for $\mathcal{M}$. To do this, we compute the representation of

$$\mathcal{B}_m v \equiv \sum_{k=1}^{m} \sum_l (v, \phi_k^l)\phi_k^l,$$

in the basis for $\mathcal{M}_m$, for $m = 1, \ldots, J$. The result at $m = J$ is of course the basis representation for $\mathcal{B}v$. For $m = 1$, the representation is already given by W_1. The representation of $\mathcal{B}_m v$, for $m > 1$ is calculated from that of $\mathcal{B}_{m-1} v$ by interpolating the $\mathcal{B}_{m-1} v$ results (i.e. expanding them in terms of the m'th basis) and adding the m'th level contribution from W_m. The work of calculating the representation of $\mathcal{B}_m v$ given that for $\mathcal{B}_{m-1} v$ is on the order of the number of unknowns in $\mathcal{M}_m$ and thus the total work for this algorithm is bounded by a constant times the number of unknowns on the finest grid.

Remark 3.3: The serial implementation of the operator $\mathcal{B}$ is closely related to the multigrid V-cycle algorithm. The step of computing W_{k-1} from W_k is nothing more that the step which "transfers the residuals" from grid level k to $k - 1$ in a multigrid V-cycle algorithm. However, the multigrid algorithm requires extra computation since it must smooth and then compute new residuals on the k'th level before transferring. The second step in the serial algorithm for $\mathcal{B}$ is also duplicated in the "coarser to finer interpolation" step in the multigrid V-cycle algorithm. The symmetric multigrid V-cycle requires extra computation since it requires additional smoothing on each grid level. Thus the serial $\mathcal{B}$ algorithm, in terms of complexity, is similar to a multigrid V-cycle algorithm without smoothing.

We next consider parallel implementation of the preconditioner $\mathcal{B}$. The execution of (1.4) can obviously be made parallel in many ways by breaking up the terms into various numbers of parallel tasks. The optimal splitting of the sum is clearly dependent on characteristics of the individual parallel computer, for example, memory management considerations, task initialization overhead, the number of parallel processors, etc. We note, however, the simplicity of the form of (1.4) allows for almost complete freedom for parallel splitting.

It is of theoretical interest to consider the algorithm on a shared memory machine with an unlimited number of processors. As above, the implementation $\mathcal{B}v$ involves two steps, the calculation of the coefficients W_k^l and the computation of the representation of $\mathcal{B}v$ in the basis for $\mathcal{M}$. Each coefficient can be computed independently and involves a linear combination (not necessarily local) of the values of W_J. With enough processors, a linear combination of m numbers can be computed in $\log_2(m)$ time. Hence the coefficient vectors $\{W_k\}$ can be computed in $\log_2(N)$ time where N is the dimension of $\mathcal{M}$. Each coefficient of $\mathcal{B}v$ involves a linear combination of $M_n J$ contributions from the J grid levels (here M_n is the maximum number of neighbors for any given level. Thus, computation of $\mathcal{B}v$ can in done in time bounded by CJ.

4. A local refinement application.

In this section, we shall discuss the application of the parallel multilevel algorithm to the finite element equations corresponding to a problem with mesh refinement. Such mesh refinements are necessary for accurate modeling of problems with various type of singular behavior. For simplicity, we shall make no attempt at generality. Instead, we shall illustrate the technique by considering an example from which many obvious generalizations are possible. For this example, the domain Ω will be the unit square and we shall approximate the solution to (1.1). The form $A(\cdot, \cdot)$ and the inner product

$(\cdot,\cdot)$ will be as in Section 3. The sequence of grids which we shall consider will be progressively more refined as we approach the corner $(1,1)$. Such a mesh would be effective if, for example, the function f in (1.1) behaved like a δ function distribution at the point $(1,1)$.

To define the mesh, we first start with a sequence of subspaces $\mathcal{M}_1,\ldots,\mathcal{M}_j$ defined using uniform grids of size $h_k = 2^{-k}$, $k = 1,\ldots,j$ as described in the quasi-uniform case (See Section 1). The $j+1$'st triangulation is then defined by refining only those triangles in the upper quarter, $[1/2,1]\times[1/2,1]$. Similarly, the $j+2$'nd triangulation is defined by refining only those triangles in the $j+1$'st grid which are in the region $[3/4,1]\times[3/4,1]$, etc. The spaces $\mathcal{M}_k$ for $k = j+1,\ldots,J$ are defined to be the continuous functions on Ω which are piecewise linear with respect to the k'th grid. Note that this introduces slave nodes into the computation, i.e. the vertices of the triangles on the boundary of the k'th refinement region which are not nodes for the $k-1$'st subspace. These nodes are slaves since the values of functions on these nodes are determined by the values of neighboring nodes and the continuity condition on the subspace. Thus, they do not represent degrees of freedom in the subspace. It is shown in [**11**], that (A.1) is satisfied for this sequence of subspaces.

We next define a sequence of operators $\{R_k\}$ satisfying (A.2). For $k \leq j$, R_k is given by (3.2). Let $\{x_k^l\}$ denote the nodes of the k'th grid and let $\{\phi_k^l\}$ denote the corresponding nodal basis functions. For each node x_k^l with $k > j$ we define

$$
h_{kl} = \begin{cases} h_k & \text{if } x_k^l \in \bar{\Omega}_k, \\ h_m & \text{if } x_k^l \in \bar{\Omega}_m/\bar{\Omega}_{m+1}, \; j \leq m < k. \end{cases}
$$

Note that if $x_k^l \in \bar{\Omega}_k/\bar{\Omega}_{k+1}$ then x_k^l is a node for each finer subspace and gets assigned the same value h_k. We then define

$$
R_k u = h_k^2 \sum_l h_{kl}^{-2} (u, \phi_k^l) \phi_k^l. \tag{4.1}
$$

We can apply Corollary 3 to show that $K(\mathcal{B}A) \leq CJ^2$ where $\mathcal{B}$ is defined by (2.13) with R_k and $\mathcal{M}_k$ as above. For this application, we have not been able to prove the regularity and approximation assumption (A.3).

For the purpose of implementation, it is more efficient to reorder the terms defining $\mathcal{B}$. For $k = j,\ldots,J$ let $\mathcal{N}_k$ be the nodes of $\mathcal{M}_k$ in $\bar{\Omega}_k$ and for $k < J$ let $\mathcal{N}_k^1$ be the nodes of $\mathcal{M}_k$ in $\bar{\Omega}_k/\bar{\Omega}_{k+1}$. For a function $u \in \mathcal{M}$, it is not difficult to see by induction on J that

$$
\mathcal{B}u = \sum_{k=1}^{j-1} R_k Q_k u + \sum_{x_j^l \in \mathcal{N}_J} (u, \phi_J^l)\phi_J^l
$$

$$
+ \sum_{k=j}^{J-1} \left[\sum_{x_k^l \in \mathcal{N}_k^1} \gamma_k^J (u, \phi_k^l)\phi_k^l + \sum_{x_k^l \in \mathcal{N}_k/\mathcal{N}_k^1} (u, \phi_k^l)\phi_k^l \right]
$$

(4.2)

where $\gamma_k^J = h_k^{-2}\sum_{m=k}^J h_m^2$. Note that the R_k terms in the first sum of (4.2) involves the same sums which appear in the uniform case of Section 3. In addition, the calculation corresponding to the k'th mesh in (4.2) for $k = j, \ldots, J$ only involves nodal basis functions on $\bar{\Omega}_k$.

Finally we define a simpler preconditioner $\hat{B}$ by replacing γ_k^J by one in (4.2), i.e.

$$\hat{B}u = \sum_{k=1}^{j}\sum_{l}(u, \phi_k^l)\phi_k^l + \sum_{k=j+1}^{J}\sum_{x_k^l \in N_k}(u, \phi_k^l)\phi_k^l. \tag{4.3}$$

Note that in (4.3), the k'th refinement grid only adds a sum over the nodes in $\bar{\Omega}_k$. We note that for $u \in M$, by (4.2)

$$(Bu, u) = \sum_{k=1}^{j-1}\sum_{l}(u, \phi_k^l)^2 + \sum_{x_J^l \in N_J}(u, \phi_J^l)^2$$

$$+ \sum_{k=j}^{J-1}\left[\sum_{x_k^l \in N_k^1}\gamma_k^J(u, \phi_k^l)^2 + \sum_{x_k^l \in N_k/N_k^1}(u, \phi_k^l)^2\right]$$

with an analogous expression for $\hat{B}$. Clearly, $1 \leq \gamma_k^J \leq 4/3$ from which it follows that

$$(\hat{B}u, u) \leq (Bu, u) \leq \frac{4}{3}(\hat{B}u, u), \qquad \text{for all } u \in M.$$

From the discussion in Section 3, it is clear that the first sum in (4.3) is a preconditioner for the problem on M_j, i.e. the finest uniform grid. As we shall see, this sum can also be replaced by any uniform preconditioner for A_j without adversely effecting the asymptotic behavior of the overall condition number. Indeed, let the operator R_j be a preconditioner for A_j (satisfying (2.16) and the second inequality of (A.2)) and define for $u \in M$,

$$\hat{B}u = R_j Q_j u + \sum_{k=j+1}^{J}\sum_{x_k^l \in N_k}(u, \phi_k^l)\phi_k^l. \tag{4.4}$$

Note that by Remark 2.3, the operator

$$\tilde{B}u = \sum_{k=j}^{J}R_k Q_k u$$

satisfies $K(\tilde{B}A) \leq C(J - j)^2$. It is shown in [11] that $\tilde{B}$ is uniformly equivalent to $\hat{B}$. Thus, $K(\hat{B}A) \leq C(J - j)^2$.

Remark 4.1: Clearly, we could generalize this example to include much more general refinements for problems in R^2 as well as higher dimensional space. Note that the refinement only changes the preconditioner $\hat{B}$ (resp. $\hat{B}$) by adding additional terms in (4.3) (resp. (4.4)) involving nodes from the refinement region. Thus, this approach is well suited to dynamic adaptive refinement techniques. New refinement regions add terms to the sum whereas the "de-refinement" of existing regions only takes away terms from the sum. The operator $\hat{B}$ is even more useful in this context since it allows the easy inclusion of this refinement preconditioner into existing large scale uniform grid codes. Preconditioners for the uniform grid already available in the existing code can be used, supplemented with additional routines implementing the terms due to the refinement.

5. Numerical results

In this section, we provide the results of numerical examples illustrating the theory discussed in the earlier sections. To demonstrate the performance of the proposed algorithms, we shall provide numerical results for a two dimensional problem with full elliptic regularity and one with less than full elliptic regularity, a two dimensional example with a geometric mesh refinement and a three dimensional example. In all of the reported results, the experimentally observed behavior of the condition number of the preconditioned system was in agreement with the theorems presented earlier. In the first example, we also compare the results of the new method with those obtained using the hierarchical preconditioning method [21] and a classical V-cycle multigrid preconditioner[4].

For our first example, we consider Problem (1.1) when $L = -\Delta \equiv -\partial^2/\partial x_1^2 - \partial^2/\partial x_2^2$ and Ω is the unit square. This example satisfies the regularity and approximation assumption (A.3) for $\alpha = 1$ as well as (A.1).

We will use a finite element discretization of (1.1) and develop a sequence of grids in a standard way. To define the coarsest grid, we start by breaking the square into four smaller squares of side length $1/2$ and then dividing each smaller square into two triangles by connecting the lower left hand corner with the upper right hand corner. Subsequently finer grids are developed as in the introduction, i.e., by dividing each triangle into the four triangles formed by the edges of the original triangle and the lines connecting the centers of these edges. The space $\mathcal{M}_i$ is defined to be the set of continuous functions on Ω which are piecewise linear on the i'th triangulation and vanish on $\partial\Omega$.

We shall compare three preconditioners for (1.2). The first preconditioner $\mathcal{B}$ is defined by the multilevel algorithm (2.13) with R_k given by (3.2) and fits into the framework considered in Section 3. For comparison, we also provide results for the hierarchical preconditioner B_H [21] and a preconditioner B_M defined by a standard symmetric V-cycle of multigrid [4]. The multigrid algorithm uses one sweep of Jacobi smoothing whenever a grid level is visited and hence results in two smoothing steps on each grid for each evaluation of the preconditioner. The multigrid uses $h_0 = 1/4$ for the coarsest grid while both the hierarchical and the parallel multilevel algorithms uses $h_0 = 1/2$.

Table 5.1 gives the condition numbers K of the preconditioned systems $B_H A$, $\mathcal{B} A$ and $B_M A$ corresponding, respectively, to the hierarchical preconditioner, the preconditioner defined by (2.13), and the V-cycle multigrid preconditioner. We note that for these examples, a preconditioned conjugate gradient algorithm using the new preconditioner would be expected to take twice as many iterations as the corresponding algorithm using the V-cycle of multigrid. However, even in a serial implementation, the multigrid algorithm involves substantially more computational effort per step. The new method outperforms the hierarchical preconditioner.

Table 5.1

Condition numbers when Ω is the square.

h_J	$K(B_H A)$	$K(\mathcal{B} A)$	$K(B_M A)$
1/16	19	7.0	2.3
1/32	31	8.1	2.4
1/64	43	9.0	2.4
1/128	58	9.8	2.4

This test problem illustrates an example where all three methods work reasonably well. However, we note that $\mathcal{B}$ is preferred over standard multigrid when the parallel aspects of the algorithm are important. In addition, $\mathcal{B}$ generalizes to higher dimensional problems without convergence rate deterioration (see Table 5.5) and hence would be preferred to the hierarchical method in three dimensional computations.

We next consider the above preconditioners on a problem with less than full elliptic regularity. We again consider (1.1) with L given by the Laplacian and Ω equal to the "slit domain", i.e. Ω is the set of points in the interior of the unit square excluding the line $\{(1/2, y) | y \in [1/2, 1)\}$. This example does not satisfy the *a priori* estimates used in the proof of the regularity and approximation assumption (A.3) for $\alpha \geq 1/2$. However, assumption (A.1) is satisfied.

Table 5.2 gives the condition numbers K of the preconditioned systems $B_H A$, $\mathcal{B} A$ and $B_M A$ corresponding, respectively, to the hierarchical preconditioner, the preconditioner defined by (2.13), and the V-cycle multigrid preconditioner. The results are in general agreement with the theoretical estimates

$$K(B_H A) \leq C \ln^2(1/h_J),$$
$$K(\mathcal{B} A) \leq C \ln^2(1/h_J),$$

for the respective methods.

Table 5.2

Condition numbers when Ω is the slit domain.

h_J	$K(B_H A)$	$K(\mathcal{B} A)$	$K(B_M A)$
1/16	14.6	7.9	2.6
1/32	25.17	10.0	2.9
1/64	38.2	12.6	3.1
1/128	53.8	14.9	3.4

We next provide numerical results for the refinement example of Section 4. We once again, consider the solution of (1.1) with L the Laplacian and Ω the unit square. The sequence of spaces $\mathcal{M}_1 \subset \cdots \subset \mathcal{M}_J$ are as developed in Section 4 and provide results for the preconditioner $\hat{\mathcal{B}}$ defined by (4.3). As noted in Section 4, some such refinement would be necessary if, for example, the function f had a δ-function behavior at the point (1,1). Table 5.3 gives the condition number of the preconditioned system $\hat{B}A$ as a function of the mesh size of the uniform grid h_j and the number of refinement levels l. The size of the finest triangle can be computed by dividing the uniform mesh size by 2^l. In all of the runs, the coarsest grid level corresponded to $h_0 = 1/2$. The numerical results seem to indicate that an increase in the number of uniform levels has a greater effect on the condition number than an increase in the number of refinement levels.

Table 5.3

Condition numbers for the refinement example.

h_j	$l = 1$	$l = 2$	$l = 3$	$l = 4$
1/8	6.3	6.5	6.7	6.9
1/16	7.7	7.9	8.05	8.1
1/32	8.8	9.0	9.1	9.2
1/64	9.6	9.7	9.8	9.9

We next present results for the refinement operator defined by (4.4). The problem and sequence of subspaces are as just described but only the subspaces $\mathcal{M}_k$, $k \geq j$ are used. In (4.4), we use a multigrid preconditioner (cf. [4]) scaled by 4 to define R_j, the operator on the finest uniform grid. The scaling was introduced to balance the size of the two terms in (4.4). Table 5.4 gives the condition number of the preconditioned system $\hat{B}A$ as a function of the mesh size of the uniform grid h_j and the number of refinement levels l.

Table 5.4

Condition numbers for $\hat{B}A$ using multigrid preconditioning on level j.

h_j	$l = 1$	$l = 2$	$l = 3$	$l = 4$
1/8	4.3	6.0	6.4	6.6
1/16	4.7	6.7	7.6	8.1
1/32	4.9	7.0	8.4	9.2
1/64	5.0	7.1	8.5	9.6

As a final example, we illustrate the preconditioning technique on a three dimensional problem. We consider a Galerkin approximation to the Laplace equation

$$-\Delta u = f \text{ in } \Omega,$$
$$u = 0 \text{ on } \partial\Omega,$$

(5.1)

where $\Delta = \partial^2/\partial x^2 + \partial^2/\partial y^2 + \partial^2/\partial z^2$ and Ω is the unit cube. We define the coarse mesh by dividing Ω into eight smaller cubes of size $h_0 = 1/2$. Successively finer meshes are formed by dividing each cube of a coarser mesh into eight smaller cubes. The finite element space $\mathcal{M}_k$ is defined to be the set of continuous functions on Ω which are trilinear with respect to the k'th mesh and vanish on $\partial\Omega$.

Table 5.5 gives the condition number K of the preconditioned system $\mathcal{B}A$ where $\mathcal{B}$ is defined by (3.3). This example satisfies full elliptic regularity and the regularity and approximation assumption (A.3) holds with $\alpha = 1$. Thus, the theory predicts only a logarithmic growth in the condition number which is in agreement with the reported results. Note the finite element spaces are of rather large dimension, in fact, the $h_J = 1/64$ example has over a quarter of a million unknowns.

Table 5.5

Condition numbers for the three dimensional example.

h_j	$K(\mathcal{B}A)$
1/8	4.1
1/16	5.2
1/32	6.0
1/64	6.6

References

1   R.E. Bank and T.F. Dupont, *An optimal order process for solving elliptic finite element equations*, Math. Comp. **36** (1981), 35–51.

2   R.E. Bank, T.F. Dupont, and H. Yserentant, *The hierarchical basis multigrid method*, Num. Math. **52** (1988), 427–458.

3   G. Birkhoff and A. Schoenstadt, eds, "Elliptic Problem Solvers II," Academic Press, New York, 1984.

4   J.H. Bramble and J.E. Pasciak, *New convergence estimates for multigrid algorithms*, Math. Comp. **49** (1987), 311–329.

5   J.H. Bramble, J.E. Pasciak and A.H. Schatz, *The construction of preconditioners for elliptic problems by substructuring, I*, Math. Comp. **47** (1986), 103–134.

6   J.H. Bramble, J.E. Pasciak and A.H. Schatz, *The construction of preconditioners for elliptic problems by substructuring, II*, Math. Comp. **49** (1987), 1–16.

7   J.H. Bramble, J.E. Pasciak and A.H. Schatz, *The construction of preconditioners for elliptic problems by substructuring, III*, Math. Comp. **51** (1988), 415–430.

8   J.H. Bramble, J.E. Pasciak and A.H. Schatz, *The construction of preconditioners for elliptic problems by substructuring, IV*, Math. Comp. (to appear).

9   J.H. Bramble, J.E. Pasciak and J. Xu, *The analysis of multigrid algorithms with non-nested spaces or non-inherited quadratic forms*, Math. Comp., (submitted).

10   J.H. Bramble, J.E. Pasciak and J. Xu, *A multilevel preconditioner for domain decomposition boundary systems*, (in preparation).

11   J.H. Bramble, J.E. Pasciak and J. Xu, *Parallel multilevel preconditioners*, Math. Comp., (submitted).

12   P. Concus, G.H. Golub and G. Meurant, *Block preconditioning for the conjugate gradient method*, SIAM J. Sci. Stat. Comput. **6** (1985), 220–252.

13   Dupont, T., Kendall, R.P., and Rachford, H.H.,, *An approximate factorization procedure for solving self-adjoint elliptic difference equations*, SIAM J. Numer. Anal. **5** (1968), 559–573.

14   R. Glowinski, G.H. Golub, G.A. Meurant, J. Périaux, eds, "Proceedings, 1'st Inter. Conf. on Domain Decomposition Methods," SIAM, Philadelphia, 1988.

15   Hackbusch, W., "Multi-Grid Methods and Applications," Springer-Verlag, New York, 1985.

16   S. F. McCormick, *Multigrid Methods for Variational Problems: Further Results*, SIAM J. Numer. Anal. **21** (1984), 255 – 263.

17   S. F. McCormick, *Multigrid Methods for Variational Problems: General Theory for the V-Cycle*, SIAM J. Numer. Anal. **22** (1985), 634–643.

18   J. Mandel, S. McCormick and R. Bank, *Variational multigrid theory*, in "Multigrid Methods," Ed. S. McCormick, SIAM, Philadelphia, Penn., pp. 131–178.

19   J.A. Meyerink and H.A. van der Vorst, *Iterative methods for the solution of linear systems of which the coefficient matrix is a symmetric M-matrix*, Math. Comp. **31** (1977), 148–162.

20   J. Xu, "Theory of Multilevel Methods," Cornell Univ. (Thesis), 1988.

21   H. Yserentant, *On the multi-level splitting of finite element spaces*, Numerische Mathematik **49** (1986), 379–412.

Author's Addresses

Cornell University
Ithaca, N.Y. 14853
E-mail: bramble@mssun7.msi.cornell.edu
Brookhaven National Laboratory
Upton, N.Y. 11973
E-mail: pasciak@bnl.gov
Pennsylvania State University
University Park, Penn. 16802
E-mail: xu@rayleigh.psu.edu

K. BURRAGE
The Hunting of the SNARK, or Why the Trapezodial Rule is Algebraically Stable

They sought it with thimbles, they sought it with care;
They pursued it with forks and hope;
They threatened its life with a railway-share;
They charmed it with smiles and soap.

They hunted till darkness came on, but they found
Not a button, or feather, or mark,
By which they could tell that they stood on the ground
Where the Baker had met with the Snark.

In the midst of the word he was trying to say,
In the midst of his laughter and glee,
He had softly and suddenly vanished away-
For the Snark was a Boojum you see.

From the hunting of the Snark by Charles Dodgson

1. Introduction

Two important recent advances in the study of numerical methods for solving ordinary differential equations are the concepts of algebraic stability and B-convergence. There is a very close relationship between these two ideas but, unfortunately, they do not appear to be equivalent because of the example of the trapezoidal rule which is B-convergent of order 2 but not algebraically stable. This has been the source of much puzzlement and even annoyance to many numerical analysts. In this paper we show that this angst is unnecessary. The trapezoidal rule is algebraically stable if interpreted in the right way - as a Nordsieck method rather than a Runge-Kutta method.

We extend this approach and examine the algebraic stability of a general class of methods, known as multivalue methods, and show that other apparent contradictions that arise when methods are interpreted as Runge-Kutta methods evaporate when they are looked at from a different and more appropriate viewpoint.

40

In determining the suitability of a numerical method for solving stiff problems a nonlinear test problem that is often used is the so-called dissipative or monotonic problem, characterized by

$$y'(x) = f(y(x)), \quad y(0) = y_0, \quad f : \mathbb{R}^N \to \mathbb{R}^N, \tag{1}$$

with

$$< f(y) - f(z), y - z > \leq 0, \quad \forall \, y, z \in \mathbb{R}^N. \tag{2}$$

If the Runge-Kutta method

$$Y_i = y_n + h \sum_{j=1}^{s} a_{ij} f(Y_j), \quad i = 1, \ldots, s \tag{3a}$$

$$y_{n+1} = y_n + h \sum_{j=1}^{s} b_j f(Y_j), \tag{3b}$$

characterized by

$$\begin{array}{c|c} c & A \\ \hline & b^{\mathsf{T}} \end{array},$$

where

$$b = (b_1, \ldots, b_s)^{\mathsf{T}}, \quad A = (a_{ij})_{i,j=1}^{s},$$

$$c = (c_1, \ldots, c_s)^{\mathsf{T}} = Ae, \quad e = (1, \ldots, 1)^{\mathsf{T}},$$

is applied to this problem then it is said to be B-stable if any two solution sequences $\{y_n\}$ and $\{z_n\}$ satisfy

$$\|y_n - z_n\| \leq \|y_{n-1} - z_{n-1}\|,$$

where $\|.\|$ denotes the standard inner product norm.

Of course any method that is B-stable is also A-stable and, in addition, there is a simple algebraic condition, called algebraic stability, characterized by

$$b_i > 0, \quad i = 1, \ldots, s, \quad M = (b_i a_{ij} + b_j a_{ji} - b_i b_j)_{i,j=1}^{s} \geq 0, \tag{4}$$

which is a sufficient condition for B-stability. Under a very mild restriction, that the c_j be distinct, this is also a necessary condition for B-stability (see Burrage and Butcher [6] and Crouzeix [10]).

It can be shown (see Burrage [2]) that if a method is of order p then rank(M) = $s - [\frac{p}{2}]$ so that it is very simple to analyse the algebraic stability of high order Runge-Kutta methods. In particular, Burrage [1] has characterized all the algebraically stable

Runge-Kutta methods of order $2s - 2$ by performing a congruence transformation on the stability matrix M based on the $s \times s$ Vandermonde matrix whose jth column is c^{j-1}. This analysis has been extended by Hairer and Wanner [17] through the technique of W-transformations based on a congruence transformation by a generalized Vandermonde matrix which allows a deeper analysis of other classes of methods such as singly implicit methods.

An obvious generalizion of dissipativity is to consider problems which satisfy a one-sided Lipschitz condition

$$< f(y) - f(z), y - z > \leq \nu ||y - z||^2, \quad \forall y, z \in \mathbb{R}^N. \tag{5}$$

A reason for studying such problems is that stiffness implies the existence of a large Lipschitz constant which, because of its size, is of little value for perturbation analysis. Nevertheless, for many problems one-sided Lipschitz constants ν can be found which are small in magnitude compared with L and may even be negative. In particular, it should be noted that if $y(x)$ and $z(x)$ represent two different solutions of (1) with different initial conditions and where f satisfies (5), then it can be shown that

$$||y(x_2) - z(x_2)|| \leq e^{(x_2 - x_1)\nu} ||y(x_1) - z(x_1)||, \quad x_2 \geq x_1.$$

Hence for problems satisfying (5) a numerical method should mirror the behaviour of the differential solution. Consequently, Burrage and Butcher [7] have generalized the concept of algebraic stability to (k, l)-algebraic stability in which, for problems of the form given by (5), we require any two solution sequences computed by a Runge-Kutta method to satisfy

$$||y_n - z_n|| \leq k(h\nu) ||y_{n-1} - z_{n-1}||, \quad \forall h\nu \leq l.$$

The function k is called the growth function and clearly $k(0) = 1$ implies algebraic stability. Burrage and Butcher [7] have shown that a method is (k, l)-algebraically stable if there exists a positive diagonal matrix D such that $M(k, l) \geq 0$, where

$$M(k, l) = \begin{pmatrix} k - 1 - 2le^T De & e^T D - b^T - 2le^T DA \\ De - b - 2lA^T De & DA + A^T D - bb^T - 2lA^T DA \end{pmatrix}. \tag{6}$$

One of the first papers on the B-stability of Runge-Kutta methods was by Wanner [24] in which he proved that the trapezoidal method

$$y_{n+1} = y_n + \frac{h}{2}(f(y_n) + f(y_{n+1}))$$

was not B-stable, by considering the problem

$$f(y) = \begin{cases} -y/\epsilon, & y \le 0 \\ -y, & y > 0, \end{cases}$$

with ϵ small and

$$y_{n-1} = 0, \ y_n = 0, \ z_{n-1} = -\epsilon, \ z_n = (1 - 2\epsilon)/3, \ h = 1.$$

This can be verified by writing the trapezoidal method in the usual formulation

$$
\begin{array}{c|cc}
0 & 0 & 0 \\
1 & \frac{1}{2} & \frac{1}{2} \\
\hline
 & \frac{1}{2} & \frac{1}{2}
\end{array}
$$

and noting that the stability matrix M given in (4) is

$$\begin{pmatrix} -\frac{1}{4} & 0 \\ 0 & \frac{1}{4} \end{pmatrix}.$$

This property of the trapezoidal rule has been the source of much confusion and even angst to many numerical analysts when the property of B-convergence is considered. As an example of this angst I quote from Frank, Schneid and Ueberhuber [14]: "The implicit trapezoidal rule is B-convergent although it can be shown not to be B-stable."

It was first noted by Prothero and Robinson [21], that when certain classes of Runge-Kutta methods are applied to the problem

$$y'(x) = \lambda\big(y(x) - g(x)\big) + g'(x), \quad y(0) = y_0, \quad \lambda << 0, \tag{7}$$

the observed order is much less than the classical order of consistency (that obtained by comparing the numerical approximations with a Taylor series expansion of the true solution) and more like the stage order. This observation was formalized by Frank, Schneid and Ueberhuber [14] and generalized to nonlinear problems. Thus a Runge-Kutta method is B-consistent of order q for problems satisfying (5) if the local error is bounded such that

$$\|l_{n+1}\| \le C_1 h^{q+1}, \quad \forall h \in (0, h_1]$$

where C_1 and h_1 are independent of stiffness but can depend on ν and on bounds for certain derivatives of the exact solution. Similarly, a method is said to be B-convergent of order q if the global error is bounded in some norm by $C_2 h^q, \forall h \in (0, h_2]$, where once again C_2 and h_2 are independent of stiffness.

Frank et al. [15] (by bounding the local error and then adding via a stability argument) have shown that certain high order methods such as the Gauss, Radau IA and Radau IIA methods are B-convergent with B-convergence order equal to the stage order. However, this is in fact not the best possible result. Burrage et al. [8] have shown that for a class of semi-linear problems many Runge-Kutta methods have a B-convergence order of one more then the stage order, while Kraaijevanger [18] has shown that the implicit midpoint rule with stage order 1 is B-convergent with order 2.

Akin to this, Spijker [23] has recently established relationships between algebraic stability and B-convergence. For a problem satisfying (5) with $\nu < 0$, Spijker [22] has shown that for any Runge-Kutta method of stage order w with nonsingular matrix and irreducible in the sense of Dahlquist and Jeltsch [12], algebraic stability implies B-convergence of order $w - \frac{1}{2}$. Spijker also shows that no similar result holds for problems with $\nu = 0$, since the Lobatto IIIC method is algebraically stable but not B-convergent.

Dekker et al. [13] have generalized the work of Spijker and have shown that for problems with $\nu < 0$, algebraic stability is equivalent to B-convergence with order equal to the stage order. However, this order reduction may not be as severe as theory predicts because these results depend on the construction of a problem with an oscillatory stepsize strategy of period two and a rapidly oscillating test problem, based on the non-autonomous version of (7). Indeed, it has recently been shown by Hairer Lubich and Roche [16] that for certain classes of stiff problems, such as singular perturbation problems, there is no order reduction for some classes of Runge-Kutta methods (such as the Radau IIA method).

The reason that Spijker [23] avoids Runge-Kutta methods with singular matrices is typified by the trapezoidal rule. Kraaijevanger has shown that the trapezoidal rule is B-convergent of order 2 even though it is not algebraically stable (as a Runge-Kutta method).

The purpose of this paper is to show that this afore-mentioned angst is unnecessary. The trapezoidal rule is algebraically stable if looked at in the correct way, that is, in the way that it should be implemented in practice. In order to do this we must first present the concept of a multivalue method and show how algebraic stability applies to this general class of methods. This we will do in section 2. In section 3 we will examine the trapezoidal rule and its stability properties in this light. In section 4 we will examine whether there are other so-called Runge-Kutta methods, such as the Lobatto IIIA methods which are not algebraically stable in the Runge-Kutta formulation but which are algebraically stable in the multivalue formulation. Finally, in section 5 we will examine a property introduced by Spijker [22] called boundedness and show how some difficulties that arise when methods are considered as Runge-Kutta methods can be overcome if they are written as multivalue methods.

2. Multivalue methods and algebraic stability

The general s-stage, r-value multivalue method for solving (1) is given by

$$Y_i = \sum_{j=1}^{r} a_{ij}^{(1)} y_j^{(n-1)} + h \sum_{j=1}^{s} b_{ij}^{(1)} f(Y_j), \quad i = 1, \ldots, s$$

$$y_i^{(n)} = \sum_{j=1}^{r} a_{ij}^{(2)} y_j^{(n-1)} + h \sum_{j=1}^{s} b_{ij}^{(2)} f(Y_j), \quad i = 1, \ldots, r$$

and can be written in the tableau form

$$\begin{array}{c|c} A_1 & B_1 \\ \hline A_2 & B_2 \end{array}. \tag{8}$$

In the above formulation the Y_i ($i = 1, \ldots, s$) are internal to each step and represent approximations to the solution at the off-step points $x_{n-1} + c_i h$ while the $y_i^{(n)}$ carry all the information from step to step. In order to start a multivalue method, r starting procedures such as r Runge-Kutta methods ($Q_1, \ldots, Q_r$, say) are needed to compute $y_1^{(0)}, \ldots, y_r^{(0)}$ from the initial value y_0. If we call this starting procedure S and call the finishing procedure F, then an application of the above multivalue method over n steps can be written as

$$SM^n F,$$

where M takes method (8) forward one step.

The order of a multivalue method is defined with respect to the r starting procedures (see Burrage and Moss [9], for example) and is said to be of order w if, starting from y_0, the Taylor series expansion of the r external solutions $(y_1^{(1)}, \ldots, y_r^{(1)})$ after one step coincide, up to order w, with the numerical results $(z_1, \ldots, z_r)$ applied to $y_1 = y(x_0 + h)$. Hence a method is of order w if

$$y_j^{(1)} - z_j = O(h^{w+1}), \quad j = 1, \ldots, r.$$

45

Two important classes of multivalue methods are multistep Runge-Kutta methods (MRKs), which include linear multistep and Runge-Kutta methods as special cases, and Nordsieck methods. In the case of MRKs the $y_i^{(n)}$ are approximations to the solution at the r previous node points. Burrage [4] and Lie [20] have shown that these methods can attain an order of $2s + r - 1$ with a stage order of $s + r - 1$. On the other hand, Nordsieck methods, in which the $y_i^{(n)}$ represent approximations to the scaled derivatives $h^{i-1}y^{(i-1)}(x_n)/(i-1)!$, do not possess as high an attainable order as MRKs. Burrage [4] has shown that if $r > 2$ the maximum order is $s + r - 1$ (with stage order $s + r - 1$) while if $r \leq 2$ the maximum order is $2s$.

Dahlquist [11] was the first to examine (through the concept of G-stability) the behaviour of linear multistep methods (or more correctly their correponding one-leg k-step counterpart) to monotonic problems and the generalization of G-stability (for one-leg methods) and B-stability (for Runge-Kutta methods) to the family of multivalue methods was given in Burrage and Butcher [7].

Burrage and Butcher [7] introduce an $r \times r$ symmetric matrix $G \geq 0$ and define a semi-inner product on $\mathbb{R}^r$ by

$$< u, v > = \sum_{i,j=1}^{r} g_{ij} u_i v_j.$$

Similarly, if $U = (u_1^\mathsf{T}, \ldots, u_r^\mathsf{T})^\mathsf{T}$ and $V = (v_1^\mathsf{T}, \ldots, v_r^\mathsf{T})^\mathsf{T}$ are vectors in $\mathbb{R}^{mr}$ with $u_i, v_i \in \mathbb{R}^m$ then

$$< U, V >_G = \sum_{i,j=1}^{r} g_{ij} < u_i, v_j > .$$

A method is said to be monotonic if for any dissipative problem there exists $G \geq 0$ such that

$$||y^{(n)}||_G \leq ||y^{(n-1)}||_G, \quad y^{(n)} = (y_1^{(n)}, \ldots, y_r^{(n)})^\mathsf{T}.$$

There is an algebraic condition that is sufficient for monotonicity. A method is said to be algebraically stable if there exists $G \geq 0$ and a diagonal matrix $D \geq 0$ such that the matrix $M \geq 0$, where

$$M = \begin{pmatrix} G - A_2^\mathsf{T} G A_2 & A_1^\mathsf{T} D - A_2^\mathsf{T} G B_2 \\ D A_1 - B_2^\mathsf{T} G A_2 & B_1^\mathsf{T} D + D B_1 - B_2^\mathsf{T} G B_2 \end{pmatrix}. \tag{9}$$

Burrage and Butcher [7] have shown that algebraic stability implies monotonicity. Thus from (9) we can write

$$||y^{(n)}||_G^2 - ||y^{(n-1)}||_G^2 = 2 \sum_{i=1}^{s} d_i < Y_i, hf(Y_i) > - \sum_{i,j=1}^{s} m_{ij} < \alpha_i, \alpha_j >, \tag{10}$$

where

$$\alpha = (y_1^{(n-1)}, \ldots, y_r^{(n-1)}, hf(Y_1), \ldots, hf(Y_s))^{\top}.$$

We note that the stability of the multivalue method, given in (8), says nothing about the stability behaviour of the starting or finishing method. It had been thought that the algebraic stability of multivalue methods was too difficult to study in a rigorous way because of the neccessity of finding an appropriate nonnegative matrix G, but recently Burrage [3] has shown that some of the properties that hold for algebraically stable Runge-Kutta methods also hold for multivalue methods. In particular, he has constructed families of MRKs (with $r - 2$ free parameters) with classical order $2s$ and stage order s and shown that they are algebraically stable with G a diagonal matrix. For such methods the rank of M is reduced from $s + r - 1$ to $r - 1$, which generalizes similar results for Runge-Kutta methods.

In the case of Nordsieck methods Burrage [5] has shown that if a Nordsieck method is of order r then a necessary condition for algebraic stability is that G is a partial Hankel matrix whose elements satisfy

$$g_{ij} = c_{i+j}, \quad \forall i, j, \ i + j \leq r + 1.$$

As a consequence of this Burrage [5] has shown that the maximum order of an irreducible algebraically stable Nordsieck method with $r > 1$ is $\min\{r+1, 2s\}$ or $\min\{r, 2s\}$ depending on whether r is odd or even, respectively.

This result is surprising in light of the results obtained in [3], where it was shown that there always exist algebraically stable MRKs of order $2s$. Evidently, the stability properties of Nordsieck methods and multistep Runge-Kutta methods are very different.

The generalization of (k, l)-algebraic stability to multivalue methods is simple and is given in Burrage and Butcher [7]. A method is said to be (k, l)-algebraically stable if there exists $G \geq 0$ and a diagonal matrix $D \geq 0$ such that the matrix $M(k, l)$, given by

$$\begin{pmatrix} kG - A_2^{\top} G A_2 - 2l A_1^{\top} D A_1 & A_1^{\top} D - A_2^{\top} G B_2 - 2l A_1^{\top} D B_1 \\ D A_1 - B_2^{\top} G A_2 - 2l B_1^{\top} D A_1 & B_1^{\top} D + D B_1 - B_2^{\top} G B_2 - 2l B_1^{\top} D B_1 \end{pmatrix} \qquad (11)$$

is nonnegative definite.

We are now in a position to examine the stability of certain classes of methods as multivalue rather than Runge-Kutta methods.

3. The trapezoidal rule

In this section we examine the nonlinear stability behaviour of the trapezoidal rule when examined as an example of a special class of multivalue methods, namely a one-stage two-value Nordsieck method. Let us first, however, examine why the trapezoidal rule should be considered in this fashion and not as a Runge-Kutta method.

Consider the trapezoidal rule when written in the usual Runge-Kutta format

$$Y_1 = y_n$$
$$Y_2 = y_n + \frac{h}{2}\left(f(Y_1) + f(Y_2)\right)$$
$$y_{n+1} = y_n + \frac{h}{2}\left(f(Y_1) + f(Y_2)\right). \tag{12}$$

At the beginning of step $n+1$, having just computed y_n, we set $Y_1 = y_n$ and compute $f(Y_1)$ which is stored. We then solve for Y_2 by some iterative method. At the end of these iterations we do a final function evaluation $f(Y_2)$ and then compute y_{n+1} using the value of $f(Y_1)$ stored as a temporary value. The problem with this implementation is that at the beginning of each step we perform an unnecessary function evaluation of, essentially, $f(y_n)$. This is unnecessary because at the end of a step we do a final function evaluation of $f(Y_2)$ and we should use this value to represent $f(y_n)$ at the next step. In this alternative formulation the trapezoidal rule can be written as a 1-stage 2-value Nordsieck method

$$Y = y_n + \frac{1}{2}z_n + \frac{h}{2}f(Y)$$
$$y_{n+1} = y_n + \frac{1}{2}z_n + \frac{h}{2}f(Y)$$
$$z_{n+1} = hf(Y), \tag{13}$$

or in multivalue format

$$
\begin{array}{cc|c}
1 & \frac{1}{2} & \frac{1}{2} \\
\hline
1 & \frac{1}{2} & \frac{1}{2} \\
0 & 0 & 1
\end{array}
.$$

In order to see that (12) and (13) are substantially different, we consider their behaviour when applied to the linear non-autonomous problem

$$y'(x) = q(x)y(x), \quad Re(q(x)) \le 0, \forall x.$$

In the case of (12) applied to this problem we have

$$y_{n+1} = R(z_1, z_2)y_n, \quad z_1 = hq(x_n), \quad z_2 = hq(x_{n+1}), \quad R(z_1, z_2) = \frac{2 + z_1}{2 + z_2},$$

and R can become unbounded.

On the other hand, an application of (13) to this problem gives

$$\begin{pmatrix} y_{n+1} \\ z_{n+1} \end{pmatrix} = R(z_2)\begin{pmatrix} y_n \\ z_n \end{pmatrix}, \quad R(z) = \frac{1}{1 - \frac{z}{2}}\begin{pmatrix} 1 & \frac{1}{2} \\ z & \frac{z}{2} \end{pmatrix}$$

and the eigenvalues of R are

$$0, \quad \frac{2 + z_2}{2 - z_2},$$

so that R is power-bounded.

We now analyse the behaviour of (12) and (13) when applied to the more general nonlinear problem given by (1) and (2). From (9) the stability matrix associated with (13) is given by

$$M = \begin{pmatrix} 0 & g_{12} - \frac{1}{2}g_{11} & d - (\frac{1}{2}g_{11} + g_{12}) \\ g_{12} - \frac{1}{2}g_{11} & g_{22} - \frac{1}{4}g_{11} & \frac{1}{2}(d - (\frac{1}{2}g_{11} + g_{12})) \\ d - (\frac{1}{2}g_{11} + g_{12}) & \frac{1}{2}(d - (\frac{1}{2}g_{11} + g_{12})) & d - (\frac{1}{4}g_{11} + g_{12} + g_{22}) \end{pmatrix},$$

and is nonnegative iff

$$g_{12} = \frac{1}{2}g_{11}, \quad g_{22} = \frac{1}{4}g_{11}, \quad d = g_{11},$$

in which case M is the zero matrix. Hence (13) is algebraically stable with

$$d = g_{11} > 0, \quad G = g_{11}\begin{pmatrix} 1 & \frac{1}{2} \\ \frac{1}{2} & \frac{1}{4} \end{pmatrix}.$$

We note that G is singular, hence

$$\|y^{(n)}\|_G^2 = \|y_n + \frac{h}{2}f(y_n)\|^2,$$

and we can write (13) as

$$\|y_{n+1} + \frac{h}{2}f(y_{n+1})\|^2 = \|y_n + \frac{h}{2}f(y_n)\|^2 + 2 < y_{n+1}, hf(y_{n+1}) > .$$

Thus, in the above sense, the trapezoidal rule given by (13) can be considered to be reducible since it is computationally equivalent to a multivalue method with $s = r = 1$. To see this, let u_n represent $y_n + \frac{h}{2}f(y_n)$, and then we can rewrite (13) as

$$Y = u_n + \frac{1}{2}hf(Y)$$
$$u_{n+1} = u_n + hf(Y).$$

This is essentially the midpoint rule except that a starting procedure is needed to compute u_0 from y_0. However this formulation is an inappropriate one if we are considering variable stepsizes in which case (13) is then the appropriate formulation.

We can strengthen this argument by considering the (k, l)-algebraic stability of the implicit midpoint rule, and the trapezoidal rule written both as a Runge-Kutta method and as a Nordsieck method. In the case of the midpoint rule Burrage and Butcher [7] have shown that the method is (k, l)-algebraically stable with $\sqrt{k} = \phi(l)$, where

$$d = \phi(l), \quad \phi(l) = \begin{cases} 1, & l \leq 0 \\ \frac{1+\frac{l}{2}}{1-\frac{l}{2}}, & 0 < l < 2. \end{cases}$$

In the case of the trapezoidal rule, written as a Runge-Kutta method, we can show that

$$M(k, l) = \begin{pmatrix} k - 1 - 2l(d_1 + d_2) & d_1 - \frac{1}{2} - ld_2 & d_2 - \frac{1}{2} - ld_2 \\ d_1 - \frac{1}{2} - ld_2 & -\frac{1}{4} - \frac{l}{2}d_2 & \frac{1}{2}(d_2 - \frac{1}{2} - ld_2) \\ d_2 - \frac{1}{2} - ld_2 & \frac{1}{2}(d_2 - \frac{1}{2} - ld_2) & d_2 - \frac{1}{4} - \frac{l}{2}d_2 \end{pmatrix}.$$

Now we note that

$$x^{\mathsf{T}} M(k, l)x = 0, \quad x^{\mathsf{T}} = (0, -1, 1)$$

so that if M is to be nonnegative definite we must have $Mx = 0$, which implies $d_2 = 0$. Hence the (2,2) element of $M(k, l)$ is negative and so (12) is not algebraically stable for any value of l.

On the other hand for the trapezoidal rule, as a Nordsieck method, we find that $M(k, l)$ is given by

$$\begin{pmatrix} (k - 1)g_{11} - 2ld & kg_{12} - \frac{1}{2}g_{11} - ld & d - (\frac{1}{2}g_{11} + g_{12}) - ld \\ kg_{12} - \frac{1}{2}g_{11} - ld & kg_{22} - \frac{1}{4}g_{11} - \frac{1}{2}d & \frac{1}{2}(d - (\frac{1}{2}g_{11} + g_{12}) - ld) \\ d - (\frac{1}{2}g_{11} + g_{12}) - ld & \frac{1}{2}(d - (\frac{1}{2}g_{11} + g_{12}) - ld) & d - (\frac{1}{4}g_{11} + g_{12} + g_{22}) - \frac{1}{2}d \end{pmatrix}.$$

Now we note that

$$x^{\mathsf{T}} M(k, l)x = (k - 1)g_{22},$$

which implies that $k \geq 1$ is a necessary condition for (k, l)-algebraic stability. But since this method is algebraically stable we must have

$$k = 1, \quad l \leq 0, \quad G = g_{11} \begin{pmatrix} 1 & \frac{1}{2} \\ \frac{1}{2} & \frac{1}{4} \end{pmatrix}.$$

But it can also be shown that

$$y^\mathsf{T} M(k,l)y = \frac{k}{2} z^\mathsf{T} Gz, \quad y^\mathsf{T} = (\tfrac{1}{2},-1,0), \quad z^\mathsf{T}=(\tfrac{1}{2},-1),$$

and with G as above we see that $M(k,l)$ is singular and some simple analysis shows that the (k,l)-algebraic stability properties are exactly the same as for the implicit midpoint rule.

4. Algebraic stability of Lobatto IIIA methods

We saw in section 3 that the trapezoidal rule considered as a Nordsieck, rather than a Runge-Kutta, method is algebraically stable. An interesting question to address is whether there are other so-called Runge-Kutta methods which can become algebraically stable when considered in a different light.

The natural class of methods to consider in this case are those methods in which

$$a_{1j} = 0, \quad a_{sj} = b_j, \quad j = 1,\ldots,s, \tag{14}$$

which include the family of Lobatto IIIA methods. Considered as Runge-Kutta methods it is clear that this family cannot be algebraically stable, since from (4) and (14) $M_{11} = -b_1^2$ and hence a necessary condition for M to be nonnegative is

$$b_1 = 0, \quad a_{j1} = 0, \quad j = 1,\ldots,s,$$

and such a method is reducible.

On the other hand, the reservations about interpreting the trapezoidal rule as a Runge-Kutta method applies to the family of methods characterized by (14). In fact (14) can be written as an $(s-1)$-stage 2-value Nordsieck method with

$$A_1 = \begin{pmatrix} 1 & a_{21} \\ \vdots & \vdots \\ 1 & a_{s1} \end{pmatrix}, \quad B_1 = \begin{pmatrix} a_{22} & \cdots & a_{2s} \\ \vdots & & \vdots \\ a_{s2} & \cdots & a_{ss} \end{pmatrix}$$

$$A_2 = \begin{pmatrix} 1 & b_1 \\ 0 & 0 \end{pmatrix}, \quad B_2 = \begin{pmatrix} b_2 & \cdots & b_{s-1} & b_s \\ 0 & \cdots & 0 & 1 \end{pmatrix}.$$

If we now form (9) with

$$G = \begin{pmatrix} g_{11} & g_{12} \\ g_{12} & g_{22} \end{pmatrix}, \quad D = \operatorname{diag}(d_2,\ldots,d_s)$$

we can write M as

$$\begin{pmatrix} M_{11} & M_{12}^T \\ M_{12} & M_{22} \end{pmatrix},$$

where

$$M_{11} = \begin{pmatrix} 0 & g_{12} - b_1 g_{11} \\ g_{12} - b_1 g_{11} & g_{22} - b_1^2 g_{11} \end{pmatrix}$$

and

$$M_{12} = \begin{pmatrix} d_2 - b_2 g_{11} & d_2 a_{21} - b_2 b_1 g_{11} \\ \vdots & \vdots \\ d_{s-1} - b_{s-1} g_{11} & d_{s-1} a_{s-11} - b_{s-1} b_1 g_{11} \\ d_s - (b_s + b_1) g_{11} & d_s a_{s1} - (b_s + b_1) b_1 g_{11} \end{pmatrix}.$$

Since we require $M \geq 0$ we must have

$$g_{12} = b_1 g_{11}, \quad d_j = b_j g_{11}, \, j = 2, \ldots, s-1, \, d_s = (b_s + b_1) g_{11}. \tag{15}$$

We can now write

$$M = \begin{pmatrix} 0 & 0 \\ 0 & \hat{M} \end{pmatrix},$$

where $\hat{M}$ is an $s \times s$ matrix. Using (9), (14) and (15), and without loss of generality setting $g_{11} = 1$, we have

$$\hat{M} = N + M_{RK},$$

where

$$(M_{RK})_{ij} = b_i a_{ij} + b_j a_{ji} - b_i b_j, \quad N = g_{22} \, \text{diag} \, (1, 0, \ldots, 0, -1).$$

As an example of the above analysis we have for the trapezoidal rule

$$\hat{M} = \begin{pmatrix} -\frac{1}{4} & 0 \\ 0 & \frac{1}{4} \end{pmatrix} + \begin{pmatrix} g_{22} & 0 \\ 0 & -g_{22} \end{pmatrix},$$

which is nonnegative (in fact $\hat{M}$ is the zero matrix) iff $g_{22} = \frac{1}{4}$, as was shown in section 3.

Now with $c_1 = 0$ and $c_s = 1$, let V be the Vandermonde matrix given by

$$V = \begin{pmatrix} 1 & 0 & \ldots & 0 \\ 1 & c_2 & \ldots & c_2^{s-1} \\ \vdots & \vdots & \vdots & \vdots \\ 1 & c_{s-1} & \ldots & c_{s-1}^{s-1} \\ 1 & 1 & \ldots & 1 \end{pmatrix},$$

then

$$V^T \hat{M} V = V^T M_{RK} V + V^T N V$$
$$= R + g_{22}(e_1^T e_1 - e^T e),$$

where $e_1^T = (1, 0, \ldots, 0)$ and $R = V^T M_{RK} V$.

It is known (see Burrage [2]) that if a Runge-Kutta method is of order 3 then $r_{11} = r_{21} = 0$. Hence the only way that $V^T \hat{M} V$ can be nonnegative is if $g_{22} = 0$. But we have already shown that M_{RK} is not nonnegative and so (14) can only be algebraically stable if the order is 2 or less.

As another example of a method which is not algebraically stable as a Runge-Kutta method but is as a Nordsieck method, consider the family of 3-stage Runge-Kutta methods

$$
\begin{array}{c|ccc}
0 & 0 & & \\
c & c - b_3 & b_3 & \\
1 & b_1 & b_2 & b_3 \\
\hline
 & b_1 & b_2 & b_3
\end{array} \quad .
$$

Clearly this method is not algebraically stable. But it can also be written as a 2-stage 2-value Nordsieck method

$$
\begin{array}{cc|cc}
1 & c - b_3 & b_3 & 0 \\
1 & b_1 & b_2 & b_3 \\
\hline
1 & b_1 & b_2 & b_3 \\
0 & 0 & 0 & 1
\end{array} \quad .
$$

If this method is to be of order 2 and if we choose $b_1 = c - b_3$ then it can be shown that this method is algebraically stable iff

$$c = \frac{1}{2}, \quad G = \begin{pmatrix} 1 & \frac{1}{4} \\ \frac{1}{4} & \frac{1}{16} \end{pmatrix}.$$

In this case the method is

$$
\begin{array}{cc|cc}
1 & \frac{1}{4} & \frac{1}{4} & 0 \\
1 & \frac{1}{4} & \frac{1}{2} & \frac{1}{4} \\
\hline
1 & \frac{1}{4} & \frac{1}{2} & \frac{1}{4} \\
0 & 0 & 0 & 1
\end{array}
$$

and is, in fact, reducible in the sense that if we let

$$u_n = y_n + \frac{h}{4} f(y_n),$$

then the method can be written as

$$Y_1 = u_n + \frac{h}{4}f(Y_1)$$

$$Y_2 = u_n + \frac{h}{2}f(Y_1) + \frac{h}{4}f(Y_2)$$

$$u_{n+1} = u_n + \frac{h}{2}(f(Y_1) + f(Y_2)). \tag{16}$$

This method has a smaller truncation error than the trapezoidal rule and has stage order 2. In fact over n steps with constant stepsize we can write (16) as

$$S^n = \left(\tfrac{1}{4}E\right)\left(\tfrac{1}{2}M\right)^n\left(\tfrac{1}{4}\hat{E}\right),$$

where this represents starting the method with the explicit Euler and finishing with the implicit Euler both with a quarter of the stepsize and doing each intermediate step with the composition of the midpoint rule with itself each with half the stepsize. Ignoring the starting and finishing methods we see that the stability properties of (16) are essentially the same as the midpoint rule.

If $b_1 + b_3 \neq c$ then we can construct a family of methods which are algebraically stable and not reducible (so that G is of full rank). An example of this family of methods is

$$\begin{array}{cc|cc}
1 & -\frac{1}{8} & \frac{3}{8} & 0 \\
1 & \frac{1}{8} & \frac{1}{2} & \frac{3}{8} \\
\hline
1 & \frac{1}{8} & \frac{1}{2} & \frac{3}{8} \\
0 & 0 & 0 & 1
\end{array} \quad ,$$

with

$$G = \begin{pmatrix} 1 & b_1 \\ b_1 & b_3^2 \end{pmatrix} = \begin{pmatrix} 1 & \frac{1}{8} \\ \frac{1}{8} & \frac{9}{64} \end{pmatrix}$$

and

$$\hat{M} = \begin{pmatrix} \frac{1}{8} & -\frac{1}{8} & 0 \\ -\frac{1}{8} & \frac{1}{8} & 0 \\ 0 & 0 & 0 \end{pmatrix}.$$

5. Boundedness

We have seen in section 3, and indeed it is well known, that the trapezoidal rule applied over n steps can be considered as a multivalue method with a starting procedure to compute $y_{\frac{1}{2}} = y_0 + \frac{h}{2}f(y_0)$ followed by $n-1$ steps of a certain method followed by

54

a finishing method. In the case of constant stepsizes we have the well-known relation between the composite trapezoidal rule and the composite midpoint rule, while in the case of variable stepsizes the base method consists of $n - 1$ steps of the one-leg method

$$u_k = u_{k-1} + h_k f\big((1 - \theta)u_{k-1} + \theta u_k\big),$$

where in the kth step $\theta = \frac{h_k}{h_k + h_{k+1}}$ and the stepsize is $\frac{h_k + h_{k+1}}{2}$. This is the approach that Kraaijevanger [18] uses to show that the trapezoidal rule is B-convergent of order 2 and is another example of the multivalue approach considered in section 3.

Recently, Spijker [22] has introduced the property of boundedness for Runge-Kutta methods. A Runge-Kutta method is said to be bounded if there exists γ such that

$$\|y_n\| \le \gamma \|y_0\|$$

for dissipative problems. Spijker notes that this is a useful approach for deriving B-convergence estimates as long as γ depends only on the method and not on h, f, or the initial conditions. Spijker proves that as long as a Runge-Kutta method is irreducible and that a method does not have the property (14), then algebraic stability is equivalent to boundedness if the possibility of allowing γ to depend on h, f, y_0 is permitted. The reason for Spijker not allowing methods satisfying (14) is because of the trapezoidal rule which is known to be bounded with $\gamma = \|y_0 + \frac{h}{2}f(y_0)\| \cdot \|y_0\|^{-1}$ but is not algebraically stable as a Runge-Kutta method. However, when viewed as a multivalue method there is no real problem here.

This work is extended by Kraaijevanger and Spijker [19] to consider another property called S-stability. A Runge-Kutta method is said to be S-stable with repect to h_n, f and the inner product if the effect of an arbitrary perturbation $\hat{y}_0 - y_0$ can be bounded by

$$\|\hat{y}_n - y_n\| \le \gamma \|\hat{y}_0 - y_0\|.$$

They prove here that algebraic stability and S-stability, with respect to all $\{h_n\}$, monotonic f and $<, >$, are equivalent assuming that the method is irreducible in the sense of Hundsdorfer and Spijker. This equivalence assumes variable stepsizes, but is not true with constant stepsizes. This result is clearly not satisfactory and occurs because of the problem with the trapezoidal rule when not viewed in an appropriate way. Kraaijevanger and Spijker note that the trapezoidal rule with constant stepsize is S-stable with

$$\gamma = \|\hat{y}_0 - y_0\|^{-1} \|\hat{y}_0 - y_0 + \frac{h}{2}\big(f(\hat{y}_0) - f(y_0)\big)\|,$$

55

but is not algebraically stable. Again there is no difficulty here if the trapezoidal rule is viewed in an appropriate way.

They also offer another counterexample to the above non-equivalence based on the method

$$
\begin{array}{c|cc}
\frac{1}{2} & \frac{1}{2} & \\
\frac{3}{2} & -\frac{1}{2} & 2 \\
\hline
 & -\frac{1}{2} & \frac{3}{2}
\end{array} \ .
$$

Clearly this method is not algebraically stable if condition (4) is applied since one of the weights is negative but then Kraaijevanger and Spijker show that it is S-stable with $\gamma = 2$. At first sight this result seems remarkable but on careful examination it is not so because of a reducibility that occurs when the stepsize is constant. To best see this, let the above method be denoted by S_0 then we can write

$$
S_0 = S_1 S_3,
$$

so that

$$
S_0^n = S_1 (S_3 S_1)^{n-1} S_3
$$
$$
= S_1 S_2^{n-1} S_3,
$$

where S_1, S_2 and S_3 are respectively

$$
\begin{array}{c|cc}
\frac{1}{2} & \frac{1}{2} & \\
\hline
 & -\frac{1}{2} &
\end{array} \ ,
$$

$$
\begin{array}{c|cc}
2 & 2 & \\
2 & \frac{3}{2} & \frac{1}{2} \\
\hline
 & \frac{3}{2} & -\frac{1}{2}
\end{array} \ ,
$$

$$
\begin{array}{c|c}
2 & 2 \\
\hline
 & \frac{3}{2}
\end{array} \ .
$$

Since S_2 is reducible to

$$
\begin{array}{c|c}
2 & 2 \\
\hline
 & 1
\end{array} \ ,
$$

we have a situation similar to (16) for which, apart from the starting and finishing procedures, the stability properties of S_0 and S_2 are the same. Thus, since in the interpretation of the algebraic stability of multivalue methods we ignore the behaviour of the starting and finishing method, we can interpret S_0 as being algebraically stable because S_2 is algebraically stable - even though S_0 does not satisfy condition (4).

We can generalize this analysis. Suppose an arbitrary method S is the composition of two methods S_1 and S_2 then we can write

$$S^n = S_1(S_2S_1)^{n-1}S_2$$
$$S^n = S_1 S_3^{n-1} S_2.$$

We can interpret S_1 as the starting method and S_2 as the finishing method, so that in the multivalue approach S will be algebraically stable if S_3 is. Now, if with constant stepsizes, S_3 is reducible then a situation can arise similar to that discussed above where S_1 does not satisfy the algebraic stability condition given by (4) but that S_3, in its reduced form, does. Now a necessary condition for reducibility in this case is that one of the abscissae for method S_2 is one more than one of the abscissae for method S_1. Of course more complicated behaviour than this can arise which can lead to reducibility but this is the essence of the matter.

As a final example we present a three stage method written as the composition of a two stage and a one stage method which is not algebraically stable but which, when implemented with constant stepsize and reducibility is taken into account, is algebraically stable in the multivalue sense: Let M_1, M_2 and M_3 be the methods

$$
\begin{array}{c|cc}
c & c & \\
1+c & \frac{1}{2}+c & \frac{1}{2} \\
\hline
 & \frac{1}{2}+c & b
\end{array} \quad ,
$$

$$
\begin{array}{c|c}
\frac{1}{2}-b & \frac{1}{2}-b \\
\hline
 & \frac{1}{2}-b-c
\end{array} \quad ,
$$

$$
\begin{array}{c|c}
\frac{1}{2} & \frac{1}{2} \\
\hline
 & 1
\end{array} \quad ,
$$

where we note that M_1 is itself decomposable. Then if $M_0 = M_1 M_2$ we can show $M_0^n = M_1 M_3^{n-1} M_2$ where M_3 is just the midpoint rule and again we can consider M_0 to be algebraically stable, since the midpoint rule is, even though we can choose c and b so that M_0 does not satisfy (4).

In conclusion, we have seen in this paper that a great deal of insight can be gained into the study of the algebraic stability of numerical methods, and Runge-Kutta methods in particular, by analysing them in a more general setting as multivalue methods. We have resolved some apparent difficulties that occur with the trapezoidal rule and other so-called Runge-Kutta methods by examining them from a different viewpoint. Finally we have seen that the application of condition (4) blindly to Runge-Kutta methods can lead to erroneous conclusions. This was seen in section 5 where methods which were decomposable and whose abscissae were separated by integer values were not algebraically stable in terms of (4) but when they were written in a multivalue format and reduced were in fact algebraically stable.

References

1. K. Burrage, "High order algebraically stable Runge-Kutta methods," *BIT* 18, 373-383, 1978.

2. K. Burrage, "Efficiently implementable algebraically stable Runge-Kutta methods," *SIAM J. Num. Anal.* 19, 245-258, 1982.

3. K. Burrage, "High order algebraically stable multistep Runge-Kutta methods," *SIAM J. Num. Anal.*, 24, 106-115, 1987.

4. K. Burrage, "Order properties of implicit multivalue methods for ordinary differential equations," *IMA J. Num. Anal.*, 8, 43-69, 1988.

5. K. Burrage, "A study of the stability properties of a general class of differential equation methods," to appear in *IMACS Transactions on Scientific Computing 88*.

6. K. Burrage and J.C. Butcher, "Stability criteria for implicit Runge-Kutta methods," *SIAM J. Num. Anal.* 16, 46-57, 1979.

7. K. Burrage and J.C. Butcher, "Non-linear stability of a general class of differential equation methods," *BIT* 20, 185-203, 1980.

8. K. Burrage, W.H. Hundsdorfer and J.G. Verwer, "A study of B-convergence of Runge-Kutta methods," *Computing* 36, 17-34, 1986.

9. K. Burrage and P.M. Moss, "Simplifying assumptions for the order of partitioned multivalue methods," *BIT* 20, 452-465, 1980.

10. M. Crouzeix, "Sur la B-stabilité des méthodes de Runge-Kutta," *Numer. Math.* 32, 75-82, 1979.

11. G. Dahlquist, "On stability and error analysis for stiff non-linear problems," Rep. TRITA-NA-7508, Royal Inst. Tech., Stockholm, 1975.

12. G. Dahlquist and R. Jeltsch, "Generalized disks of contractivity for explicit and implicit Runge-Kutta methods," Rpt. TRITA-NA-7906, Dept. Comp. Sci., Royal Inst. Tech., Stockholm, 1979.

13. K. Dekker, J.F.B.M. Kraaijevanger and J. Schneid, "On the relation between algebraic stability and B-convergence for Runge-Kutta methods," Rep. 88-39 TUDelft.

14. R. Frank, J. Schneid, and C.W. Ueberhuber, "The concept of B-convergence," *SIAM J. Num. Anal.* 18, 753-780, 1981.

15. R. Frank, J. Schneid, and C.W. Ueberhuber, "Order results for implicit Runge-Kutta methods applied to stiff systems," *SIAM J. Num. Anal.* 22, 515-534, 1985.

16. E. Hairer, Ch. Lubich and M. Roche, "Error of Runge-Kutta methods for stiff problems studied via differential algebraic equations," *BIT,* 1988, to appear.

17. E. Hairer and G. Wanner, "Algebraically stable and implementable Runge-Kutta methods of high order," *SIAM J. Num. Anal.* 18, 1048-1108, 1981.

18. J.F.B.M. Kraaijevanger, "B-convergence of the implicit midpoint rule and the trapezoidal rule," *BIT* 25, 652-666, 1985.

19. J.F.B.M. Kraaijevanger and M.N. Spijker, "Algebraic stability and error propagation in Runge-Kutta methods," Rep. No. 1986-06, Inst. of App. Math. and Comp. Sci., University of Leiden, 1986.

20. I. Lie, "Multistep collocation for stiff systems," Ph.D. Thesis, Dept. of Numerical Mathematics, Trondheim, 1985.

21. A. Prothero and A. Robinson, "On the stability and accuracy of one-step methods for solving stiff systems of ordinary differential equations," *Math. Comp.* 28, 145-162, 1974.

22. M.N. Spijker, "Monotonicity and boundedness in implicit Runge-Kutta methods," Rep. No. 1985-15, Inst. of App. Math. and Comp. Sci., University of Leiden, 1985.

23. M.N. Spijker, "The relevance of algebraic stability in implicit Runge-Kutta methods," in *Numerical Treatment of differential equations,* third seminar Halle, Ed. K. Strehmel, Teubner Texte zur Mathematik 82 Leipzig, 158-164, 1986.

24. G. Wanner, "A short proof on nonlinear A-stability," *BIT* 16, 226-227, 1976.

Kevin Burrage

Department of Mathematics and Statistics

University of Auckland

Auckland, New Zealand

T.F. COLEMAN and Y. LI

A Global and Quadratic Affine Scaling Method for (Augmented) Linear L_1 Problems

1 Introduction

In [1] Coleman and Li present a new algorithm for the linear l_1 problem; convergence analysis is given and numerical results are provided. Here we extend the approach proposed in [1]: the linear l_1 function is augmented with an additional linear term. An application of the augmented form is the linear programming approach of Conn [4]. In this short report we motivate and develop an algorithm for the augmented l_1 problem. Results of numerical experiments are also provided.

Let C be an n-by-m matrix with $m > n$ and consider the overdetermined system

$$C^T x \approx \beta \tag{1.1}$$

where row i of C^T is denoted by c_i^T, for $i = 1 : m$. The linear l_1 problem is to find a vector x which is a solution to

$$\min_{x \in \Re^n} \{\bar{\psi}(x) \overset{def}{=} \sum_{i=1}^{m} |c_i^T x - \beta_i|\}. \tag{1.2}$$

We define the *augmented l_1* problem:

$$\min_{x \in \Re^n} \{\psi(x) \overset{def}{=} c^T x + \sum_{i=1}^{m} |c_i^T x - \beta_i|\}. \tag{1.3}$$

Note that the choice $c = 0$ yields the linear l_1 problem.

The function $\psi(x)$ is piecewise linear; ψ is not differentiable at any point x such that $c_i^T x = \beta_i$, for any index i. Most methods for minimizing piecewise linear functions, such as (1.3), are finite algorithms, related to the simplex method. Here we propose a method that, in real arithmetic, generates an infinite sequence of approximate solutions. Our method is related to recent "Karmarkar–like" methods for linear programming – especially Meketon's [7] interior adaptation for the linear l_1 problem. However, our proposed method differs fundamentally in two ways:

1. There is no "interior"; our method is not an interior point method. Instead, lines of nondifferentiability, i.e., $c_i^T x = \beta_i$, are avoided (and sometimes crossed);

2. In addition to guaranteed convergence, our proposed method is ultimately quadratically convergent (proven in [1] for the case $c = 0$).

2 Notations and Definitions

We use the sign function $sgn(v)$, where v is a vector, in the following sense: if $w = sgn(v)$,

$$w_i = \begin{cases} 1 & \text{if } v_i \geq 0, \\ -1 & \text{otherwise.} \end{cases} \tag{2.1}$$

We use e_i to denote the ith elementary vector. We let $\mathcal{A}(x)$ denote the indices of zero residuals at any point x, i.e.,

$$\mathcal{A}(x) = \{ i \mid c_i^T x - \beta_i = 0 \} \tag{2.2}$$

and $\mathcal{A}^c(x)$ denotes the complementary set. (We will suppress the argument when it is clear from context.)

Definition 1 *We say an augmented l_1 problem is* **primal nondegenerate** *if and only if, at any point x the vectors c_i, $i \in \mathcal{A}(x)$, are linearly independent.*

Definition 2 *We call λ a* **dual basic point** *if and only if $C\lambda = c$ and $|\{\lambda_i : |\lambda_i| = 1\}| \geq m - n$. We say an augmented l_1 problem is* **dual nondegenerate** *if and only if, at any dual basic point λ, $|\{\lambda_i : |\lambda_i| = 1\}| = m - n$.*

The optimal solution can be characterized in various ways. For example, x is optimal if and only if there exists $\lambda \in R^{|\mathcal{A}|}$ such that

$$c + \sum_{i \in \mathcal{A}^c} \mathrm{sgn}(c_i^T x - \beta_i) c_i = \sum_{i \in \mathcal{A}} \lambda_i c_i$$

where

$$-1 \leq \lambda_i \leq 1, \quad \forall i \in \mathcal{A}.$$

It is easy to verify that this formulation is equivalent to the following: x is optimal if and only if there exists $\lambda \in R^m$ such that

$$(c_i^T x - \beta_i) * (\mathrm{sgn}(c_i^T x - \beta_i) - \lambda_i) = 0, \quad i = 1 : m \tag{2.3}$$
$$C\lambda = c \tag{2.4}$$
$$-1 \leq \lambda_i \leq 1, \quad \forall i \in \mathcal{A}. \tag{2.5}$$

This characterization is especially interesting because the nonlinear equation (2.3) suggests the possibility of a Newton method.

Remark: Primal and dual nondegeneracy imply at solution (x, λ),

$$|\mathcal{A}(x)| = \mathrm{rank}(\{c_i \mid i \in \mathcal{A}\}) = n, \quad \text{and}$$

$$|\mathcal{A}^c(x)| = |\{\lambda_i : |\lambda_i| = 1\}| = m - n.$$

Moreover, the augmented linear l_1 problem has a unique minimizer, (x^*, λ^*), under these nondegeneracy assumptions.

3   An Equivalent Formulation

We now consider an alternative (but equivalent) formulation of the augmented l_1 problem. Let Z denote a matrix whose rows form a basis for the null space of $A \overset{def}{=} [C, c]$. Hence Z has dimensions $((m+1) - \text{rank}(A)) \times (m+1)$, $\text{rank}(Z) = (m+1) - rank(A)$, and

$$AZ^T = 0.$$

Then, defining $r_i = \beta_i - c_i^T x$, $i = 1 : m$, $r_{m+1} = -c^T x$, and $b = [\beta^T, 0]^T$, the augmented linear l_1 problem is equivalent to the following constrained augmented l_1 problem with $m+1$ variables r:

$$\min_{r \in \Re^{m+1}} \sum_{i=1}^{m} |r_i| - r_{m+1}$$

$$\text{subject to} \quad Zr = Zb. \tag{3.1}$$

It should be noted that implementation of our method does not necessarily involve the computation of matrix Z [1].

Once again optimality conditions can be expressed in different ways. For example, r is a solution if and only if there exists $\mu \in R^{|\mathcal{A}|}$ and $w \in R^{m+1-|\mathcal{A}|}$ such that

$$\sum_{\mathcal{A}^c} \text{sgn}(r_i)e_i - e_{m+1} = \sum_{\mathcal{A}} e_i \mu_i + Z^T w \tag{3.2}$$

$$Z(r - b) = 0$$

$$-1 \leq \mu_i \leq 1, \quad \forall i \in \mathcal{A}.$$

Therefore, if we define $\lambda = Z^T w$, an alternative expression of optimality conditions is

$$r_i(g_i - \lambda_i) = 0, \quad i = 1 : m + 1 \tag{3.3}$$

$$Z(r - b) = 0 \tag{3.4}$$

$$-1 \leq \lambda_i \leq 1, \quad i \in \mathcal{A}. \tag{3.5}$$

where $g_{m+1} = -1$ and $g_i = \text{sgn}(r_i)$, $i = 1 : m$.

This formulation is important to our design of a quadratically convergent algorithm: equations (3.3) and (3.4) lead to a Newton system (introduced in Section 5).

4   An Affine Scaling Method

We now present a new affine scaling method for the augmented linear l_1 problem. Our approach maintains feasibility with respect to the equality constraints in (3.1); the scaling involved in the computation of a descent direction is determined by the distance the current point is from the lines of nondifferentiability (i.e., $r_i = 0$ for some $1 \leq i \leq m$). It is this notion that replaces the more standard " distance from infeasibility " definition used in most interior methods. However, since the differentiable region in (3.1) is not connected, and since we do not know how to immediately identify a connected differentiable region adjacent to the optimal point, our algorithm must also allow for the ability to cross lines of nondifferentiability.

Let D be a positive diagonal matrix: e.g., $D = \mathrm{diag}\{|r|^{\frac{1}{2}} + \|r\|_\infty e_{m+1}\}$.

Assuming we are in the differentiable region (i.e., $r_i \neq 0$, $i = 1:m$), we can define a descent direction by the following "trust region problem":

$$\min_{d \in R^m} g^T d$$
$$\text{subject to} \quad Zd = 0 \tag{4.1}$$
$$\|D^{-1}d\|_2 \leq \delta$$

where δ is a positive number reflecting the trust region size; $g = g(r)$ is the gradient of $\psi(r) = \sum_{i=1}^m |r_i| - r_{m+1}$; the matrix D defines the shape of the ellipsoid. Therefore, for $i = 1:m$, the ellipsoid is short in directions corresponding to components of r_i close to zero, and long in directions corresponding to relatively large $|r_i|$; the ellipsoid is always relatively long in the direction corresponding to r_{m+1}. The solution to (4.1) is of the form $d_* = \alpha d$ where

$$d = -D^2(g - Z^T(ZD^2Z^T)^{-1}ZD^2g), \tag{4.2}$$

and $\alpha = \alpha(\delta)$, for small $\delta > 0$.

Instead of choosing δ a priori, we compute d as suggested by (4.2) and then define α using a piecewise linear minimization technique along the ray d (allowing for the ability to cross lines of nondifferentiability). Hence we must determine all nonnegative breakpoints,

$$J = \{\alpha_i : \alpha_i = -\frac{r_i}{d_i}, \ r_id_i < 0, \ i = 1:m\},$$

and then determine the minimizer in direction d:

$$\alpha_* = \min_{\alpha > 0} \psi(r + \alpha d) = \min_{\alpha \in J} \psi(r + \alpha d).$$

This is done by considering each breakpoint in J in turn, adjusting the gradient to reflect a step just beyond the breakpoint and then determining if d continues to be a descent direction for ψ. For example, if α_j is the smallest positive breakpoint, then a step just beyond this point yields the following gradient:

$$g^+ = g - 2g_je_j.$$

If $(g^+)^T d < 0$ the next breakpoint is considered, etc. Of course we cannot step all the way to the minimizer (since points of nondifferentiability must be avoided) and therefore we compute $\tilde{\alpha} = \max_{J \cup \{0\}}\{\alpha_i : 0 \leq \alpha_i < \alpha_*\}$ and take as our steplength

$$\alpha = \tilde{\alpha} + \tau(\alpha_* - \tilde{\alpha}) \tag{4.3}$$

where $0 < \tau < 1$.

The linear algorithm follows. Let r^0 be an initial differentiable point satisfying $Zr^0 = Zb$; $k \leftarrow 0$.

Algorithm 1

Step 1 Define $D^k = \mathrm{diag}\{|r^k|^{\frac{1}{2}} + \|r\|_\infty e_{m+1}\}$ and $g^k = \nabla\psi(r^k)$.

Step 2 Compute

$$d^k = -(D^k)^2\{g^k - (Z^k)^T[Z^k(D^k)^2(Z^k)^T]^{-1}Z^k(D^k)^2 g^k\}.$$

Step 3 Do a line search on the piecewise linear function $\psi(\alpha)$, as described above, to determine α^k. Then

$$r^{k+1} \leftarrow r^k + \alpha^k d^k, \quad k \leftarrow k + 1.$$

Remarks:
i) Note that at any point r^k, $\psi(r)$ is differentiable and $g^k_{m+1} = -1, g^k_i = \mathrm{sgn}(r^k_i), i = 1 : m$.
ii) Note that the linesearch maintains $D^k > 0$; therefore, d^k is well-defined and the line-search guarantees $\psi(r_{k+1}) < \psi(r_k)$.

5 A Local Newton Process

A Newton process for the augmented l_1 problem can be defined by considering conditions (3.3) and (3.4). First, consider (3.3) which we rewrite as

$$\mathrm{diag}(r)(g - Z^T w) = 0. \tag{5.1}$$

In general this system is not differentiable due to the discontinuities caused when $r_i = 0$, for $i = 1 : m$. However, system (5.1) can certainly be differentiated at any point with no zero residual. Moreover, in a neighbourhood of a solution r^*, $\mathrm{sgn}(r^*_i)$ will remain constant for any $i \in \mathcal{A}^c(r^*)$. Therefore, in a neighbourhood of r^*, discontinuities will be due only to the active equations, $i \in \mathcal{A}(r^*)$. As we formally establish in [1] (for the case $c = 0$), these discontinuities do not impede the local quadratic convergence behaviour of a Newton process. Therefore, define $g = \nabla\psi(r)$, $D_r = \mathrm{diag}(r)$, and $D_\lambda = \mathrm{diag}(g - Z^T w)$; differentiate (5.1) and (3.4) to define a Newton correction,

$$\begin{bmatrix} D_\lambda & -D_r Z^T \\ Z & 0 \end{bmatrix}\begin{bmatrix} \Delta r \\ \Delta w \end{bmatrix} = \begin{bmatrix} -D_r(g - Z^T w) \\ 0 \end{bmatrix}. \tag{5.2}$$

Define $d_N = \Delta r$; it is easy to prove that the Newton step with respect to r is equal to

$$d_N = -A^T(AD_r^{-1}D_\lambda A^T)^{-1}Ag. \tag{5.3}$$

It is now clear that the Newton step is similar in form to the linear step. In particular, the linear step (4.2) can be expressed as

$$d = -A^T(AD^{-2}A^T)^{-1}Ag. \tag{5.4}$$

It is this similarity in form that yields a smooth transition from the linear algorithm to the Newton method. Before presenting this procedure, it is worth noting that the matrix $AD_r^{-1}D_\lambda A^T$ is positive definite when (r, w) is close enough to the solution (r^*, w^*) and $\prod_{i=1}^m r_i \neq 0$. We refer the interested reader to [1] for a proof of the theorem when $c = 0$. Since $\lambda_{m+1}^* = -1$, the proof can be easily extended to the more general problem discussed in this paper.

Theorem 1 *Assume (r^*, w^*) is a solution and the augmented l_1 problem is both primal and dual nondegenerate. Then there exists a neighbourhood of (r^*, w^*) such that when (r, w) is within this neighbourhood, and no component r_i is zero for $i = 1 : m$, the matrix $AD_r^{-1}D_\lambda A^T$ is positive definite.*

Theorem 1 implies that Newton directions are descent directions in a neighbourhood of the solution. However, in general, the Newton step is not a descent direction. Therefore, as proposed in the next section, a modified Newton algorithm is needed.

6   A Hybrid Method

The Newton and linear directions, introduced in the previous sections, have similar forms but differ in their definitions of diagonal matrix. Consider now:

$$D^2 = |D_\gamma D_\theta^{-1}| \tag{6.1}$$

where $D_\gamma = \mathrm{diag}(|r| + \|r\|_\infty e_{m+1})$, $D_\theta = \theta \cdot \mathrm{diag}(g) + (1-\theta)D_\lambda$, and $0 \leq \theta \leq 1$. Clearly if $\theta = 1$ then $D^2 = |D_\gamma|$ and d is the linear step proposed in Section 4, i.e., Algorithm 1. On the other hand, as $\theta \to 0$,

$$A^T D^{-2} A^T \to AD_r^{-1}D_\lambda A^T,$$

(which, by Theorem 1, is positive definite in a neighbourhood of the solution) and so the Newton direction is approached asymptotically (i.e., as $\theta \to 0$).

Our remaining task is to define $\theta \in [0, 1]$ so that $\theta \to 0$ if and only if $(r, \lambda) \to (r^*, \lambda^*)$. Two possible choices are:

$$\theta = \frac{\frac{\sqrt{\sum_{i=1}^{m+1}(r_i(g_i - \lambda_i))^2}}{\|r^0\|_\infty} + e^T \max\{|\lambda| - e, 0\}}{\rho + \frac{\sqrt{\sum_{i=1}^{m+1}(r_i(g_i - \lambda_i))^2}}{\|r^0\|_\infty} + e^T \max\{|\lambda| - e, 0\}}, \tag{6.2}$$

$$\theta = \frac{\max\{\max(\frac{|r_i(g_i - \lambda_i)|}{\|r^0\|_\infty}), \max\{\max\{|\lambda| - e, 0\}\}\}}{\rho + \max\{\max(\frac{|r_i(g_i - \lambda_i)|}{\|r^0\|_\infty})), \max\{\max\{|\lambda| - e, 0\}\}\}}, \tag{6.3}$$

where $0 < \rho < 1$. Clearly θ is bounded above by 1; assuming $Z(r - b) = 0$ then $\theta = 0$ if and only if $(r, \lambda) = (r^*, \lambda^*)$. The following result indicates that, unless optimal, D_θ has a zero-free diagonal. Consequently, when $\theta \neq 0$ and no component r_i is zero for $i = 1 : m$, D^2 is a *positive* diagonal matrix and so d is a descent direction. A proof for this result when $c = 0$ can be found in [1].

6

Theorem 2 *Suppose $0 < \rho < 1$. Assume θ is defined by either (6.2) or (6.3). Then D_θ satisfies*

$$|D_\theta| \geq \min\{\theta, (1 - \rho)\theta\}I.$$

Therefore, D_θ is nonsingular when $\theta \neq 0$; consequently, either definition of θ leads to a descent direction. Moreover, θ induces a smooth transition from the linear step to the Newton step.

7 Computation of the Hybrid Step

We briefly consider three different methods for computing the hybrid steps.

The hybrid step can be viewed as an approximate Newton step. Assuming (r, w) is our current guess, where $Zr = Zb$, the approximate Newton step, in analogy with (5.2), can be expressed as

$$\begin{bmatrix} \text{diag}(g)|D_\theta| & -D_\gamma Z^T \\ Z & 0 \end{bmatrix} \begin{bmatrix} d \\ d_w \end{bmatrix} = \begin{bmatrix} -D_\gamma(g - Z^T w) \\ 0 \end{bmatrix}.$$

Note that d is the hybrid direction defined above. The dimension of this system is quite large, $(2(m + 1) - n)$-by-$(2(m + 1) - n)$, and can be reduced by implicitly satisfying $Zd = 0$:

$$Solve \ [\text{diag}(g)|D_\theta|A^T, -D_\gamma Z^T] \begin{bmatrix} d_x \\ d_w \end{bmatrix} = -D_\gamma(g - Z^T w) \tag{7.1}$$

and then

$$d \leftarrow A^T d_x. \tag{7.2}$$

After solving this full-space system and obtaining (d_x, d_w), the dual variables can be updated $w^+ \leftarrow w + D_\gamma Z^T d_w$; the primal variables are updated using a linesearch along direction d, $r^+ \leftarrow r + \alpha d$.

Note that as $\theta \to 0$, i.e., as we converge to the solution, the matrix in (7.1) approaches $(D_{\lambda^*} A^T, -D_{\gamma^*} Z^T)$ which, under nondegeneracy assumptions, is full rank (and bounded). This is the advantage of using the full-space implementation: the limit matrix is well-behaved. The major disadvantage is that the system is still large, $(m + 1)$-by-$(m + 1)$, and requires the computation of Z.

It is possible to reduce the dimension of the system to be solved (and still compute the same correction) by realizing that what is needed is either d_x and $Z^T d_w$ or w and $A^T d_x$ but not both d_x and d_w. We give the formulae for the computation here; detail on their derivation is given in [1]. One possibility is the range-space computation:

$$Solve \ D^{-1} A^T d_x \overset{\text{l.s.}}{=} Dg \tag{7.3}$$

and

$$d \leftarrow -A^T d_x, \quad \lambda^+ \leftarrow g + D^{-2}d, \tag{7.4}$$

where D is defined by (6.1). The main computational work is the solution of the linear least squares problem of order $(m+1)$-by-n. Note that Z is not required; this approach is particularly attractive when n is small or A is sparse. The primal variables are updated via a linesearch: $r^+ \leftarrow r + \alpha d$.

A second possibility is the null space implementation: first, compute Z and then

$$Solve \quad DZ^T w^+ \overset{\text{l.s.}}{=} Dg \tag{7.5}$$

and

$$A^T d_x \overset{def}{=} d \leftarrow -D^2(g - Z^T w^+). \tag{7.6}$$

In this case an $(m+1)$-by-$((m+1)-n)$ least squares system is solved at each iteration; hence, this approach is attractive when n is almost as large as m. Note that Z is computed only once; moreover, it is often possible to compute a sparse Z given that A is sparse (e.g., [2,3]).

We conclude this section with a presentation of the simple hybrid algorithm. Bear in mind that there are a number of numerical concerns, such as unequal row scaling and stopping criteria, that must be taken care of before this algorithm can be reliably used.

In order to ensure quadratic convergence, we require $\alpha^k \to 1$ [1]. Therefore, the linesearch algorithm differs slightly from the procedure presented in Section 2: update (4.3) is replaced with

$$\alpha = \tilde{\alpha} + \max\{\tau, 1 - \theta\}(\alpha_* - \tilde{\alpha}). \tag{7.7}$$

Let r^0 be an initial differentiable point satisfying $Z(r^0 - b) = 0$; $k \leftarrow 0$; compute an initial point w^0.

Algorithm 2

Step 1 Compute θ^k from (6.2) or (6.3). Define D^k from (6.1) and $g^k \leftarrow \nabla\psi(r^k)$.

Step 2 Compute d^k and w^{k+1} using one of the three methods above .

Step 3 Do a line search on the piecewise linear function $\psi(\alpha)$ (as described in Section 4 and using (7.7)) to determine α^k,

$$r^{k+1} \leftarrow r^k + \alpha^k d^k, \quad k \leftarrow k + 1.$$

8 Numerical Testing

In this section we provide numerical results concerning Algorithm 2, the hybrid method (*New*). In particular, we first compare our range-space implementation of this algorithm - see Section 7 - with our implementation of the interior point algorithm described in [5] (for the case $c = 0$). We denote the latter method by *Dual*. Our comparisons are based on the number of iterations: since the dominant work in both implementations is the solution of a linear least squares problem at each iteration, of size $(m+1)$-by-n,the number of iterations accurately reflects the overall relative computational cost.

The dependent variable θ^k is a measure of the distance from optimality: our stopping criterion for both algorithms is:

$$\theta^k < 10^{-13},$$

where machine precision on our system is approximately 10^{-16}. In our experiments, we compute θ^k as defined by (6.3). The origin is a natural starting point for algorithm *Dual*; the starting point for *New* is computed as follows:

$$r^0 \leftarrow b - A^T x^0, \quad \text{where} \quad A^T x^0 \stackrel{\text{l.s.}}{=} b$$

$$\lambda^0 \leftarrow \frac{\tau}{\max\{|r|\}} * r^0.$$

The settings of the parameters for Algorithm 2, *New*, are:

$$\tau \leftarrow .975, \quad \rho \leftarrow .99$$

Algorithm *Dual* also uses $\tau = .975$ in the linesearch algorithm; there are no other parameters.

We have implemented the methods in PRO-MATLAB [6] using SUN 3/50 and 3/160 workstations. The linear least squares subproblems are solved with the orthogonal QR-factorization using row interchanges for greater stability (row interchanges are advisable when a least squares problem involves widely varying row scalings [8].) No account was made of sparsity in our experiments.

We have generated several kinds of test problems. First, since l_1 minimization is often used in a function approximation context [7], we have tried several such problems. We have also generated several random l_1 problems of varying dimensions. Finally, we experiment with several randomly generated augmented l_1 problems (i.e. $c \neq 0$).

Problem 1. Approximate $f(z)$, evaluated at $z = 0, \frac{1}{m}, ..., 1$, by a polynomial of degree $n - 1$:

$$\phi(z) = \sum_{j=1}^{n} \alpha_j \, z^{j-1} \, ,$$

where $x = (\alpha_1, ..., \alpha_n)$.

$n = 5, f_1(z) = \exp(z)$.

Number of Steps		
m	Dual	New
100	18	8
200	19	11
400	21	10
600	21	12

$n = 6, f_2(z) = \sin(z)$

68

Number of Steps		
m	Dual	New
100	15	9
200	16	8
400	16	9
600	17	9

Problem 2. Approximate $f(z)$, evaluated at $z = 0, \frac{1}{m}, ..., 1$, by a polynomial of degree $n = 10$:

$$\phi(z) = \sum_{j=1}^{n} \alpha_j \, z^{j-1} \, ,$$

where $x = (\alpha_1, \cdots, \alpha_n)$.

$$f_3(z) = \exp(z) + \begin{cases} 1 & \text{if } 0.1 < z \le 0.2 \\ 0 & \text{otherwise .} \end{cases}$$

Number of Steps		
m	Dual	New
40	30	13
100	27	12
200	29	12
500	33	12

Approximation problems often result in a very large number of rows, m. The number of iterations required by algorithm *New* is essentially insensitive to the growth of m, as demonstrated in Problem 3.

Problem 3. Approximate $f(z)$, evaluated at $z = 0, \frac{1}{m}, ..., 1$, by a polynomial of degree $n - 1$:

$$\phi(z) = \sum_{j=1}^{n} \alpha_j \, z^{j-1} \, ,$$

where $x = (\alpha_1, ..., \alpha_n)$.

$n = 6$.

	Number of Steps			
m	$\sqrt{1+z}$	$\sin \frac{\pi z}{2}$	$\log(1+z)$	$\sinh(z)$
100	10	8	8	9
200	9	8	9	9
300	8	7	9	9
400	8	7	9	9
500	11	8	10	9
600	9	10	10	8
700	10	10	10	9
800	12	11	10	10
900	10	9	9	9
1000	9	9	8	9
1500	11	9	9	10
2000	10	12	10	11

Problem 4. Random l_1 problems (i.e. $c = 0$): We generated the elements of matrix C and right-hand-side β in a uniform random manner.

$m = 50$

	Number of Steps	
n	Dual	New
10	25	9
20	25	12
30	25	10
40	23	9

$m = 100$

	Number of Steps	
n	Dual	New
10	25	10
20	26	13
40	26	13
50	26	11
70	25	11
90	23	10

$m = 150$

Number of Steps		
n	Dual	New
10	26	10
20	27	13
50	26	17
70	27	11
90	27	13
135	24	12
140	23	9

$m = 200$

Number of Steps		
n	Dual	New
10	26	13
20	27	17
50	27	16
70	27	19
100	26	15
140	26	14
160	25	14
190	24	9

The next experiment indicates that the performance of the algorithm (*New*) is not adversely affected when $c \neq 0$. (We do not compare to *dual* here since *dual* applies to l_1 problems only (i.e., $c = 0$). Note the relative insensitivity to n.

Problem 5. Random augmented l_1 problems (i.e. $c \neq 0$): We generated the elements of matrix C, right-hand-side β and c in a uniform random manner.

$m = 50$

Number of Steps	
n	New
10	10
20	13
30	11
40	9

$m = 100$

Number of Steps	
n	New
10	12
20	12
40	14
50	14
70	12
90	9

$m = 200$

Number of Steps	
n	New
20	15
50	17
70	17
100	16
140	12
160	14
190	12

9   Conclusions

We have presented a new algorithm for the augmented linear l_1 problem. For l_1 problems (i.e., $c = 0$), the algorithm appears to be consistently superior to the interior point approach [5], typically requiring a factor of two to three fewer iterations. (The iterations are comparable in cost and so the overall running times compare in a similar way.) Moreover, the algorithm solves discrete l_1 approximation problems efficiently: the number of iterations appears to be independent of the number of data points. The algorithm appears to work equally well when $c \neq 0$ and is not sensitive to n; therefore, the proposed algorithm may be useful for large linear programming problems.

10   Acknowledgement

The authors would like to thank their colleague, Mike Todd, for many helpful suggestions.

References

[1] Thomas F. Coleman and Yuying Li. A global and quadratic affine scaling method for linear l_1 problems. Technical Report 89–1026, Computer Science Department, Cornell University, 1989.

[2] Thomas F. Coleman and Alex Pothen. The null space problem I: Complexity. *SIAM Journal on Algebraic and Discrete Methods*, 7:527–537, 1987.

[3] Thomas F. Coleman and Alex Pothen. The null space problem II: Algorithms. *SIAM Journal on Algebraic and Discrete Methods*, 8:544–563, 1987.

[4] Andrew R. Conn. Linear programming via a non-differentiable penalty function. *SIAM Journal on Numerical Analysis*, 13:224–241, 1988.

[5] M. S. Meketon. Least absolute value regression. Technical report, AT & T Bell Laboratory, 1987.

[6] C. B. Moler, J. Little, S. Bangert, and S. Kleiman. *ProMatlab User's guide*. Math-Works, Sherborn, MA, 1987.

[7] M. R. Osborne. *Finite Algorithms in Optimization and Data Analysis*. John Wiley & Sons, 1985.

[8] Charles Van Loan. On the method of weighting for equality-constrained least squares problems. *SIAM Journal on Numerical Analysis*, 22:851–864, 1985.

Thomas F. Coleman & Yuying Li
Department of Computer Science
Cornell University
Ithaca
New York 14853
U.S.A.

P.J. DEUFLHARD

Uniqueness Theorems for Stiff ODE Initial Value Problems

Abstract. Reliable numerical algorithms are, in one way or the other, appropriate implementations of uniqueness theorems of the underlying analytic problem. In *non-stiff* numerical integration, the associated uniqueness theorem for ODE initial value problems is the well-known *Picard-Lindelöf theorem* - which characterizes the growth of the solution by means of the Lipschitz constant of the right-hand side of the ODE system. For *stiff* integration, this characterization is known to be inappropriate, but an associated analytic uniqueness theorem in terms of a different characterization seems to be missing - despite of the enormous amount of literature dealing with stiff integration. Such an analytic uniqueness theorem should then serve as a theoretical frame for investigations of *discretization methods* for stiff ODE systems. After some preliminary considerations two variants of the intended theorem are derived on the basis of affine invariant convergence theorems for *Newton's method* in function space. The emerging theoretical characterization is then used to derive comparison theorems for the implicit Euler discretization. The new theory has an impact also on the implementation of stiff integrators. A brief review of numerical experiments with semi-implicit extrapolation integrators is included.

1. Preliminary Considerations

Usually, an initial value problem (IVP) for ordinary differential equations (ODE's) is given in the form

$$y' = f(y), \quad y(0) = y_0 \ . \tag{1.1}$$

For the subsequent proofs, however, the equivalent formulation in terms of a *homotopy* is preferable:

$$F(y, \tau) := y(\tau) - y_0 - \int_0^\tau f(y(t)) dt = 0 \ . \tag{1.2}$$

Herein the interval length $\tau \geq 0$ represents the embedding parameter. Let Γ denote some neighborhood of the graph of a solution of (1.1). Then *Peano's existence theorem* requires that

$$L_0 := \sup_\Gamma \|f(y)\| < \infty \ . \tag{1.3}$$

In order to prove *uniqueness*, the standard approach is to construct the so-called *Picard iteration*

$$y^{i+1}(\tau) = y_0 + \int_0^\tau f(y^i(t)) dt \tag{1.4}$$

to be started with $y^0(t) \equiv y_0$. From this fixed point iteration, one immediately derives

$$\|y^{i+1}(\tau) - y^i(\tau)\| \le \int_0^\tau \|f(y^i(t)) - f(y^{i-1}(t))\| dt \ . \tag{1.5}$$

Hence, in order to study contraction, the most natural theoretical characterization is in terms of the Lipschitz constant L_1 defined by

$$\|f(u) - f(v)\| \le L_1 \|u - v\| \ . \tag{1.6}$$

With this definition, the sequence $\{y^i\}$ can be shown to converge to some solution y^* such that

$$\|y^*(\tau) - y_0\| \le L_0 \tau \varphi(L_1 \tau) \tag{1.7.a}$$

with

$$\varphi(s) := \begin{cases} (\exp(s) - 1)/s & s \ne 0 \\ 1 & s = 0 \ . \end{cases} \tag{1.7.b}$$

Moreover, y^* is unique in Γ. This is the main result of the well–known Picard-Lindelöf theorem.

A similar term arises in the analysis of one-step discretization methods for ODE-IVP's. Let $p \ge 1$ denote the consistency order of such a method, then the discretization error between the analytic solution y and the discrete solution y_h can be represented in the form (see, for example, the recent textbook [11]):

$$\|y_h(\tau) - y(\tau)\| \le C_p \cdot h^p \cdot \tau \cdot \varphi(L_1 \tau) \tag{1.8}$$

Herein C_p typically is a bound of a higher order derivative of f. In order to bound the discretization error, a condition of the kind

$$L_1 \tau \le C, \ \ C = \mathcal{O}(1) \tag{1.9}$$

is needed. Therefore, this characterization is only appropriate for *non-stiff* discretization methods.

In the beginning of the study of *stiff* integration, it was first thought that the use of *implicit discretization methods* would be the essential item to overcome the observed difficulties - see, for instance, the early pioneering paper by Dahlquist [4]. In the next stage of insight, it was recognized that the solution method for the thus arising algebraic equations is equally important: the early paper of Liniger/Willoughby [12] pointed out that fixed point iteration only based on f-evaluations for the algebraic equations would once more bring in condition (1.9), whereas a *Newton-like iteration* were just the method of choice. Much later, so-called semi-implicit or linearly-implicit discretization methods (e.g. Rosenbrock methods, W-methods, extrapolation methods) were constructed that only apply 1 Newton-like iteration per discretization step. Therefore, at the present stage of the development, *the essence of non-stiff integration seems to be that only f is sampled, whereas stiff integration additionally requires sampling of the Jacobian f_y or an approximation of it.*

With these preparations the natural approach towards the intended uniqueness theorem seems to be replacing the Picard iteration (1.4) by a Newton iteration in function space. For the *ordinary* Newton method one has

$$
\begin{aligned}
F_y(y^i)\Delta y^i &= -F(y^i) \\
y^{i+1} &= y^i + \Delta y^i
\end{aligned}
\tag{1.10}
$$

or, in more explicit notation:

$$
\begin{aligned}
\Delta y^i(\tau) - \int_0^\tau f_y(y^i(t))\Delta y^i(t)dt = \\
= - \left[y^i(\tau) - y_0 - \int_0^\tau f(y^i(t))dt \right].
\end{aligned}
\tag{1.10'}
$$

Obviously, the above iteration requires global information in terms of f_y rather than just pointwise information as in numerical stiff integration. Therefore, the *simplified* Newton method will be the method of choice: one just replaces

$$
F_y(y^i) \to F_y(y^0), \quad y^0(t) \equiv y_0
$$

or, equivalently,

$$
f_y(y^i(t)) \to f_y(y_0) =: A .
\tag{1.11}
$$

Insertion of (1.11) into (1.10') then leads to

$$
\begin{aligned}
y^{i+1}(\tau) - A \int_0^\tau y^{i+1}(t)dt = \\
= y_0 + \int_0^\tau [f(y^i(t)) - Ay^i(t)]dt
\end{aligned}
\tag{1.12}
$$

Upon comparing (1.12) with (1.4), one may recognize that (1.12) may be regarded as a Picard iteration for the equivalent ODE

$$
y' - Ay = \bar{f}(y) := f(y) - Ay, \quad y(0) = y_0
\tag{1.13}
$$

Starting with (1.13), a so-called *deflated* Lipschitz constant has been introduced in [2, 5]:

$$
\|\bar{f}(u) - \bar{f}(v)\| \le \bar{L}_1 \|u - v\|
\tag{1.14}
$$

It is clear that this characterization will lead to the analogon of (1.9), namely

$$
\bar{L}_1 \tau \le C , \quad C = \mathcal{O}(1) .
\tag{1.15}
$$

On the other hand, the definition of $\bar{L}_1$ contains some additional τ-dependence, compare (3.7') in [5]. Hence, condition (1.15) is theoretically unsatisfactory. The alternative is to recall the above derivation and to analyze the iteration (1.12) in terms of Newton's method.

2. Newton-Type Uniqueness Theorems

In this section, the above Newton-like iteration (1.12) will be used to prove uniqueness theorems for ODE-IVP's.

Theorem 1 *With the notation above let $f \in C^1(D)$, $D \subseteq \mathbb{R}^n$. For the Jacobian $A := f_y(y_0)$ assume a one-sided Lipschitz condition of the form*

$$\langle u, Au \rangle \leq \mu \langle u, u \rangle \equiv \mu \|u\|^2 \; , \tag{2.1.a}$$

where $\langle \cdot, \cdot \rangle$ denotes an inner product in $\mathbb{R}^n$, which induces the $\mathbb{R}^n$-norm $\|\cdot\|$. In this norm, assume that

$$\|f(y_0)\| \leq L_0 \qquad \forall y \in D \tag{2.1.b}$$

$$\|f_y(u) - f_y(v)\| \leq L_2 \|u - v\| \qquad \forall u, v \in D \tag{2.1.c}$$

Then, for D sufficiently large, existence and uniqueness of the solution of the ODE–IVP (1.1) is guaranteed in $[0, \tau]$ for

$$\tau \text{ unbounded }, \text{ if } \mu\bar{\tau} \leq -1 \tag{2.2.a}$$

$$\tau \leq \bar{\tau}\Psi(\mu\bar{\tau}) \; , \text{ if } \mu\bar{\tau} > -1 \tag{2.2.b}$$

where $\bar{\tau} := (2L_0 L_2)^{-1/2}$ and

$$\Psi(s) := \begin{cases} \ln(1+s)/s & s \neq 0 \\ 1 & s = 0 \end{cases} \tag{2.2.c}$$

Proof. Upon performing the variation of constants, one may rewrite (1.12) as

$$\Delta y^i(\tau) = \int_{t=0}^{\tau} \exp(A(\tau - t))[f(y^i(t)) - \frac{d}{dt}y^i(t)]dt \tag{2.3}$$

where $\exp(At)$ denotes the matrix exponential characterizing the solution of (1.1) for $f = Ay$. Let $|\cdot|$ denote the standard C^0-norm:

$$|u| := \max_{t \in [0,\tau]} \|u(t)\| \; .$$

In order to study convergence, Theorem 3 of Deuflhard/Heindl [8] will be applied, which essentially requires that

$$|\Delta y^0| \leq \alpha \tag{2.4.a}$$

$$|F_y(y^0)^{-1}(F_y(u) - F_y(v))| \leq \omega|u - v| \tag{2.4.b}$$

$$\alpha\omega \leq \frac{1}{2} \tag{2.4.c}$$

The rest of the assumptions holds for D sufficiently large. The task is now to estimate α, ω and to assure (2.4.c). With $y^0(t) \equiv y_0$, the first Newton correction satisfies - compare (2.3):

$$\Delta y^0(\tau) = \int_{t=0}^{\tau} \exp(A(\tau - t)) f(y_0) dt$$

Hence

$$\|\Delta y^0(\tau)\| \leq \int_{s=0}^{\tau} \|\exp(As) f(y_0)\| ds \leq$$
$$\leq L_0 \int_{s=0}^{\tau} \exp(\mu s) ds = L_0 \tau \varphi(\mu \tau) =: \alpha(\tau) \tag{2.4.d}$$

with φ as introduced in (1.7.b).

In order to estimate $\omega(\tau)$, one introduces the operator norm in (2.4.b) by

$$z := F_y(y^0)^{-1}(F_y(v + w) - F_y(v))u$$
$$|z| \leq \omega \cdot |u| \cdot |w|$$

Once more by variation of constants, one obtains

$$\|z(\tau)\| \leq \int_{t=0}^{\tau} \|\exp(A(\tau - t))[f_y(v + w) - f_y(v)]u\| dt \ ,$$

which, similar as above, yields

$$\|z(\tau)\| \leq L_2 \cdot \tau \cdot \varphi(\mu \tau) \cdot |u| \cdot |w|$$

Hence, a natural definition is

$$\omega(\tau) := L_2 \tau \varphi(\mu \tau) \ . \tag{2.4.e}$$

Insertion into the Kantorovitch condition produces

$$(\tau \varphi(\mu \tau))^2 \leq (2L_0 L_2)^{-1} =: \bar{\tau}^2 \tag{2.5.a}$$

or, equivalently,

$$\tau \varphi(\mu \tau) \leq \bar{\tau} \ . \tag{2.5.b}$$

Then (2.2) is an immediate consequence. ∎

Remark. Note that the estimate (2.4.d) can be refined by introducing

$$\mu_0 := \frac{\langle f(y_0), A f(y_0) \rangle}{\langle f(y_0), f(y_0) \rangle} \ , \tag{2.6}$$

which immediately implies

$$\|\exp(As) f(y_0)\| \leq L_0 \exp(\mu_0 s) \ ,$$

and, in turn

$$\|\Delta y^0(\tau)\| \leq L_0 \tau \varphi(\mu_0 \tau) =: \alpha(\tau) \ .$$

Unfortunately, the easily computable quantity μ_0 cannot replace the less desirable quantity μ in (2.4.e).

Theorem 1 characterizes the local continuation property of solutions of (1.1) in terms of the two quantities μ and $\bar{\tau}$ for a given exact Jacobian $f_y(y_0)$. If, however, an approximation error

$$\delta A := A - f_y(y_0) \tag{2.7}$$

must be taken into account, then a further characterizing quantity will be needed.

Theorem 2 *Notation and assumptions as in Theorem 1, but with*

$$\|\delta A\| \leq \delta_0 \ , \quad \delta_0 \geq 0 \ . \tag{2.8}$$

Then the results (2.2) hold with $\bar{\tau}$ replaced by

$$\hat{\tau} := \bar{\tau}/(1 + \delta_o \bar{\tau}) \tag{2.9}$$

Proof. Once more, Theorem 3 of [8] is applied with $F_y(y^0)$ replaced by $M_F(y^0)$, which means replacing $f_y(y_0)$ by $A \neq f_y(y_0)$. With μ now associated with the Jacobian *approximation A* , the estimates for $\alpha(\tau), \omega(\tau)$ carry over. In addition, the assumptions (2.4) must be extended by

$$\|M_F(y^0)^{-1}(F_y(y^0) - M_F(y^0))\| \leq \bar{\delta}_0 < 1 \tag{2.10.a}$$

Upon defining

$$z := M_F(y^0)^{-1}(F_y(y^0) - M_F(y^0))u \ ,$$

a similar estimate as in the proof of Theorem 1 leads to

$$\|z(\tau)\| \leq \int_0^\tau \| \exp(A(\tau - t) \cdot \delta A \cdot u\|dt \leq \delta_0 \tau \varphi(\mu\tau)|u|$$

Hence, one obtains the condition

$$\bar{\delta}_0 := \delta_0 \tau \varphi(\mu\tau) < 1 \tag{2.10.b}$$

Insertion into the modified Kantorovitch condition

$$\frac{\alpha\omega}{(1 - \bar{\delta}_0)^2} \leq \frac{1}{2} \tag{2.10.c}$$

then yields

$$\tau\varphi(\mu\tau) \leq \bar{\tau}/(1 + \delta_0\bar{\tau}) =: \hat{\tau}$$

Note that condition (2.10.b) is automatically satisfied. ■

The above characterization in terms of μ and $\bar{\tau}$ (let $\delta_0 = 0$) may lead to some refined estimates by ignoring the derivation in terms of Newton's iteration – this fact was brought to the author's knowledge by W. Walter.

Theorem 3 *[14] Assumptions and notation as in Theorem 1. Then there exists a bounded solution $y(t)$ in $[0, \tau]$ with*

$$\tau \text{ unbounded, if } \mu\bar{\tau} \le -1 \tag{2.11.a}$$

$$\tau < \bar{\tau}\bar{\Psi}(\mu\bar{\tau}), \text{ if } \mu\bar{\tau} > -1 \tag{2.11.b}$$

where

$$\bar{\Psi}(s) := \begin{cases} \dfrac{1}{\sqrt{1-s^2}}\left[\pi - 2\arctan\left(\dfrac{s}{\sqrt{1-s^2}}\right)\right], & -1 < s < 1 \\ 2, & s = 1 \\ \dfrac{1}{\sqrt{s^2-1}}\ln\dfrac{s+\sqrt{s^2-1}}{s-\sqrt{s^2-1}}, & s > 1 \end{cases}$$

Proof. Let $L_0 > 0$ without loss of generality. Define

$$\rho^2 = \langle y(t) - y_0, y(t) - y_0 \rangle \equiv \|y(t) - y_0\|^2 \ .$$

Then

$$\rho\rho' = \frac{1}{2}(\rho^2)' = \langle y(t) - y_0, f(y(t)) \rangle \tag{2.12}$$

Upon using the characterization in terms of L_0, μ, L_2 one immediately obtains:

$$\begin{aligned} f(y) &= f(y_0) + A(y - y_0) + \\ &+ \int_{s=0}^{1}\left[f_y(y_0 + s(y - y_0)) - A\right](y - y_0)ds \end{aligned} \tag{2.13}$$

Insertion into (2.12) and division by $\rho > 0$ yields

$$\rho' \le L_0 + \mu\rho + \frac{L_2}{2}\rho^2, \quad \rho(0) = 0.$$

Let the associated majorizing equation be

$$\sigma' = L_0 + \mu\sigma + \frac{L_2}{2}\sigma^2 =: g(\sigma), \quad \sigma(0) = 0. \tag{2.14}$$

Then $\rho(t) \le \sigma(t)$ is known to hold for positive right-hand side [15]. For the subsequent case study, the following reformulation is useful:

$$g(\sigma) = L_0[1 - (\mu\bar{\tau})^2 + (\mu\bar{\tau} + L_2\bar{\tau}\sigma)^2]$$

The roots of g are

$$\sigma_{1,2} = \frac{1}{L_2\bar{\tau}}\left[-\mu\bar{\tau} \pm \sqrt{(\mu\bar{\tau})^2 - 1}\right]$$

Positive roots are only possible for

$$\mu\bar{\tau} \le -1 \ . \tag{2.15.a}$$

In this case, $\sigma(t)$ is bounded by σ_2 starting from $\sigma(0) = 0$. Hence

$$\rho(t) \leq \sigma_2 \tag{2.15.b}$$

which means the global existence of a bounded solution $y(t)$. This is (2.11.a).

For $\mu\bar{\tau} > -1$ and $\sigma \geq 0$ one has $g > 0$. Hence σ may be unbounded, which implies a restriction on τ:

$$\tau < \int_0^\infty \frac{d\sigma}{g(\sigma)} \tag{2.16}$$

Upon substituting

$$s := \mu\bar{\tau} + L_2\bar{\tau}\sigma$$

(2.16) is equivalent to

$$\tau < 2\bar{\tau} \int_{\mu\bar{\tau}}^\infty \frac{ds}{1 - (\mu\bar{\tau})^2 + s^2} \tag{2.16'}$$

Distinction of the cases

$$\mu\bar{\tau} \in [-1, +1), \quad \mu\bar{\tau} = 1, \quad \mu\bar{\tau} > 1$$

and standard integration confirms (2.11.b). ∎

A comparison of the functions Ψ from Theorem 1 and $\bar{\Psi}$ from Theorem 3 is given in Figure 1. In view of possible comparison theorems in the discrete case, the above constractive Theorems 1 and 2 will be preferred.

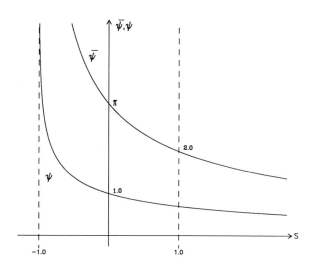

Figure 1 Comparison of Ψ (Theorem 1) and $\bar{\Psi}$ (Theorem 3).

Remark. The above condition $\mu\bar{\tau} \leq -1$, may be written as

$$\mu + \frac{1}{\bar{\tau}} \leq 0 \; , \tag{2.17}$$

a possible characterization of *nonlinear contractivity*, which comes out here in a rather natural way without further assumptions.

3. Stepsize Bounds for the Implicit Euler Discretization

The question to be studied next is to which extent the above structure of the underlying analytic problem is inherited after discretization. In order to avoid technical difficulties, the formalism will be exemplified herein by the simple case of the *implicit Euler discretization*. In each step of this discretization, one must solve the algebraic system

$$F(y, h) := y - y_0 - hf(y) = 0 \; , \tag{3.1}$$

which now represents a homotopy in $\mathbb{R}^n$ with embedding in terms of the stepsize h - say $h \geq 0$. The Newton-like iteration for solving this system is

$$
\begin{aligned}
(I - hA)\Delta y_i &= -(y_i - y_0 - hf(y_i)) \\
y_{i+1} &= y_i + \Delta y_i
\end{aligned}
\tag{3.2}
$$

where $\delta A := A - f_y(y_0) \neq 0$ will be assumed.

Theorem 4 *Assumptions and notation as in Theorem 1 and Theorem 2 above. Then the Newton-like iteration (3.2) for the implicit Euler discretization converges to a unique solution for all stepsizes.*

$$h \text{ unbounded } , \text{ if } \mu\hat{\tau} \leq -1 \tag{3.3.a}$$

$$h \leq \hat{\tau}\Psi_{\mathrm{IE}}(\mu\hat{\tau}) \; , \text{ if } \mu\hat{\tau} > -1 \tag{3.3.b}$$

where

$$\Psi_{\mathrm{IE}}(s) := (1 + s)^{-1} \tag{3.3.c}$$

Proof. Once more, Theorem 3 of [8] is applied, here to the homotopy (3.1). The Jacobian approximation $A \approx f_y(y_0)$ leads to the approximation

$$I - hA =: M_F(y_0) \approx F_y(y_0) \; ,$$

which is used in the definition of the affine invariant Lipschitz constant

$$\|M_F(y_0)^{-1}(F_y(u) - F_y(v))\| \leq \omega(h)\|u - v\| \; , \tag{3.4.a}$$

the first correction bound

$$\|\Delta y_0\| = \|M_F(y_0)^{-1}F(y_0)\| \leq \alpha(h) \; , \tag{3.4.b}$$

and the approximation measure

$$\|M_F(y_0)^{-1}(M_F(y_0) - F_y(y_0))\| \le \bar{\delta}_0(h) < 1 \qquad (3.4.c)$$

With these definitions, the modified Kantorovitch condition reads

$$\frac{\alpha\omega}{(1 - \bar{\delta}_o)^2} \le \frac{1}{2} \; . \qquad (3.4.d)$$

Upon using similar techniques as in the proofs of Theorem 1 and Theorem 2 above, one comes up with the estimates:

$$
\begin{aligned}
\alpha(h) &:= hL_0/(1 - \mu h) \\
\omega(h) &:= hL_2/(1 - \mu h) \\
\bar{\delta}_0(h) &:= h\delta_0/(1 - \mu h)
\end{aligned}
\qquad (3.5)
$$

where L_0, L_2, δ_0 are defined as in section 2. Insertion into (3.4.d) yields, for $\mu h < 1$:

$$\frac{h}{1 - \mu h} \le \hat{\tau} \; , \qquad (3.6)$$

or, equivalently,

$$h \le \hat{\tau}/(1 + \mu\hat{\tau}) \; . \qquad (3.6')$$

This is just (3.3). Finally, note that for $\mu > 0$

$$\mu h \le \mu\hat{\tau}/(1 + \mu\hat{\tau}) < 1 \; ,$$

which assures the above requirement. ∎

Upon comparing the comparable theorems 1 and 4, the essence seems to be contained in the characterizing functions Ψ and Ψ_{IE} - see Fig. 2. Common features of these functions are:

a) $\Psi(0) = \Psi_{IE}(0) = 1$,

b) Ψ and Ψ_{IE} have a pole at $s = -1$, $\qquad (3.7)$

c) Ψ and Ψ_{IE} are monotonically decreasing.

In view of Theorem 3, one might also expect a refinement of Theorem 4. The discrete analog of (2.12) is then

$$\frac{\rho^2}{h} = \langle y - y_0, f(y) \rangle, \quad \rho = \rho(h). \qquad (3.8)$$

Insertion of (2.13) leads to

$$\rho \le h(L_0 + \mu\rho + \frac{L_2}{2}\rho^2), \quad \rho(0) = 0 \; . \qquad (3.9)$$

The associated quadratic equation

$$\frac{hL_2}{2}\sigma^2 + (\mu h - 1)\sigma + hL_0 = 0$$

has the roots

$$\sigma_{1,2} = \frac{1 - \mu h}{h L_2} \left[1 \pm \sqrt{1 - \left(\frac{h}{\bar\tau(1 - \mu h)} \right)^2} \right] .$$

Positive roots occur for $\mu h < 1$ and

$$\frac{h}{\bar\tau(1 - \mu h)} \le 1$$

which is equivalent to

$$h \le \frac{\bar\tau}{1 + \mu\bar\tau} \qquad . \tag{3.10}$$

This is just the result (3.3.b) of Theorem 4 - hence no improvement comes out in this case.

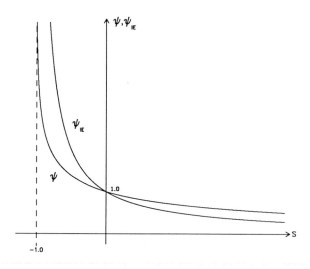

Figure 2 Comparison of functions Ψ (analytic case) and Ψ_{IE} (implicit Euler discretization).

4. Computational Estimation of Stepsize Bounds in Stiff Integrators

In view of reliable actual computation one will aim at estimating the quantities μ, $\bar\tau$, δ_0 in numerical stiff integrators. The estimation of μ will be replaced by the actual evaluation of μ_0 from (2.6) – which is an easy task in all stiff integrators. The cheap estimation of $\bar\tau$ and δ_0, however, is hard, since these quantities enter in coupled form into available computational results. This fact is now illustrated in the context of two extrapolation integrators.

In the semi-implicit Euler discretization [5] let

$$\Delta y_0 = h(I - hA)^{-1} f(y_0)$$
$$y_1 = y_0 + \Delta y_0 \tag{4.1}$$

Since $\bar{\tau}$ contains second order information, at least two Newton iterations would be required in the implicit Euler discretization. Then the second Newton correction would read:

$$\Delta y_1 := (I - hA)^{-1}(hf(y_1) - \Delta y_0). \tag{4.2}$$

The computation of Δy_1 in the semi-implicit Euler discretization requires some (small) additional work, whereas the semi-implicit mid-point rule [2] contains the information Δy_1 implicitly – compare [5], (3.28).

Assume now that Δy_1 is given. Straightforward Taylor expansion for

$$w := \frac{1}{h}(I - hA)\Delta y_1 \tag{4.3.a}$$

yields

$$w := -\delta A \Delta y_0 + \frac{1}{2} \int_{s=0}^{1} f_{yy}(y_0 + s \cdot \Delta y_0) \Delta y_0 \Delta y_0 ds . \tag{4.3.b}$$

Let

$$\Theta := \|\Delta y_1\| / \|\Delta y_0\| \tag{4.4}$$

denote the contraction factor. Then standard estimation as in the preceding sections leads to

$$\Theta \leq \delta_0 \cdot z + (z/2\bar{\tau})^2$$
$$z := h/(1 - \mu h) \tag{4.5}$$

Combining (4.5) with (3.6), one obtains the maximum contraction factor

$$\Theta_{\text{max}} = \frac{1}{4} \cdot \left(\frac{1 + 2\delta_0 \bar{\tau}}{1 + \delta_0 \bar{\tau}}\right)^2 , \tag{4.6}$$

which nicely reflects the limiting cases $\Theta_{\text{max}} = \frac{1}{4}$ (simplified Newton method) and $\Theta_{\text{max}} = 1$ (fixed point iteration). Upon requiring

$$\Theta \leq \hat{\Theta} = \frac{1}{8} \tag{4.7}$$

one comes up with the (possible) stepsize restriction

$$h \leq \hat{\tau} \Psi_{\text{IE}}(\mu \hat{\tau})$$
$$\hat{\tau} := 4\hat{\Theta}\bar{\tau}/(\alpha + \sqrt{4\hat{\Theta} + \alpha^2}) \tag{4.8}$$
$$\alpha := 2\delta_0 \bar{\tau}$$

as a sufficient contractivity condition for the implicit Euler method.

Even though extrapolation methods offer a rather promising way of iteratively estimating the quantities $\bar{\tau}$ and δ_0, these estimates are not fully satisfactory – both due to the information coupling in Δy^1 and due to the replacement of μ by μ_0. For this reason, only the requirement $\Theta \leq \hat{\Theta}$ is monitored: whenever $\Theta > 2\hat{\Theta}$, then the stepsize is reduced by an ad-hoc factor. If $\hat{\Theta} < \Theta \leq 2\hat{\Theta}$, then an increase of the extrapolation order for the next integration step is prevented. Moreover, whenever $\Theta \ll \hat{\Theta}$, then the Jacobian approximation is kept. Furthermore, the value μ_0 is used to assure that

$$\mu_0 h < 1 \qquad (4.9)$$

Numerical experiments using the above computational devices show nearly no effect over the well-known problem set STIFF DETEST [12]. In hard real life problems, however, reliability and computational speed turned out to increase – a fact, which is now illustrated by two examples.

Example 1. RNA-polymerization. A documentation and illustration of this large scale problem is given in [3] in the context of the chemical kinetics package LARKIN. Recall that LARKIN generates the sparse analytic Jacobian fairly cheaply, so that savings from keeping the Jacobian are not expected. The stiff integrator EULSIM based on the semi-implicit Euler discretization was used. In this problem, the quantity μ turns out to be positive – reflecting the "polymerization wave" [3]. The total computing time in this large scale example follows the behavior of NSUBST, the number of forward/backward substitutions.

Table 1 Comparison for example 1, EULSIM adaptation within LARKIN (TOL=10^{-3}).

	NFEV	NDEC	NSUBST
OLD	995	425	1345
NEW	784	380	1083

Example 2. Hydrogen-Oxygen Auto-Ignition. This example comprises a physical and chemical model due to [13]. The model describes the auto-ignition process of hydrogen-oxygen mixtures in space and time. The example presented here is a *1D-system of reaction-diffusion-convection type*. Method of lines preprocessing (with strongly non-uniform grid) leads to a *differential-algebraic system of index = 1*, which is treated by the extrapolation code LIMEX [9, 11].

The run of interest for the physical chemist is the one up to final time $t_{fin}=2 \cdot 10^{-2}$ – which is shortly after the auto-ignition point. At $t_{fin}=10^3$, the stationary state of the system has been reached. It may be worth noting that investigations of this kind help to quantitatively understand the phenomenon of "engine knock". Note that this problem, too, shows a non-negative quantity μ_0 – which is typical for combustion problems.

Table 2 Comparison of computing times for example 2, LIMEX adaptation within a flame code due to [13], TOL=10^{-3}.

	$t_{fin}=2 \cdot 10^{-2}$	$t_{fin}=10^3$
LIMEX-OLD	952 sec	2233 sec
LIMEX-NEW	433 sec	1525 sec

Acknowledgments. The author gratefully acknowledges detailed discussions with Chr. Lubich on an earlier draft of this paper. Moreover, he wants to thank S. Wacker for her excellent TEX–typing of this manuscript.

References

[1] R. Aiken (ed.): *Stiff Computation.* Oxford University Press, Cambridge, 1985.

[2] G. Bader, P. Deuflhard: *A Semi-Implicit Mid-Point Rule for Stiff Systems of Ordinary Differential Equations.* Numer. Math. 41 (1983), p. 373-398.

[3] G. Bader, U. Nowak, P. Deuflhard: *An Advanced Simulation Package for Large Chemical Reaction Systems.* In [1], p. 255-264 (1985).

[4] G. Dahlquist: *A special stability problem for linear multistep methods.* BIT 3 (1963), p. 27-43.

[5] P. Deuflhard: *Recent Progress in Extrapolation Methods for Ordinary Differential Equations.* SIAM Rev. 27 (1985), p. 505-535.

[6] P. Deuflhard, B. Engquist (ed.): *Large Scale Scientific Computing.* Birkhuser/Boston, Series "Progress in Scientific Computing", vol. 7 (1987).

[7] P. Deuflhard, E. Hairer, J. Zugck: *One-Step and Extrapolation Methods for Differential-Algebraic Systems.* Numer. Math., 51, p. 501-516 (1987).

[8] P. Deuflhard, G. Heindl: *Affine Invariant Convergence Theorems for Newton's Method and Extensions to Related Methods.* SIAM J. Numer. Anal. 16 (1979), p. 1-10.

[9] P. Deuflhard, U. Nowak: *Extrapolation Integrators for Quasilinear Implicit ODEs.* In [6], p. 37-50 (1987).

[10] W.H. Enright, T.E. Hull, B. Lindberg: *Comparing numerical methods for stiff systems of ODE's.* BIT 15 (1975), p. 10-48.

[11] E. Hairer, S.P. Nørsett, G. Wanner: *Solving Ordinary Differential Equations I. Nonstiff Problems.* Springer Ser. Comp. Math. 8 (1987).

[12] W. Liniger, R.A. Willoughby: *Efficient Integration Methods for Stiff Systems of Ordinary Differential Equations.* SIAM J. Numer. Anal. 7 (1970), p. 47.

[13] U. Maas, J. Warnatz: *Ignition Processes in Hydrogen-Oxygen Mixtures.* Manuscript, to be published (1987).

[14] W. Walter: *Private Communication* (1987).

[15] W. Walter: *Differential and integral inequalities.* Berlin, Heidelberg, New York: Springer (1970). (German edition 1964).

Prof. Dr. P. Deuflhard
Konrad-Zuse-Zentrum (ZIB)
Heilbronner Strasse 10
1000 Berlin 31
Fed. Rep. Germany

R. FLETCHER and J.A.J. HALL

Towards Reliable Linear Programming

1. Introduction

Difficulties with simplex-like methods for linear programming arise when degeneracy and round-off errors interact to cause cycling and other instances of non-termination. This paper develops a method [6] which guarantees termination under these circumstances. This method may truncate small numbers to zero in near-degenerate situation in order to obtain exact degeneracy, which is then treated accordingly. A brief review of the method is given later in this section, together with an illustrative example.

In certain situations it has been observed that more significant truncations can arise. In Section 2 it is shown that there are various ways in which this can occur, and some possibilities for avoiding the difficulty are discussed. One particular situation is isolated in which a pivot on a small number, whose magnitude is comparable to the round-off error, causes a large truncation to occur at a later stage. This indicates the need for a strategy in which small potential pivots are truncated so as to avoid large truncations occurring at a later stage. Such a strategy is described in Section 2 and it appears to work well.

In Section 3 various aspects in the design of a production code to implement the algorithm are discussed. An extension of the method in [6] is described, which allows upper and lower bounds on all variables and constraints, and permits the use of pseudo-constraints. All these modifications to [6] are shown to maintain the guarantee of termination. A new level-indexing scheme is described which is more efficient than that in [6]. A production code has been developed which allows the use of dense or sparse linear algebra, and includes the provision of warm and hot start facilities. A feature of this code is that it has proved effective in low accuracy single precision, as indicated by numerous trials on problems that are near-degenerate or ill-conditioned.

An unsophisticated sparse matrix code has been developed as part of the production code, and this is described in Section 4. A simple scheme, requiring no provision for fill-in, is obtained by using Tarjan's algorithm for block triangularization, with dense LU factors of the blocks. Our original intention was to use this idea in conjunction with Schur complement updates, in order to avoid factorizing large blocks, if possible. However results on a standard large-scale test set indicate that no advantage over the the simpler method is obtained. Nevertheless, one feature of interest has been the development of a stable algorithm for extending the LU factors of a matrix when it is bordered by an extra row and column, based on the use of Fletcher-Matthews updates, and this is described in Section 5.

The aim of this research has been to develop a highly reliable LP (and in the future, QP) code, particularly for solving subproblems which arise in situations such as successive LP and QP methods for nonlinear programming, and integer programming.

In these applications it is of prime importance that the routine does not break down, or cycle, or exhibit any other behaviour that requires external intervention. We have chosen to develop a recursive method which probably dates back to Balinski and Gomory [1], and was shown by Fletcher [6] to provide the basis for guaranteed termination in the presence of degeneracy, *even when the arithmetic is inexact*. The method is able to use any current technique for updating matrix factors in LP, without significant overheads. Despite the current vogue for algorithms with polynomial complexity, this method is essentially simplex-like. An important advantage for the kind of application that we have in mind is that it readily permits hot starts. In its simplest form the method solves the LP

$$\text{maximize} \quad c^T x$$
$$\text{subject to} \quad A^T x \geq b, \qquad x \geq 0 \tag{1.1}$$

where $A \in \mathbb{R}^{n \times m}$. The method is a Phase I/Phase II active set method and is described in detail in [6]. However a short illustrative example is given here: the main ideas are contained in how the method handles Phase I so we consider finding a feasible point for (1.1) when

$$A = \begin{bmatrix} -1 & 2 & 2 \\ 1 & -3 & 0 \\ 0 & 1 & -1 \end{bmatrix}, \qquad b = \begin{pmatrix} -\frac{1}{2} \\ 1 \\ 2 \end{pmatrix}. \tag{1.2}$$

The method is conveniently described in tableau form. Initially the simple bounds $x \geq 0$ are taken as the active constraints and a tableau is formed (see (a)) with $[I \mid A]$ below the line, and the constraint residuals $r^T = [0^T \mid -b^T]$ above the line (row 0). For clarity we have indicated known zero elements by a dot, and unimportant nonzero elements by a $*$, although the actual values can easily be obtained from (1.2).

(a)

.	.	.	$\frac{1}{2}$	-1	-2
1	.	.	-1	2	2
.	1	.	$*$	$*$	0
.	.	1	$*$	$*$	-1

(b)

$\frac{1}{2}$	.	.	.	0	-1
$*$	.	.	1	$*$	-2
1	1	.	.	-1	2
$*$	.	1	.	$*$	-1

To start the method, the most infeasible residual (r_q say: here $q = 6$) is regarded as the objective function, and the method aims to increase its value to zero, without allowing any currently feasible residuals to become infeasible. The elements in the column below r_q are referred to as *costs* and indicate the effect of relaxing each of the active constraints. A search is made for the largest positive cost (the *optimality test*): if no positive costs exist then r_q cannot be increased and we are optimal. Otherwise the active constraint corresponding to the largest positive cost (r_p say: here $p = 1$) is chosen to be relaxed. The corresponding row of the tableau indicates how the residuals change as this constraint is relaxed. A line search or *ratio test* is now made to determine the maximum allowed increase in r_p. The resulting value is denoted by α and is referred to as the *step*. The step is the largest multiple of row p that can be added into row 0, without making any feasible residuals go infeasible. This is done by looking at certain ratios of elements in row 0 to those in row p, and usually determines a column (q' say) whose feasible residual is driven to zero. Here $\alpha = \frac{1}{2}$ and $q' = 4$. Finally a pivot (underlined) is made on the element p, q' in the tableau, by reducing column q' to a unit column with row operations in the usual way. This corresponds to replacing p by q' in the set of active constraints. Here the new tableau is that given in (b). In general these iterations are repeated until r_q itself is driven to zero. At this stage, a new candidate for r_q is determined and the whole process is again repeated. All these iterations are said to take place at *level 1*. Finally, when there are no infeasible residuals, a feasible point has been located.

In our example the next iteration continues to try to reduce r_6, and this time r_2 is chosen to be relaxed. However r_5 would then immediately go negative, so a *degeneracy block* is said to have occurred. In order to make progress, a local duality result can be used (see [6]) which gives rise to a system of linear inequalities in the dual variables. This is regarded as a *level 2* problem and is solved by the same type of process as for level 1. Of course if a degeneracy block arises at level 2 then the method must go to level 3 to make progress, and so on. Thus a recursive type of method is obtained. Apart from a sign change, the residuals at level 2 are the costs at level 1, and the costs at level 2 are the denominators in the ratio test at level 1. Illustrating the procedure on our example, we now view tableau (b) in a different way. The nondegenerate constraint r_1 is ignored, and column 6 now represents the level 2 residuals. At level 2, residuals are regarded as infeasible if they are positive, so the most infeasible level 2 residual is that one chosen in the level 1 optimality test: here it is the 2,6 element in tableau (b). The rest of row 2 gives the level 2 costs and we seek the most negative of these in the level 2 optimality test (here there is only the one element in column 5 because we are ignoring column 1). The elements in column 5 now give the denominators for the level 2 ratio test. The situation is shown in tableau (c). A ratio test between columns 5 and 6 indicates that a multiple $\alpha = 1$ of column 5 can be added into column 6. Row 3 is critical because a larger step would make the residual in the 6,3 position become infeasible. Finally a pivot on the element in the 5,3 position is made, giving tableau (d).

(c)

*	.	.	.	0	-1
*	.	.	1	-2	-2
*	1	.	.	-1	2
*	.	1	.	$\underline{1}$	-1

(d)

*	.	0	.	.	-1
*	.	*	1	.	-3
*	1	1	.	.	1
*	.	*	.	1	-1

The next level 2 iteration continues from tableau (d), trying to decrease the infeasible level 2 residual in position 2,6 to zero. However there are now no negative level 2 costs in row 2, so we are now *optimal at level 2*. Here we see the point of the whole method, in that the degeneracy block at level 1 is no longer present. Thus we can return to level 1 and progress by increasing the level 1 objective function. In our example we now view the tableau as in (e), and a level 1 iteration now reduces r_6 to zero, giving tableau (f), which corresponds to a feasible point, and completes Phase I. In general the method can also return to the lower level if the blocking variable is zeroed (see [6]). For Phase II the 'natural' objective function is added to the tableau (as column zero: see (2.2) below) and becomes the level 1 objective function to be maximized.

(e)	$\frac{1}{2}$	·	0	·	·	-1
	$*$	·	$*$	1	·	-3
	1	1	1	·	·	1
	$*$	·	$*$	·	1	-1

(f)	$\frac{3}{2}$	1	1	·	·	·
	$*$	$*$	$*$	1	·	·
	$*$	$*$	$*$	·	·	1
	$*$	$*$	$*$	·	1	·

In practice use of a tableau representation would be inefficient and it is shown in [6] that the method can be implemented using any current technique for updating an invertible representation of the basis matrix B. (Here the basis matrix is the matrix whose columns are the normal vectors of the active constraints.) This representation is then used to calculate a selected row or column from the tableau as required. Other than this, the main storage requirement is for three vectors of length $m + n$. One of these, denoted by r, stores all the residuals in the problems. This can be done in a very compact manner. Reference to tableau (a) shows that there are three nontrivial residuals (r_4, r_5, r_6), whereas residuals (r_1, r_2, r_3) are known to be zero. These zero locations provide just enough space to store the level 1 costs from column 6, which are equivalently level 2 residuals. When the method moves to level 2 in tableau (c), there is a known zero in r_5 which is a degenerate level 1 residual. This provides just enough space to store the level 2 cost in position 2,5 in the tableau, and this can equivalently be thought of as a level 3 residual. Thus there is sufficient room in r to store all the current residuals and costs in a compact way. Another vector, denoted by ls, is required to indicate the level to which each entry in r corresponds, and a third vector w is used to store the denominators that are required in the ratio test. This compact storage scheme makes the recursive method practicable since only one scalar (the address of the cost function at that level) needs to be stacked when going to a higher level.

2. A truncation scheme for error control

It is straightforward to prove termination of the algorithm in exact arithmetic, by applying standard arguments about finiteness of the simplex algorithm at each level of recursion. (There is an upper limit to the level of recursion since at least one variable (the objective function) is removed every time that the level is increased.) However it is shown in [6] that a guarantee of termination for inexact arithmetic can also be obtained by an extension of the algorithm. The idea depends on the fact that there is an objective function at each level of recursion, and the proof of termination requires that its stored value is strictly increasing. With inexact arithmetic it is possible that a small but nonzero step would not increase the stored value. By some simple error analysis it is possible to determine whether or not the step is large enough to guarantee an increase in the objective function. If not then a degeneracy block is deemed to have occurred and the algorithm proceeds to a higher level. It is also necessary to argue that the algorithm cannot repeatedly increase and reduce levels infinitely, without increasing the objective function. More details of the argument are given in [6].

It is a feature of the compact storage scheme that higher level residuals overwrite known zero values of lower level residuals. With this new definition of a degeneracy block we must now be prepared to overwrite nonzero (but hopefully small) residuals. In this respect there is an interesting difference between this algorithm and others: this method handles degeneracy by making perturbations (truncations) which create exact degeneracy from near-degeneracy. Other methods, e.g. [8], make perturbations which destroy degeneracy so that it need not be considered. To our knowledge there is no termination result for such an algorithm when the arithmetic is inexact.

In our algorithm, making truncations is a consequence of the need to retain the compact storage scheme. It is therefore important to try to ensure that only small truncations are made by the algorithm, otherwise we might expect serious loss of accuracy. (Of course, with inexact arithmetic, the stored values of residuals will indicate a solution when the algorithm terminates, but there is no guarantee that this solution is exact.) Nonetheless it is usually the case that the truncations are small. Consider an iteration in which the objective function is r_q, and an inactive constraint with residual r_i becomes active. The step calculated by the ratio test is $\alpha = r_i/w_i$ (give or take a sign), and the objective function would be updated by $r_q := r_q + \alpha w_q$. If this calculation produces no change in the stored value of r_q, then it is usually because r_i and hence α is negligible. Therefore truncating r_i to zero is a negligible perturbation.

It was recognised in [6] that significant truncations could arise in one situation. If w_q were small then a negligible change to r_q could result, even though α were large. Truncating r_i could then give a significant perturbation. This situation can arise for example when the solution is almost non-unique, and the slope (w_q) along the line joining two solutions is negligible. Here the most appropriate remedy would be to truncate w_q to zero, rather than r_i. In practice a *least truncation rule* has been adopted, in which either r_i or w_q is truncated, whichever is smallest in modulus, when a degeneracy block arises. If it is w_q that is truncated, then the algorithm must back-track and re-do the optimality test.

Since that time we have gained much more experience with the algorithm and other cases have occurred in which large truncations are made. These situations arise as a consequence of making a pivot on a negligible element, ϵ say, in the tableau. One immediate problem can occur if the element is a zero element that has become nonzero due to round-off error. Then the new basis matrix can be exactly singular so that the method fails completely. The guarantee of termination could be maintained by perturbing the singular matrix to become near-singular. However this is unsatisfactory because it leads to tableau elements of $\sim \epsilon^{-1}$ which then cause other numerical difficulties. Sometimes the least truncation rule avoids the difficulty by truncating a w_q which otherwise would subsequently have been chosen as a pivot. This is not always the case however, and we have constructed an example, simplified from practical experience, in which a substantial truncation is inevitable. Consider the problem

$$\text{maximize} \quad x_2 + 1$$
$$\text{subject to} \quad x_2 \leq \epsilon x_1 \tag{2.1}$$
$$x_1 \leq \tfrac{1}{2}, \qquad x_1, \ x_2 \geq 0$$

in which the value of ϵ is equal to the relative precision. The tableaux for this example are given in (2.2) below. There is a degeneracy block both at level 1 and level 2, and a level 3 step of $\alpha = \epsilon^{-1}$ is made, followed by a pivot on the ϵ. Since this step zeroes the blocking variable the method returns to level 2, where it is optimal, and then to level 1. A step of $\alpha = \epsilon/2$ is calculated and in exact arithmetic this step would increase r_0 to $1 + \epsilon/2$. However in floating point arithmetic $\mathrm{fl}(1 + \epsilon/2) = 1$ and so there is no increase in the stored value of r_0. Thus a degeneracy block is deemed to exist. The candidates for truncation are $w_0 = 1$ and $r_4 = \tfrac{1}{2}$ and the latter is actually chosen. However both of these values are nontrivial and the algorithm cannot avoid making a significant truncation.

1	·	·	0	$\frac{1}{2}$		1	0	·	·	$\frac{1}{2}$	
0	1	·	$\underline{\epsilon}$	-1		0	ϵ^{-1}	·	1	$-\epsilon^{-1}$	(2.2)
1	·	1	-1	0		1	ϵ^{-1}	1	·	$-\epsilon^{-1}$	

The difficulty in this example occurs as a result of a pivot on a near-zero number and an obvious remedy is to truncate all such potential pivots to zero. Since the least truncation rule already allows us to truncate a near-zero pivot, this is not significantly different to what we do already. In fact LP problems generally become ill-conditioned when near-zero elements in the tableau come into play. For example, negligible perturbations to (2.1) exist which move the solution to the origin, or even render the problem infeasible. Thus, although the truncation of a near-zero element can cause a significant change to the solution, it can be argued that such a problem is inherently ill-conditioned, and it is not possible to distinguish between two apparently dissimilar solutions. Another example of this is the apparently trivial example of finding a feasible point of the constraint set

$$\epsilon x_1 - x_2 \geq 1, \qquad x_1, \ x_2 \geq 0. \tag{2.3}$$

If $\epsilon > 0$ then $(\epsilon^{-1}, 0)$ is a feasible point, whereas if $\epsilon \leq 0$ then the problem is infeasible. Clearly there is a significant difference between these extremes. In practice however, we much prefer to truncate ϵ to zero and indicate that the problem is infeasible, rather than to return the feasible point $(\epsilon^{-1}, 0)$ which could be entirely erroneous from the user's point of view.

As a consequence of these observations it was decided to incorporate a truncation scheme into the algorithm, and a quite simple scheme has been found to work well. All numbers that are derived from the tableau are truncated to zero if they are within a tolerance t of zero. The user is required to initialize t and a value $\epsilon^{\frac{1}{2}}$ times the overall size of A is suggested, where ϵ is the relative machine precision. (If A is badly scaled, some pre-scaling might be appropriate, but it is not easy to automate this process.)

The factor of $\epsilon^{\frac{1}{2}}$ is larger than might be necessary, to allow for build up of round-off, and to try to ensure that the necessary truncations are made. Subsequently the algorithm attempts to improve the accuracy of the solution, if necessary, by having an outer iteration in which the value of t is reduced. However it is not desirable to let t go to zero, because of situations like (2.3). Therefore the user is also asked to provide a smaller tolerance t_{min} which is a lower bound on t. A value of about 100ϵ times the overall size of A is suggested. The following heurisics control the outer iteration. Each iterate of the outer iteration is carried out with a fixed value of t and consists of a solution of the LP, with t being used to truncate tableau elements. A record of the maximum actual truncation τ during the solution process is kept. On termination of the inner iteration, if $\tau \leq t_{min}$ or if τ is not less than its value on the previous outer iterate, then the outer iteration is terminated. Otherwise the value of $t = \max(t_{min}, \tau/10)$ is chosen for the next outer iterate. The next iteration is initialized by a hot start from the previous solution, using iterative refinement to recalculate x, so as to obtain accurate values of the inactive constraint residuals. Because the outer iteration terminates if τ is not reduced, and because the process takes place in finite arithmetic, termination of the outer iteration is guaranteed. In practice it is very rare for more than one extra outer iterate to be required.

3. An LP production code

A production code has been developed for solving the bounded LP problem

$$
\begin{aligned}
\text{maximize} \quad & c^T x \\
\text{subject to} \quad & l \leq \begin{bmatrix} A^T x \\ x \end{bmatrix} \leq u.
\end{aligned}
\tag{3.1}
$$

Because of the extra generality of (3.1), additional features have been required to maintain the guarantee of termination for inexact arithmetic, and these are described in this section. The code has a more efficient level indexing scheme than that given in [6], and this is also described. The code is written so that it can be incorporated with either a dense matrix package for matrix/vector operations, or a sparse matrix package. We have implemented a fairly simple sparse matrix scheme which is described in Section 4. However further matrix packages which implement more complex schemes could readily be incorporated. Our dense matrix scheme is based on Fletcher-Matthews updates. The code permits cold-warm-hot starts and is written in Fortran 77. Our general experience with the code as a whole is described at the end of this section, and that with the sparse matrix code in the next section.

The presence of simultaneous upper and lower bounds on each constraint necessitates some additional features. In its most simple mode of operation (*mode 0*), the method only includes an equality constraint ($u_i = l_i$) in the active set when such a constraint becomes active during Phase I. However there is also an option (*mode 1*) in which as many equality constraints as possible are included in the initial active set. The initial basis is then completed by including a selection of unit columns chosen on the grounds of numerical stability. For the dense matrix package, this is quite straightforward and usually all the equality constraints can be included. For the sparse matrix package in mode 1, an algorithm derived from the P^5 algorithm of Erisman et al. [4] is used. It forms a block triangular matrix using as many equality constraints as possible, although the lack of a pivotal strategy at this stage has the effect that it may not be safely possible to include all the equality constraints. Again the basis is completed with a selection of unit columns. In practice the algorithm is very effective: in 10 out of the 13 test problems in the SOL test set (see below), all the equality constraints are included, and in the remaining cases a total of only 13 equalities out of 533 are not included immediately.

During the course of the algorithm the following rules apply to equality constraints

i) If an equality constraint becomes active in the ratio test, a degeneracy block is not flagged.

ii) Once an equality constraint is in the active set, it is not subsequently eligible to be removed (i.e. equality constraints at level 2 do not contribute to the optimality test).

iii) When a degeneracy block is flagged at level 1 or level 2, an equality constraint is not allowed to go to a higher level.

These rules do not affect the guarantee of termination because equality constraints can only become active a finite number of times, after which their status remains the same (usually a level 2 residual), so that they cannot contribute to any cycling. In addition to these rules, consider an iteration at level 1 in which the algorithm moves between the lower and upper bounds (or vice versa) for one particular inequality constraint, without changing the objective function. The near-degeneracy procedure of Section 2 would require a truncation to take place which would make the constraint into an equality constraint. However changing the data of the problem is undesirable, so an alternative solution is adopted. If this situation arises then the constraint is regarded as being a *near-equality constraint* and is subsequently treated like an equality constraint. As above, such a change of status can only occur a finite number of times so does not affect the termination guarantee. Also any constraint for which $u_i - l_i \leq t$ is regarded as being a near-equality constraint. The status of near-equality constraints is reconsidered before starting the next outer iteration, because of the reduction in t.

The code uses the feature of *pseudo-bounds* and this also requires some special attention. A pseudo-bound (e.g. see [5]) is an active simple bound for which $l_i < x_i < u_i$ and so is always eligible for removal from the active set. The feature is useful in various ways, for example it provides a simple and stable means of handling unbounded variables. The following rules apply to pseudo-bounds.

iv) A pseudo-bound is always eligible to be chosen in the optimality test.

v) Once a pseudo-bound is removed from the active set it becomes a level 1 residual and no longer has any special status.

vi) A degeneracy block is not flagged if the constraint selected by the optimality test is a pseudo-bound.

vii) When a degeneracy block is flagged at level 2, a pseudo-bound is not eligible to go to level 4.

Rule vi) is similar to rule i) in that it can only occur a finite number of times and so cannot affect the termination guarantee. After a finite number of iterations the number of pseudo-bounds is constant (usually zero) and their status is fixed as level 2 residuals, so they also cannot contribute to cycling.

An improved feature of the code is a new way of handling the level indexing vector ls and the pointer stack lp which is more efficient than that in [6] whilst requiring the same amount of storage. In [6] $ls(j)$ is simply the level number of $r(j)$, whilst the element $lp(level)$ points to the residual in r that is the objective function for that level. A disadvantage of this scheme is that all the $m + n$ elements of ls have to be referenced when searching through r for residuals at a given level. In the new scheme the vector ls is broken up into blocks of adjacent elements, one block for each level, and each block contains pointers to the elements in r that are at that level. The leftmost element in each block points to the objective function for that level. The element $lp(level)$ now points to the start of the block in ls for that particular level. The most convenient ordering of blocks in ls is $1, 3, \ldots$ followed by $2, 4, \ldots$. It is then trivial to return higher level residuals to the lower level when reducing the level. A search through the residuals at any one level is then more efficiently accomplished: a look-up in lp indicates the initial and final elements in the block for the given level, and then a pass through the block in ls gives the required residuals in r. The presence of near-equality constraints and pseudo-bounds is an extra complication. The level 2 block in ls is broken up into three sub-blocks, one for near-equality constraints, one for pseudo-bounds, and one for all other level 2 residuals. Extra pointers for the start of these blocks are required. Another additional complication for the problem with two-sided bounds is that it is necessary to indicate whether a residual in $r(j)$ is computed from the upper bound u_j or the lower bound l_j. The sign of the element in ls which points to $r(j)$ is used for this purpose.

We have gained experience with the code on a variety of LP problems. These have been solved on a SUN 3/50 workstation in single precision with relative precision of $6.0_{10} - 8$. It is worth observing that most LP production codes use double precision: it is our view that if an algorithm is intrinsically reliable it should be possible to use it in single precision and this has been one of our aims. This results of course in significant savings, particularly in storage requirements. The code has solved the test set of small degenerate/ill-conditioned problems given in [6] without difficulty. The code has also been used as a subprogram in an implementation of Kocis and Grossman's method [9] for integer nonlinear programming. Here some LP problems are generated which are highly degenerate due to the use of certain integer cuts. We have solved a large number of such problems of up to about 100 variables and 250 general constraints in size, without any failures. Finally we have applied the code to the test problems in the SOL test problem set, and details of this are reported in the next section. Again no failures have as yet been detected. Some comments about the relative efficiency of the method are also made in the next section.

4. A simple sparse matrix scheme

The aim of this section has been to develop a relatively simple sparse matrix package for use with the LP algorithm. The aim has been to avoid the need for vast amounts of code, so that for instance the code would be easily transportable for use on workstations.

We were hoping to get a substantial increase in speed compared with that obtained by the dense matrix package, and also to solve much larger problems. Our plan was based on the optimal reduction of the basis matrix B to block triangular form by Tarjan's algorithm, together with the use of a Schur complement update scheme to avoid factorizing large blocks. This scheme allows a simple data structure and code because we do not need to provide for fill-in within matrix factors. The algorithm is optimal (and very efficient) for network problems, although it could be prohibitively expensive for problems in which B is almost irreducible for many iterations. Another advantage is that structural singularity of B is always detected. In some cases this can back up the truncation scheme when the latter fails to truncate a pivot that is a zero element which has been corrupted by round-off error.

Consider therefore the problem of permuting B to an optimum block upper triangular form, typically illustrated by

$$
PBQ = \begin{bmatrix}
\bullet & \bullet & * & * & * & * & * & * & * & * & * \\
\bullet & \bullet & * & * & * & * & * & * & * & * & * \\
 & & \bullet & * & * & * & * & * & * & * & * \\
 & & & \bullet & \bullet & \bullet & * & * & * & * & * \\
 & & & \bullet & \bullet & \bullet & * & * & * & * & * \\
 & & & \bullet & \bullet & \bullet & * & * & * & * & * \\
 & & & & & & \bullet & * & * & * & * \\
 & & & & & & & \bullet & \bullet & * & * \\
 & & & & & & & \bullet & \bullet & * & * \\
 & & & & & & & & & \bullet & \bullet \\
 & & & & & & & & & \bullet & \bullet
\end{bmatrix}, \tag{4.1}
$$

where P and Q are permutation matrices. The diagonal blocks (elements denoted by $\bullet$) are irreducible, even though some of these elements may be structurally zero. (A structural zero is one that arises due to the absence of an entry in the sparse matrix representation.) The elements above the block diagonal, denoted by $*$, include the remaining nonzero elements of PBQ, but will usually include a substantial proportion of structurally zero elements. Very efficient methods for determining this optimum structure have become available in recent years, see for example Duff, Erisman and Reid, [3]. For an arbitrary matrix the first stage is to find a *transversal*, that is to say a permutation of the matrix in which the diagonal elements are structurally nonzero. Then a symmetric permutation which transforms this matrix to the optimum block triangular form is found by the use of *Tarjan's algorithm*. Tarjan's algorithm only requires a single reference to each structural nonzero in B, so is very efficient. The complexity of current algorithms for finding a transversal is somewhat worse but it is observed in [3] that an algorithm based on depth first search is also very efficient in practice.

In our application we are more interested in updating the optimal form when one column of B is replaced by another. In this case the search for a new transversal is fairly straightforward. If the incoming column can directly replace the outgoing one and maintain a transversal then that is done. Otherwise the incoming column must displace a different column in B and the only possibilities are those for which the incoming column has nonzero entries. Now we see if the displaced column can replace the outgoing column. If so we are done, if not we must make a further displacement. Positions where a replacement cannot take place, or which have already been considered, are marked to eliminate some options. Making a depth first search through this tree of possible replacements gives a very effective algorithm whose order of complexity is the same as for Tarjan's algorithm.

In the simplest form of our sparse matrix code, after a column update on B we find a transversal as described and then use Tarjan's algorithm to determine the optimal block matrix. To factorize this matrix we simply calculate dense LU factors of the blocks by Gaussian elimination with partial pivoting. No operations on the super-diagonal elements ($*$ in (4.1)) are required. In doing this we rely on the blocks being moderate in size, and we return to this point later in the section. We could of course use sparse LU factors for each diagonal block, e.g. by using the Harwell library subroutine MA28, and this would be much more effective when the diagonal blocks are large but sparse. However this would involve considerable extra code and complication. In fact we only need to refactorize a subset of the blocks. For example if column 5 of (4.1) is replaced by the vector $(*, *, *, *, *, *, *, *, 0, 0, 0)^T$, where the $*$ elements may or may not be structurally nonzero, then we only need to do our transformations (transversal search and Tarjan's algorithm) in columns 4 to 9 inclusive and the first two and the last blocks are unchanged.

Our original intention was to use this block factorization scheme in conjunction with the idea of Schur complement updates, given by Bisschop and Meeraus [2]. This technique can be explained as follows. Let B_0 be a basis matrix for which the optimum block matrix and the block factors are known. In subsequent LP iterates, let

$$\begin{array}{llllll}
v_1 & \text{replace column} & i_1 & \text{of } B_0 & =: & B_1 \\
v_2 & \text{replace column} & i_2 & \text{of } B_1 & =: & B_2 \\
\vdots & & \vdots & \vdots & & \vdots \\
v_k & \text{replace column} & i_k & \text{of } B_{k-1} & =: & B_k.
\end{array}$$

The solution of systems of linear equations involving B_k can be obtained by solving a system involving the matrix

$$\begin{bmatrix} B_0 & V_k \\ I_k & 0 \end{bmatrix} \qquad \text{where} \qquad \begin{array}{ll} V_k &= [v_1 \ v_2 \ \ldots \ v_k] \\ I_k &= [e_{i_1} \ e_{i_2} \ \ldots \ e_{i_k}]^T. \end{array}$$

For example the system $B_k x = b$ can be expressed as the system

$$\begin{bmatrix} B_0 & V_k \\ I_k & 0 \end{bmatrix} \begin{pmatrix} y \\ z \end{pmatrix} = \begin{pmatrix} b \\ 0 \end{pmatrix}. \tag{4.2}$$

The equations $I_k y = 0$ set k elements of y to zero, and the remaining equations determine the nontrivial elements of y and z such that $x = y + I_k^T z$. A similar construction holds for solving $B_k^T x = b$ [7]. Block LU factors of the extended matrix are given by

$$\begin{bmatrix} B_0 & V_k \\ I_k & 0 \end{bmatrix} = \begin{bmatrix} I & \\ I_k B_0^{-1} & I \end{bmatrix} \begin{bmatrix} B_0 & V_k \\ & C_k \end{bmatrix}, \tag{4.3}$$

test problems	numbers of			largest LU storage required	non-trivial blocks in B (largest underlined)
	variables n	general constraints m	nonzero elements in A		
afiro	32	27	88	0	0
adlittle	97	56	465	729	$\underline{27}$
share2b	79	96	730	1127	$2, 3, 5, \underline{33}$
share1b	225	117	1182	2791	$3*2, 3*5, \underline{52}$
beaconfd	262	173	3476	0	0
israel	142	174	2358	4761	$\underline{69}$
brandy	249	220	2150	8649	$\underline{93}$
e226	282	223	2767	5059	$2*3, \underline{71}$
capri	353	271	1786	3570	$2*2, 9*3, \underline{59}$
bandm	472	305	2659	14675	$2*3, 4, \underline{121}$
stair	467	356	3857	117649	$\underline{343}$
etamacro	688	400	2489	4237	$2*2, 5, 6, 18, \underline{62}$
pilot	3652	1441	43220	see text	

Table 1. Results for SOL test problems

where $C_k = -I_k B_0^{-1} V_k$ is known as the *Schur complement*. Given LU factors of B_0, and if LU factors of the Schur complement are known, then (4.3) can be used to solve systems like (4.2). Each replacement of a column of B_0 causes V_k and I_k^T to be extended by an extra column, and this corresponds to adding an extra row and column to the Schur complement. It is possible to update factors of the Schur complement in $O(k^2)$ operations and we show how this can be done in a stable way for LU factors in Section 5. An additional case of interest is when the incoming constraint v_k replaces a column of V_k that was not in B_0. In this case it is easily verified that one column of the Schur complement is replaced by another column, and for such a change the LU factors are readily updated by the Fletcher-Matthews method.

The motivation behind our idea for using Schur complement updates can be explained as follows. Suppose at some stage of the process that the basis matrix, B_0 say, has an optimal block form in which the blocks are quite small and have been factorized. Suppose also that the optimal form for B_1 contains a very large block which would be expensive to factorize. In this case it would be preferable to use the Schur complement approach which makes use of the more simple factors of B_0, and hence avoids factorizing the blocks of B_1. If at a later stage the large block is broken up then the algorithm could return to the optimal block form and would refactorize the current basis matrix. This algorithm requires the structure of the optimal block form to be monitored, which involves applying the transversal/Tarjan process at each iteration. However these algorithms are so efficient that this cost is relatively small. Once the Schur complement option is initiated, then an inverse operation with B_k involves two solves with B_0, as against one otherwise, and this is the main penalty for using the Schur complement option. Also required are tests for when to initiate the Schur complement option, and when to refactorize. We have considered the use of two possible tests: one of these makes the choice which minimizes the storage required for the next iteration. The alternative test makes the choice which minimizes an estimate of the number of computer operations

test	test on storage space			test on operations		
problems	max block	max S.C.	max storage	max block	max S.C.	max storage
afiro	1	0	0	1	0	0
adlittle	13	17	351	26	3	694
share2b	13	13	261	18	2	531
share1b	23	14	1056	51	10	2688
beaconfd	1	0	0	1	0	0
israel	56	55	3136	56	51	3136
brandy	43	37	3545	90	38	8100
e226	61	52	4022	74	4	5485
capri	26	28	931	52	39	2713
bandm	54	51	3362	119	17	14187
stair	30	159	26202	343	6	117649
etamacro	32	36	1587	48	25	2693

Table 2. Comparison of Schur complement tests

for the next iteration, with an override to use the Schur complement if refactorizing would overfill the available storage.

Numerical results, again with single precision on a SUN 3/50 workstation, have been obtained for the SOL test problem set. Table 1 gives details of the problems and indicates the worst case storage requirement in the case that the optimal block form is refactorized on every occasion. These results are obtained with a mode 1 start (see Section 3) although similar results are obtained with a mode 0 start. One particularly poor case is the stair problem for which the largest block has dimension 343, which is a substantial proportion of n. (In fact, since there are only 356 general constraints, this would provide an upper limit on the largest block, so this outcome is not far short of the worst possible.) However for most problems the largest block is quite manageable and is substantially smaller than the lesser of n and m. The afiro and beaconfd problems are very favourable and show the best possible outcome: this is typical of what would happen if the problem were a network flow problem. In general these results show that the simple technique will usually enable problems of up to at least about 1000 variables to be solved using a basic workstation configuration, whilst much larger problems can be tackled if the basis matrices retain a favourable structure. These observations are generally consistent with what has been observed elsewhere (e.g. [3]) about the typical structure of LP bases. It is also worth pointing out that no difficulties due to the use of single precision arithmetic were encountered on this size of problem. An attempt was also made to solve the problem called pilot. After 4 weeks computing (!) and 4000 pivots being performed, the calculation is still running and the current iterate is still some way from being feasible. However the number of infeasibilities is continually reducing and there is no evidence of any malfunction. The current largest block has a dimension of over 400. We therefore attribute our present failure to solve pilot to the lack of computing power of the SUN 3/50. This outcome is not unexpected since MINOS 5.0 is expected to take about 20000 pivots and 2 hours on an IBM 3081K to solve the problem.

test	our method				MINOS 5.0	
problems	mode 0		mode 1		unscaled	scaled
afiro	16	(6)	10	(1)*	6	6
adlittle	113	(18)	131	(17)	123	98
share2b	116	(75)	107	(71)	91	121
share1b	389	(249)	267	(106)	296	260
beaconfd	109	(89)	34	(0)	38	39
israel	376	(105)	376	(105)	345	231
brandy	1100**	(329)	226	(812)	292	377
e226	652	(152)	603	(95)	570	471
capri	715	(591)	578	(472)	273	235
bandm	1458	(1266)	566	(269)	362	534
stair	756	(498)	514	(271)	519	389
etamacro	832	(448)	697	(399)	904	927

* bracketed figure is number of pivots in Phase I
** a run with different round-off gave values of 608 and (501)

Table 3. Comparison of total numbers of pivots required

The effect of modifying the method to include Schur complement updates with the aim of avoiding the factorization of large blocks is shown in Table 2. These results are obtained using the mode 1 start (see Section 3) although similar conclusions are obtained if a mode 0 start is used. The test on storage space is certainly seen to be effective in keeping down the size of the largest block. However, with the exception of the stair problem, the test on computer operations gives runs which are noticeably faster overall. For the stair problem the test on storage space gives better results for both criteria. Unfortunately the test on operations gives block sizes that are not much different to those obtained when the basis matrix is refactorized on every iteration (see Table 1). Given that the number of solves is doubled when using the Schur complement option, it is clear that these figures do not indicate an improvement in efficiency, and this observation is backed up by computer timings. Because of this we have discontinued using the Schur complement option.

A comparison of our code with other LP codes is of interest, but it is difficult to give meaningful statistics. The relative efficiency of our matrix scheme will depend on such factors as whether vectorization is in use, and on how efficiently the underlying BLAS modules are coded. We have gone for simplicity and convenience in our sparse matrix scheme and there is no doubt that a more complex scheme would show improvements on many problems. Another criterion that can be measured is the number of pivots required to solve a particular problem. Even this is not as transparent as it might seem. For example, small changes to a program which only affect round-off propagation can give significantly different outcomes, once the vertex sequence is changed. Often this gives a change in the number of pivots of order $\pm 10\%$, and occasionally a much larger variation, as illustrated by the brandy problem. However, even allowing for this variation, it is clear from the results in Table 3 that the use of the mode 1 start gives a significant improvement when there are many equalities, particularly if n is large. We also have some statistics regarding the performance of MINOS 5.0 on these problems.

These show some minor differences according to whether or not the problems were pre-scaled. These results are also shown in Table 3. It is not possible to read too much into these figures because MINOS uses a crash start which is different from our crash start in mode 1. However the comparison between our method in mode 1 and MINOS is possibly marginally in favour of the latter, and it may be that this can be ascribed to our use of a technique which drives one residual to zero at a time rather than minimizing an l_1 sum of residuals. Some ideas for using the latter approach to get guaranteed termination have already been studied in [6] and we may consider improving this aspect of the code in the future.

5. Stable extension of LU factors

In this section we consider the problem of extending the LU factors of a matrix when it is bordered by an extra row and column. It is assumed that we can make both row and column permutations but to simplify the presentation we shall omit the column permutation from the notation. Thus we assume that A is nonsingular with factors $PA = LU$, where P is a permutation matrix, and we wish to calculate factors of the matrix

$$A^{\#} = \begin{bmatrix} A & b \\ c^T & d \end{bmatrix}. \tag{5.1}$$

Of course this is straightforward if A is well-conditioned because we can write

$$A^{\#} = \begin{bmatrix} P^T & \\ & 1 \end{bmatrix} \begin{bmatrix} L & \\ l^T & 1 \end{bmatrix} \begin{bmatrix} U & u \\ & \mu \end{bmatrix} \tag{5.2}$$

where $u = L^{-1}Pb$ and $l = U^{-T}c$ are calculated by forward substitution, and $\mu = d - l^T u$. However this process is unsatisfactory when A and hence U is ill-conditioned, but $A^{\#}$ is well-conditioned. Then the operation $l = U^{-T}c$ causes growth in l which gives rise to numerical instability.

The current wisdom is to use alternative factorizations which can be updated in a stable manner, and QR factors are the most obvious choice. It is also possible to update LU factors in which L is a full matrix (not lower triangular). References to both these ideas are given by Gill et al. [7]. However both these ideas require $\frac{3}{2}k^2 + O(k)$ storage, and there is also a time penalty when using the factors. In what follows we describe a method in which the regular LU triangular factors are updated in a stable way, using $k^2 + O(k)$ storage and $O(k^2)$ arithmetic operations.

The idea is to allow an initial column permutation in (5.1) so as to give a well-conditioned leading submatrix. Let column i of A be replaced by b giving the submatrix

$$A' = A - (a_i - b)e_i^T. \tag{5.3}$$

It follows using the identity $\det(I - uv^T) = 1 - v^T u$ that

$$\det A' = e_i^T A^{-1} b \det A. \tag{5.4}$$

We use $\det A'$ as a measure of the conditioning of the new matrix. Implicitly this assumes that the scaling of columns of $[A \,|\, b]$ is similar, and if this is not so then the following test should be modified by implicitly prescaling the columns of $[A\,|\,b]$. Given that no prescaling is required, then we choose the option which maximizes $|\det A'|$. This gives the following algorithm.

   i) Select the best interchange:
      a) no interchange required if $\|A^{-1}b\|_\infty \le 1$
      b) otherwise denote $r = \text{argmax}_i\,|(A^{-1}b)_i|$ and replace a_r by b, using the Fletcher-Matthews method to update the LU factors of A (this interchange is a standard simplex update).
 ii) Extend the updated factors as in (5.2).

This algorithm has been used as part of our Schur complement code and no evidence of numerical instabitily has been detected.

6. References

[1] Balinski M.L. and Gomory R.E. (1963). A mutual primal-dual simplex method, in *Recent Advances in Mathematical Programming*, R.L.Graves and P.Wolfe (Eds.), McGraw Hill, New York.

[2] Bisschop J. and Meeraus A. (1977). Matrix augmentation and partitioning in the updating of the basis inverse. *Math. Programming*, **18**, 7-15.

[3] Duff I.S., Erisman A.M. and Reid J.K. (1986). *Direct Methods for Sparse Matrices*. Oxford Science Publications, Oxford.

[4] Erisman A.M., Grimes R.G., Lewis J.G. and Poole W.G. (1985). A structurally stable modification of Hellerman-Rarick's P^4 algorithm for reordering unsymmetric sparse matrices. *S.I.A.M. J. Numer. Anal.*, **22**, 369-385.

[5] Fletcher R. (1987). *Practical Methods of Optimization, 2nd. Edition*, Wiley, Chichester.

[6] Fletcher R. (1988). Degeneracy in the presence of roundoff errors. *Linear Algebra Appl.*, **106**, 149-183.

[7] Gill P.E., Murray W., Saunders M.A. and Wright M.A. (1984). Sparse matrix methods in optimization. *S.I.A.M. J. Sci. Stat. Comput.*, **5**, 562-589.

[8] Harris P.M.J. (1973). Pivot selection methods of the Devex LP code. *Math. Programming*, **5**, 1-28.

[9] Kocis G.R. and Grossman I.E. (1988). Global optimization of nonconvex MINLP problems in process synthesis. *Ind. Engng. Chem. Res.*, **27**, 1407-1421.

R.Fletcher and J.A.J.Hall
Deptartment of Mathematics and Computer Science
University of Dundee
Dundee DD1 4HN
Scotland, U.K.

B. FORNBERG
Rapid Generation of Weights in Finite Difference Formulas

1. INTRODUCTION

In the context of finite difference approximations of differential equations, one encounters the problem of determining the weights in the most accurate approximations to d^m/dx^m extending over prescribed numbers of points. In the first section, we review a classical approach, which is computationally efficient and easy to code, but limited to equi–spaced grids. This is followed by the description of a recently discovered algorithm for the general case: Given $M \geq 0$ (the order of the highest derivative we wish to approximate) and $N+1$ grid points (at x–coordinates $\alpha_0, \alpha_1, \ldots, \alpha_N$), the new algorithm finds all the weights $\delta^m_{n,\nu}$ for $m=0,1, \ldots, M$, $n = m, m+1, \ldots, N$ such that the approximations

$$\frac{d^m f}{d x^m} \bigg|_{x=x_0} \approx \sum_{\nu=0}^{n} \delta^m_{n,\nu} f(\alpha_\nu) \tag{1}$$

all achieve optimal accuracy. This algorithm (listed, with a very brief description, in Fornberg (1988)) is very short (less than 20 lines long) and only requires about 4 arithmetic operations per calculated weight.

2. ALGORITHMS FOR EQUI–SPACED GRIDS

Tables 1, 2 and 3 give some examples of weights on equi–spaced grids. For example, the first two lines for the first derivative in Table 1 should be interpreted as

$$f'(x) = [\qquad\qquad -\tfrac{1}{2}\,f(x{-}h) + 0\,f(x) + \tfrac{1}{2}\,f(x{+}h) \qquad\qquad]/h + O(h^2), \tag{2}$$

$$= [\tfrac{1}{12}\,f(x{-}2h) - \tfrac{2}{3}\,f(x{-}h) + 0\,f(x) + \tfrac{2}{3}\,f(x{+}h) - \tfrac{1}{12}\,f(x{+}2h)]/h + O(h^4),$$

Approximations at x = 0 ; x - coordinates at nodes:

Order of derivative	Order of accuracy	-4	-3	-2	-1	0	1	2	3	4
0	∞					1				
1	2				$-\tfrac{1}{2}$	0	$\tfrac{1}{2}$			
1	4			$\tfrac{1}{12}$	$-\tfrac{2}{3}$	0	$\tfrac{2}{3}$	$-\tfrac{1}{12}$		
1	6		$-\tfrac{1}{60}$	$\tfrac{3}{20}$	$-\tfrac{3}{4}$	0	$\tfrac{3}{4}$	$-\tfrac{3}{20}$	$\tfrac{1}{60}$	
1	8	$-\tfrac{1}{280}$	$-\tfrac{4}{105}$	$\tfrac{1}{5}$	$-\tfrac{4}{5}$	0	$\tfrac{4}{5}$	$-\tfrac{1}{5}$	$\tfrac{4}{105}$	$-\tfrac{1}{280}$
2	2				1	-2	1			
2	4			$-\tfrac{1}{12}$	$\tfrac{4}{3}$	$-\tfrac{5}{2}$	$\tfrac{4}{3}$	$-\tfrac{1}{12}$		
2	6		$\tfrac{1}{90}$	$-\tfrac{3}{20}$	$\tfrac{3}{2}$	$-\tfrac{49}{18}$	$\tfrac{3}{2}$	$-\tfrac{3}{20}$	$\tfrac{1}{90}$	
2	8	$-\tfrac{1}{560}$	$\tfrac{8}{315}$	$-\tfrac{1}{5}$	$\tfrac{8}{5}$	$-\tfrac{205}{72}$	$\tfrac{8}{5}$	$-\tfrac{1}{5}$	$\tfrac{8}{315}$	$-\tfrac{1}{560}$
3	2			$-\tfrac{1}{2}$	1	0	-1	$\tfrac{1}{2}$		
3	4		$\tfrac{1}{8}$	-1	$\tfrac{13}{8}$	0	$-\tfrac{13}{8}$	1	$-\tfrac{1}{8}$	
3	6	$-\tfrac{7}{240}$	$\tfrac{3}{10}$	$-\tfrac{169}{120}$	$\tfrac{61}{30}$	0	$-\tfrac{61}{30}$	$\tfrac{169}{120}$	$-\tfrac{3}{10}$	$\tfrac{7}{240}$
4	2			1	-4	6	-4	1		
4	4		$-\tfrac{1}{6}$	2	$-\tfrac{13}{2}$	$\tfrac{28}{3}$	$-\tfrac{13}{2}$	2	$-\tfrac{1}{6}$	
4	6	$-\tfrac{7}{240}$	$-\tfrac{2}{5}$	$\tfrac{169}{60}$	$-\tfrac{122}{15}$	$\tfrac{91}{8}$	$-\tfrac{122}{15}$	$\tfrac{169}{60}$	$-\tfrac{2}{5}$	$-\tfrac{7}{240}$

Table 1.　Some weights for approximations centered at a grid point

the top two lines in Table 2 as

$$f(x) = [\; \tfrac{1}{2}\, f(x-\tfrac{h}{2}) + \tfrac{1}{2}\, f(x+\tfrac{h}{2})\;] + O(h^2),$$

$$= [\; \tfrac{1}{16}\, f(x-\tfrac{3h}{2}) + \tfrac{9}{16}\, f(x-\tfrac{h}{2}) + \tfrac{9}{16}\, f(x+\tfrac{h}{2}) - \tfrac{1}{16}\, f(x+\tfrac{3h}{2})\;] + O(h^4),$$

(3)

etc. (for derivatives of order m, the sum should be divided by h^m).

Approximations at $x = 0$; x - coordinates at nodes:

Order of derivative	Order of accuracy	-7/2	-5/2	-3/2	-1/2	1/2	3/2	5/2	7/2
0	2				$\frac{1}{2}$	$\frac{1}{2}$			
	4			$\frac{-1}{16}$	$\frac{9}{16}$	$\frac{9}{16}$	$\frac{-1}{16}$		
	6		$\frac{3}{256}$	$\frac{-25}{256}$	$\frac{75}{128}$	$\frac{75}{128}$	$\frac{-25}{256}$	$\frac{3}{256}$	
	8	$\frac{-5}{2048}$	$\frac{49}{2048}$	$\frac{-245}{2048}$	$\frac{1225}{2048}$	$\frac{1225}{2048}$	$\frac{-245}{2048}$	$\frac{49}{2048}$	$\frac{-5}{2048}$
1	2				-1	1			
	4			$\frac{1}{24}$	$\frac{-9}{8}$	$\frac{9}{8}$	$\frac{-1}{24}$		
	6		$\frac{-3}{640}$	$\frac{25}{384}$	$\frac{-75}{64}$	$\frac{75}{64}$	$\frac{-25}{384}$	$\frac{3}{640}$	
	8	$\frac{5}{7168}$	$\frac{-49}{5120}$	$\frac{245}{3072}$	$\frac{-1225}{1024}$	$\frac{1225}{1024}$	$\frac{-245}{3072}$	$\frac{49}{5120}$	$\frac{-5}{7168}$
2	2			$\frac{1}{2}$	$\frac{-1}{2}$	$\frac{-1}{2}$	$\frac{1}{2}$		
	4		$\frac{-5}{48}$	$\frac{13}{16}$	$\frac{-17}{24}$	$\frac{-17}{24}$	$\frac{13}{16}$	$\frac{-5}{48}$	
	6	$\frac{259}{11520}$	$\frac{-499}{2304}$	$\frac{1299}{1280}$	$\frac{-1891}{2304}$	$\frac{-1891}{2304}$	$\frac{1299}{1280}$	$\frac{-499}{2304}$	$\frac{259}{11520}$
3	2			-1	3	-3	1		
	4		$\frac{1}{8}$	$\frac{-13}{8}$	$\frac{17}{4}$	$\frac{-17}{4}$	$\frac{13}{8}$	$\frac{-1}{8}$	
	6	$\frac{-37}{1920}$	$\frac{499}{1920}$	$\frac{-1299}{640}$	$\frac{1891}{384}$	$\frac{-1891}{384}$	$\frac{1299}{640}$	$\frac{-499}{1920}$	$\frac{37}{1920}$
4	2		$\frac{1}{2}$	$\frac{-3}{2}$	1	1	$\frac{-3}{2}$	$\frac{1}{2}$	
	4	$\frac{-7}{48}$	$\frac{59}{48}$	$\frac{-45}{16}$	$\frac{83}{48}$	$\frac{83}{48}$	$\frac{-45}{16}$	$\frac{59}{48}$	$\frac{-7}{48}$

Table 2. Some weights for approximations centered half–way between grid points.

Order of derivative	Order of accuracy	Approximations at x = 0; x-coordinates at nodes:								
		0	1	2	3	4	5	6	7	8
0	∞	1								
1	1	-1	1							
	2	$-\frac{3}{2}$	2	$-\frac{1}{2}$						
	3	$-\frac{11}{6}$	3	$-\frac{3}{2}$	$\frac{1}{3}$					
	4	$-\frac{25}{12}$	4	-3	$\frac{4}{3}$	$-\frac{1}{4}$				
	5	$-\frac{137}{60}$	5	-5	$\frac{10}{3}$	$-\frac{5}{4}$	$\frac{1}{5}$			
	6	$-\frac{49}{20}$	6	$-\frac{15}{2}$	$\frac{20}{3}$	$-\frac{15}{4}$	$\frac{6}{5}$	$-\frac{1}{6}$		
	7	$-\frac{363}{140}$	7	$-\frac{21}{2}$	$\frac{35}{3}$	$-\frac{35}{4}$	$\frac{21}{5}$	$-\frac{7}{6}$	$\frac{1}{7}$	
	8	$-\frac{761}{280}$	8	-14	$\frac{56}{3}$	$-\frac{35}{2}$	$\frac{56}{5}$	$-\frac{14}{3}$	$\frac{8}{7}$	$-\frac{1}{8}$
2	1	1	-2	1						
	2	2	-5	4	-1					
	3	$\frac{35}{12}$	$-\frac{26}{3}$	$\frac{19}{2}$	$-\frac{14}{3}$	$\frac{11}{12}$				
	4	$\frac{15}{4}$	$-\frac{77}{6}$	$\frac{107}{6}$	-13	$\frac{61}{12}$	$-\frac{5}{6}$			
	5	$\frac{203}{45}$	$-\frac{87}{5}$	$\frac{117}{4}$	$-\frac{254}{9}$	$\frac{33}{2}$	$-\frac{27}{5}$	$\frac{137}{180}$		
	6	$\frac{469}{90}$	$-\frac{223}{10}$	$\frac{879}{20}$	$-\frac{949}{18}$	41	$-\frac{201}{10}$	$\frac{1019}{180}$	$-\frac{7}{10}$	
	7	$\frac{29531}{5040}$	$-\frac{962}{35}$	$\frac{621}{10}$	$-\frac{4006}{45}$	$\frac{691}{8}$	$-\frac{282}{5}$	$\frac{2143}{90}$	$-\frac{206}{35}$	$\frac{363}{560}$
3	1	-1	3	-3	1					
	2	$-\frac{5}{2}$	9	-12	7	$-\frac{3}{2}$				
	3	$-\frac{17}{4}$	$\frac{71}{4}$	$-\frac{59}{2}$	$\frac{49}{2}$	$-\frac{41}{4}$	$\frac{7}{4}$			
	4	$-\frac{49}{8}$	29	$-\frac{461}{8}$	62	$-\frac{307}{8}$	13	$-\frac{15}{8}$		
	5	$-\frac{967}{120}$	$\frac{638}{15}$	$-\frac{3929}{40}$	$\frac{389}{3}$	$-\frac{2545}{24}$	$\frac{268}{5}$	$-\frac{1849}{120}$	$\frac{29}{15}$	
	6	$-\frac{801}{80}$	$\frac{349}{6}$	$-\frac{18353}{120}$	$\frac{2391}{10}$	$-\frac{1457}{6}$	$\frac{4891}{30}$	$-\frac{561}{8}$	$\frac{527}{30}$	$-\frac{469}{240}$
4	1	1	-4	6	-4	1				
	2	3	-14	26	-24	11	-2			
	3	$\frac{35}{6}$	-31	$\frac{137}{2}$	$-\frac{242}{3}$	$\frac{107}{2}$	-19	$\frac{17}{6}$		
	4	$\frac{28}{3}$	$-\frac{111}{2}$	142	$-\frac{1219}{6}$	176	$-\frac{185}{2}$	$\frac{82}{3}$	$-\frac{7}{2}$	
	5	$\frac{1069}{80}$	$-\frac{1316}{15}$	$\frac{15289}{60}$	$-\frac{2144}{5}$	$\frac{10993}{24}$	$-\frac{4772}{15}$	$\frac{2803}{20}$	$-\frac{536}{15}$	$\frac{967}{240}$

Table 3. Some weights for one–sided approximations at a grid point.

For derivatives of low orders, closed form expressions can be found for individual weights. Some examples (accuracy p, order of derivative m) are:

Table 1: p even, at x-coordinates $x = k$, $k = 0, \pm1, \pm2, \pm3, \ldots, \pm\frac{p}{2}$:

$$m = 1: \quad \text{weight} = \begin{cases} 0 & , \quad k = 0 \\ \dfrac{[(p/2)!]^2(-1)^{k+1}}{k(p/2+k)!(p/2-k)!} & , \quad k \neq 0 \end{cases} \tag{4}$$

$$m = 2: \quad \text{weight} = \begin{cases} -2\sum\limits_{\nu=1}^{p/2}\dfrac{1}{\nu^2} & , \quad k = 0 \\ \dfrac{2}{k}\cdot(\text{weight for } m{=}1) & , \quad k \neq 0 \end{cases} \tag{5}$$

Table 2: p even, at x-coordinates $x = k$, $k = \pm\frac{1}{2}, \pm\frac{3}{2}, \pm\frac{5}{2}, \ldots, \pm\frac{p-1}{2}$:

$$m = 0: \quad \text{weight} = \frac{(p!)^2\,(-1)^{k+\frac{1}{2}}}{k\cdot 2^{2p}[(p/2)!]^2[(p-1)/2+k]![(p-1)/2-k]!} \tag{6}$$

$$m = 1: \quad \text{weight} = \frac{2}{k}\cdot(\text{weight for } m{=}0)$$

Table 3: p any positive integer, at x-coordinates $x = k$, $k = 0, 1, 2, \ldots, p$:

$$m = 1: \quad \text{weight} = \begin{cases} -\sum\limits_{\nu=1}^{p}\dfrac{1}{\nu} & , \quad k = 0 \\ \dfrac{p!\,(-1)^{k+1}}{k\cdot k!\,(p-k)!} & , \quad k > 0 \end{cases} \tag{7}$$

For higher derivatives, explicit formulas become prohibitively complicated. However, that is of no significance as regards practical computation of the weights since very effective recursions are available.

The 'classical' recursions (limited to equi–spaced grids) are best described with the help of some operator notation. Define

$$
\begin{aligned}
D &: & D\,f(x) &= f'(x) \\
E &: & E\,f(x) &= f(x+h) \\
\delta &:= & E^{1/2} &- E^{-1/2} \\
\mu &:= & \tfrac{1}{2}\big(E^{1/2} &+ E^{-1/2}\big) \ .
\end{aligned}
\tag{8}
$$

The most compact approximations to D^m of order $2p$ can be written in the forms

	Gives approximations when	
	m odd	m even
$D^m \approx \dfrac{\mu\,\delta^m}{h^m} \displaystyle\sum_{\nu=0}^{p-1} a_\nu^m\,\delta^{2\nu}$	at grid point	at half–way pt
$D^m \approx \dfrac{\delta^m}{h^m} \displaystyle\sum_{\nu=0}^{p-1} b_\nu^m\,\delta^{2\nu}$	at half–way pt	at grid point

$$\tag{9}$$

These formulations (in power series of δ) are particularly convenient because

1. Like for Taylor expansions, the coefficients a_ν^m and b_ν^m are independent of p, i.e. approximations of successively higher orders of accuracy are obtained by simply adding more terms.

2. The coefficients a_ν^m and b_ν^m can be expressed in closed form in terms of generalized Bernoulli numbers (Milne–Thomson, 1933, Chapter VII). For numerical purposes, the following recursion is far simpler (Fornberg, 1989):

for $m = 0$ to $mmax$

$$a_0^m := 1$$

$$b_0^m := 1$$

for $\nu = 1$ to $p{-}1$

$$a_\nu^m := \frac{-(m{+}2\nu{-}1)\ a_{\nu-1}^m\ /4 \ + \ m\ b_\nu^{m-1}}{m\ +\ 2\nu} \tag{10}$$

$$b_\nu^m := \frac{m\ a_\nu^{m-1}}{m\ +\ 2\nu}$$

next ν

next m

Note: $m\ b_\nu^{m-1}$ and $m\ a_\nu^{m-1}$ are to be taken as zero when $m = 0$ (and b_ν^{m-1} and a_ν^{m-1} are undefined).

By converting powers of δ into sums over powers of E (binomial theorem; again best performed recursively), the weights follow (for any order derivative, approximated to any order of accuracy, at a grid point or half–way in–between grid points).

If boundaries are present, one might need to use one–sided approximations in their neighborhood. One example suffices to illustrate how any such formula can be derived. Suppose we need the weights for the 2nd derivative, approximated to 4th order of accuracy. We have (centered approximation, for example from the recursions above; cf. Table 1):

$$f\ '' = \{ \quad -\tfrac{1}{12} \quad \tfrac{4}{3} \quad -\tfrac{5}{2} \quad \tfrac{4}{3} - \tfrac{1}{12} \quad \} f\ /\ h^2 \qquad\qquad + \ O(h^4) \tag{11}$$

$\Uparrow$ accurate here

Since $\delta^6 f = O(h^6)$

$$0 = \{ \quad 1 \quad -6 \quad 15 \quad -20 \quad 15 \quad -6 \quad 1 \quad \} f\ /\ h^2 \quad + \ O(h^4), \tag{12}$$

accurate at any position

we add $1/12$th of (12) to (11) to get

$$f'' = \{ \quad \tfrac{5}{6} \quad -\tfrac{5}{4} \quad -\tfrac{1}{3} \quad \tfrac{7}{6} \quad -\tfrac{1}{2} \quad \tfrac{1}{12} \quad \} \, f \, / \, h^2 \quad + \, O(h^4). \tag{13}$$

⇑ accurate here

The same idea applied once more gives (cf. Table 3)

$$f'' = \{ \quad \tfrac{15}{4} \quad -\tfrac{77}{6} \quad \tfrac{107}{6} \; - 13 \quad \tfrac{61}{12} \quad -\tfrac{5}{6} \quad \} \, f \, / \, h^2 \quad + \, O(h^4). \tag{14}$$

⇑ accurate here

This procedure gives the most compact approximations possible for any derivative and any required accuracy. (This is not the case if one generates formulas for high derivatives by repeated use of formulas for lower ones).

Bickley (1941) gives tables for centered to one–sided formulas up to $m = 4$ (fourth derivative) and up to 10 points wide (generated in a different way than described here).

3. ALGORITHM FOR ARBITRARY GRID SPACING

With the grid points α_ν, $\nu = 0,1,2, \dots n,$ we can define

$$\omega_n(x) \; := \; \prod_{k=0}^{n} (\, x - \alpha_k \,) \quad . \tag{15}$$

Then

$$F_{n,\nu}(x) \; := \; \frac{\omega_n(x)}{\omega_n'(\alpha_\nu)(x - \alpha_\nu)} \tag{16}$$

is the polynomial of degree n which takes the values

$$F_{n,\nu}(\alpha_k) \; = \; \left\{ \begin{array}{ll} 0 \; , & k \neq \nu \\ 1 \; , & k = \nu \end{array} \right. . \tag{17}$$

Lagrange's interpolation formula becomes

$$p(x) = \sum_{\nu=0}^{n} F_{n,\nu}(x) \, f(\alpha_\nu) \qquad . \tag{18}$$

To simplify the notation, assume from now on that we wish to approximate derivatives at $x_0 = 0$. Using the $n+1$ points $\alpha_0, \dots, \alpha_n$, the best approximation we can obtain for $D^m f(x) \big|_{x=0}$ is $D^m p(x) \big|_{x=0}$. The weights $\delta_{n,\nu}^m$ express how much this approximation to the derivative changes if the function value $f(\alpha_\nu)$ is changed. Since such changes affect only one term in (18), it follows

$$\delta_{n,\nu}^m = \left[\frac{d^m}{dx^m} F_{n,\nu}(x) \right]_{x=0} \qquad , \tag{19}$$

implying

$$F_{n,\nu}(x) = \sum_{m=0}^{n} \frac{\delta_{n,\nu}^m}{m!} \, x^m \tag{20}$$

Next, note that (11) can be written

$$F_{n,\nu}(x) = \frac{(x-\alpha_0)\cdot(x-\alpha_1)\cdot \ \dots \ \cdot(x-\alpha_{\nu-1})\cdot(x-\alpha_{\nu+1})\cdot \ \dots \ \cdot(x-\alpha_n)}{(\alpha_\nu-\alpha_0)\cdot(\alpha_\nu-\alpha_1)\cdot \ \dots \ \cdot(\alpha_\nu-\alpha_{\nu-1})\cdot(\alpha_\nu-\alpha_{\nu+1})\cdot \ \dots \ \cdot(\alpha_\nu-\alpha_n)} \tag{21}$$

From this, several recursion relations can be obtained; the two we will use are

$$F_{n,\nu}(x) = \frac{x-\alpha_n}{\alpha_\nu - \alpha_n} \, F_{n-1,\nu}(x) \tag{22}$$

and

$$F_{n,n}(x) = \frac{\omega_{n-2}(\alpha_{n-1})}{\omega_{n-1}(\alpha_n)} \, (x-\alpha_{n-1}) \, F_{n-1,n-1}(x) \qquad . \tag{23}$$

Substituting (20) into (22) and (23) and equating coefficients of equal powers give two recursion relations for the $\delta_{n,\nu}^m$:

$$\delta^m_{n,\nu} = \frac{1}{\alpha_n - \alpha_\nu} \left(\alpha_n \, \delta^m_{n-1,\nu} - m \, \delta^{m-1}_{n-1,\nu} \right) , \; \nu = 0,1,..., \; n-1 , \tag{24}$$

$$\delta^m_{n,n} = \frac{\omega_{n-2}(\alpha_{n-1})}{\omega_{n-1}(\alpha_n)} \left(m \, \delta^{m-1}_{n-1,n-1} - \alpha_{n-1} \, \delta^m_{n-1,n-1} \right) \tag{25}$$

The stencils to the right of the formulas in (24) and (25) indicate how they connect different weights $\delta^m_{n,\nu}$, seen from the same perspective as is used in Figure 1.

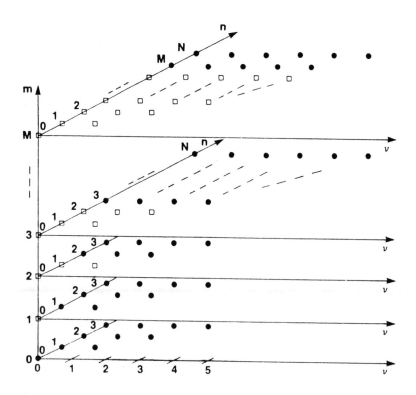

Figure 1.   Overwiew of the m,n,ν–space (solid black circles marking the positions at which $\delta^m_{n,\nu}$ need to be calculated.

114

To obtain all the $\delta_{n,\nu}^m$, we first note that for $m = 0$, the vertical 'descenders' in the small stencils vanish. Therefore, starting with the trivial $\delta_{0,0}^0 = 1$, (24) gives $\delta_{0,n}^0$ for $n = 1,2, \ldots , N$. Then we can repeat for $\nu = 1,2, \ldots ,N$ the two steps { (25) gives $\delta_{\nu,\nu}^0$, (24) gives $\delta_{\nu,n}^0$ for $n = \nu+1, \ldots N$ }. That completes the plane $m = 0$. For the further planes $(m = 1,2, \ldots ,M)$, the descenders are present, but the lower planes are known. After initializing by setting $\delta_{m-1,\nu}^m = 0,$ $\nu = 0,1, \ldots ,m-1$, (25) and (24) can again be used to complete each successive m–level.

The description above corresponds to a nest of three loops:

$$
\begin{aligned}
&\text{for} \quad m = 1 \quad \text{to} \quad M \\
&\qquad \text{for} \quad \nu = 0 \quad \text{to} \quad N \\
&\qquad\qquad \text{for} \quad n = \max\,(m,\nu) \quad \text{to} \quad N \qquad\qquad\qquad (26) \\
&\qquad\qquad\qquad \vdots \\
&\qquad\qquad\qquad \vdots
\end{aligned}
$$

There are $3! = 6$ different ways to order these three loops. They are all acceptable (i.e. whichever ordering is used, each new $\delta_{n,\nu}^m$ will depend only on weights already calculated). However, the calculation of $\dfrac{\omega_{n-2}(\alpha_{n-1})}{\omega_{n-1}(\alpha_n)}$ in (25) becomes particularly convenient if we choose the loop ordering:

$$
\begin{aligned}
&\text{for} \quad n = 1 \quad \text{to} \quad N \\
&\qquad \text{for} \quad \nu = 0 \quad \text{to} \quad n \\
&\qquad\qquad \text{for} \quad m = 0 \quad \text{to} \quad \min\,(n,M) \qquad\qquad\qquad (27) \\
&\qquad\qquad\qquad \vdots \\
&\qquad\qquad\qquad \vdots
\end{aligned}
$$

This leads to the algorithm:

Enter M, N, x_0, α_0, α_1, α_2, ... , α_N

$\delta^0_{0,0}$:= 1

$c1$:= 1

for n := 1 to N

    $c2$:= 1

    for ν := 0 to $n-1$

        $c3$:= $\alpha_n - \alpha_\nu$

        $c2$:= $c2 \cdot c3$

        if $n \le M$ then $\delta^n_{n-1,\nu} := 0$ (28)

        for m := 0 to min (n,M)

$$\delta^m_{n,\nu} := ((\alpha_n - x_0)\, \delta^m_{n-1,\nu} - m\, \delta^{m-1}_{n-1,\nu}) \,/\, c3$$

        next m

    next ν

    for m := 0 to min (n,M)

$$\delta^m_{n,n} := \frac{c1}{c2} \left(m\, \delta^{m-1}_{n-1,n-1} - (\alpha_{n-1} - x_0)\, \delta^m_{n-1,n-1} \right)$$

    next m

  $c1$:= $c2$

next n

Note 1: $m\, \delta^{m-1}_{n-1,\nu}$ and $m\, \delta^{m-1}_{n-1,n-1}$ are to be taken as zero when $m = 0$ (and $\delta^{m-1}_{n-1,\nu}$ and $\delta^{m-1}_{n-1,n-1}$ are undefined).

Note 2: If N is very large and floating point is used, the calculation of $\omega_{n-1}(\alpha_n)$ (in the variable $c2$) might cause overflow (or underflow).

For example, in generating extensions of Tables 1–3, this problem arises when N ! exceeds the largest possible number (i.e. $N > 34$ in typical 32 bit precision with $3 \cdot 10^{38}$ as the largest number, $N > 965$ in CRAY single precision (64 bit word length, 15 bit exponent, largest number approximately 10^{2465})). In such cases, scaling of $c1$ and $c2$ (only used in forming the ratio $c1/c2$) should be added to the code.

Note 3: If we only want the weights for formulas which use all the given points (rather than also for formulas of lower accuracies, using only leading subsets of the points α_0, α_1, ... , α_N), the code can easily be modified so that each n–plane overwrites the previous one (i.e. we need only assign memory for a 2–D array δ^m_ν of weights rather than for a 3–D one $\delta^m_{n,\nu}$).

4. VERIFICATION OF NUMERICAL STABILITY

Tables 1, 2 and 3 were each obtained by single calls to the routine (28) (implemented in REDUCE, which provides exact rational arithmetic). The relation

$$\sum_{\nu=0}^{n} \delta^m_{n,\nu} \;=\; \delta_m \;=\; \begin{cases} 1 & , \quad m = 0 \\ 0 & , \quad m > 0 \end{cases} \tag{29}$$

offers a good test of numerical stability if floating point arithmetic is used (since the different ν–planes are calculated independently because of the structure of (19)). The coefficients in Table 1 (or Table 2) form a particularly sensitive test because of two circumstances often associated with instabilities:

1. The algorithm is capable of generating weights which grow very rapidly with both m and n. However, in this case, there is only a comparatively weak growth with m and none with n.

2. For m fixed and n and ν increasing (in the case of Table 1; reversed in Table 2, cf. Fornberg (1989)):

$$\delta^m_{n,\nu} \;=\; \begin{cases} O(1/\nu^2) & , \quad m \text{ even} \\ O(1/\nu) & , \quad m \text{ odd} \end{cases} \tag{30}$$

For m even, comparatively small numbers are generated with help of larger ones from the level $m{-}1$.

To separate the growth in errors from changes in the sizes of the weights, we evaluate

$$S(m,n) = \frac{\left| \sum_{\nu=0}^{n} \delta_{n,\nu}^{m} - \delta_{m} \right|}{\sum_{\nu=0}^{n} |\delta_{n,\nu}^{m}|} \tag{31}$$

for some m and n in the case of (an extended) Table 1. Table 4 shows the results of such a calculation, performed in single precision (32 bit floating point) on an IBM PC compaible computer. In no case (including this one) have we noticed any signs of instabilities.

order of
derivative

m	$n=2$	$n=4$	$n=10$	$n=20$	$n=50$	$n=100$	$n=200$
200							$.67 \cdot 10^{-7}$
100						$.16 \cdot 10^{-7}$	$0.$
50					$.44 \cdot 10^{-7}$	$.16 \cdot 10^{-7}$	$0.$
20				$.12 \cdot 10^{-7}$	$.14 \cdot 10^{-7}$	$0.$	$0.$
10			$0.$	$.19 \cdot 10^{-7}$	$.40 \cdot 10^{-7}$	$.91 \cdot 10^{-8}$	$0.$
4		$0.$	$0.$	$0.$	$.17 \cdot 10^{-6}$	$.26 \cdot 10^{-6}$	$.46 \cdot 10^{-6}$
2	$.30 \cdot 10^{-7}$	$0.$	$.39 \cdot 10^{-7}$	$.35 \cdot 10^{-7}$	$.95 \cdot 10^{-7}$	$.24 \cdot 10^{-6}$	$.50 \cdot 10^{-6}$
1	$.34 \cdot 10^{-7}$	$.99 \cdot 10^{-8}$	$.14 \cdot 10^{-7}$	$.17 \cdot 10^{-7}$	$.14 \cdot 10^{-7}$	$.52 \cdot 10^{-7}$	$.75 \cdot 10^{-7}$

Formulas possible only if $n \geq m$.

$n+1$ = number of grid points

Table 4.   Error function $S(m,n)$ for some m and n in the case of (an extended) Table 1. Results from a calculation in 32 bit floating point.

4. SOME APPLICATIONS OF THE ALGORITHM

Generation of Tables

Tables 1 – 3 are typical examples of tables which can be generated by a single application of the algorithm. They were obtained by choosing $M = 4$, $x_0 = 0$ and: Table 1: $N = 8$, $\alpha_\nu = \{0,1,-1,2,-2,3,-3,4,-4\}$, Table 2: $N = 7$, $\alpha_\nu = \{1/2,-1/2,3/2, -3/2,5/2,-5/2,7/2,-7/2\}$, Table 3: $N = 8$, $\alpha_\nu = \{0,1,2,3,4,5,6,7,8\}$.

Irregular and dynamically changing grids.

If a 1–D grid is irregular, a full set of weights must be calculated at each grid point. Since the present algorithm contains no data dependent tests, these tasks can be vectorized for maximal efficiency on modern supercomputers. The cost of calculating the weights, both on sequential and on vector computers, is therefore comparable to applying the weights only a few times. This makes it affordable to change the grid at almost every time step, if desired, in applications such as the tracking of rapidly moving fronts.

Interpolation

Suppose the values of a function $f(x)$ at $x = \alpha_\nu$, $\nu = 0,1, \ldots n$, are stored in a vector f_ν and we want to find $f(x_0)$ by interpolation. The algorithm by Aitken is particularly convenient (as is a related version by Neville; cf. Press et al (1986)). The key steps are:

$$
\begin{aligned}
&\text{for} \quad j = 0 \text{ to } n{-}1 \\
&\qquad \text{for} \quad \nu = j{+}1 \text{ to } n \\
&\qquad\qquad f_\nu := ((\alpha_\nu - x_0)\, f_j - (\alpha_j - x_0)\, f_\nu) \,/\, (\alpha_\nu - \alpha_j) \qquad (32)\\
&\qquad \text{next} \quad \nu \\
&\qquad \text{next} \quad j
\end{aligned}
$$

After the completion of these steps, the contents of f_ν, $\nu = 0,1,2, \ldots n$, have been changed into the interpolated results at $x = x_0$, based on the old f_0, $f_1, \ldots, f_\nu$. Number of operations required (to leading order):

COST: $0.5\ n^2$ independent of f_ν (pre–compute $\alpha_\nu - x_0$, $\alpha_j - x_0$, and $\alpha_\nu - \alpha_j$)

$2\ n^2$ dependent on f_ν (i.e. need to be repeated for each new function on the same grid).

The incremental cost for additional functions on the same grid can be decreased at the expense of more preparations:

COST: $1.5\ n^2$ independent of f_ν (pre–compute $\dfrac{\alpha_\nu - x_0}{\alpha_\nu - \alpha_j}$ and $\dfrac{\alpha_j - x_0}{\alpha_\nu - \alpha_j}$)

$1.5\ n^2$ dependent on f_ν.

Alternatively, we can use the present algorithm in the special case of $M = 0$ (i.e. ignoring its ability to find formulas for derivatives). Then

COST: $2\ n^2$ independent of f_ν (only x_0 and α_ν enter into algorithm)

$2\ n$ dependent on f_ν (dot product of weights and function values)

The initial cost, $2\ n^2$ operations after the first set of function values, is lower than for either version of Aitken's method. If additional sets of function values are to be interpolated on the same grid, the incremental cost is only $2\ n$ operations vs. $1.5\ n^2$ (at best) for Aitken's (or Neville's) method.

In analogy with Aitken's method, which generates interpolated results based on $f_0, \ldots, f_\nu$, $\nu = 0,1, \ldots n$ (and not only the last one of these), the present method provides the full sets of weights for all these cases.

120

REFERENCES

Bickley, W.G. (1941), *Formulae for Numerical Differentiation*, Math. Gazette, **25**, pp 19–27.

Fornberg, B. (1988), *Generation of Finite Difference Formulas on Arbitrarily Spaced Grids*, Math. Comput. **51**, pp 699–706.

Fornberg, B. (1989), *High Order Finite Differences and the Pseudospectral Method on Staggered Grids*. Submitted to SIAM J. Numer. Anal.

Milne–Thomson, L.M. (1933), *The Calculus of Finite Differences*. MacMillan and Co. Ltd.

Press, W.H., Flannery, B.P., Teukolsky,S.A. & Vetterling, W.T. (1986), *Numerical Recipes*. Cambridge University Press.

Bengt Fornberg
Exxon Research and Engineering Company
Annandale, NJ 08801, USA.

E–mail: bfornbe @ erenj.bitnet

I.G. GRAHAM and Y.YAN

Boundary Integral Methods for Laplace's Equation

1. INTRODUCTION

In this paper we present a brief review of recent results on the direct boundary integral method for solving Laplace's equation

$$\Delta \varphi = 0 \tag{1}$$

in the subset of $\mathbf{R}^2$ either exterior or interior to a closed polygonal contour Γ. We will be concerned here with collocation-type methods, rather than with the more thoroughly analysed (but less practical) Galerkin schemes. We do not attempt a rigorous treatment, but refer to various other papers for the details. Some emphasis will be placed on a simple collocation scheme for first kind equations, but we also review what is known for second kind equations.

The following notation will be used throughout. We assume Γ has corners x_0, x_2, $\cdots$, $x_{2r} = x_0$, and that the exterior angle at each x_{2l} is $(1 + \chi_l)\pi$. Let x_{2l-1} be the mid-point of the straight line joining x_{2l-2} to x_{2l}. We use Γ_j to denote the portion of Γ joining x_{j-1} to x_j. For $l = 1, \cdots, r$ and $v: \Gamma \to \mathbf{R}$, we let v^l denote the restriction of v to $\Gamma^{2l} \cup \Gamma^{2l+1}$. Then we choose fixed integers m_j, $j = 1, \cdots, 2r$ and, for any $N \in \mathbf{N}$, we divide Γ_j up into $m_j N$ segments. This defines a mesh on Γ with $n = (m_1 + \cdots + m_{2r})N$ segments, which will be used to define the discretizations of our boundary integral equations. We will be concerned with convergence as $N \to \infty$.

2. BOUNDARY INTEGRAL EQUATIONS

A useful reference for basic potential theory in polygonal regions is [8] and the references therein.

Let Ω^i and Ω^e denote, respectively, the regions interior and exterior to Γ. For $\xi \in \Gamma$, and $x \neq \xi$, define

$$G(x,\xi) = \frac{1}{2\pi} \log \frac{1}{|x-\xi|} \quad , \quad G'(x,\xi) = \frac{\partial}{\partial n(\xi)} G(x,\xi),$$

where $\partial/\partial n(\xi)$ denotes the normal derivative at $\xi \in \Gamma$. (All normal derivatives are outward from Ω^i). Then, for any $v: \Gamma \to \mathbf{R}$, define the *single* and *double layer potentials*, respectively, by

$$\mathcal{V}v(x) = \int_\Gamma G(x,\cdot)v(\cdot) \quad , \quad \mathcal{W}v(x) = \int_\Gamma G'(x,\cdot)v(\cdot), \quad x \in \mathbf{R}^2.$$

If φ satisfies (1) in Ω^i, and if φ and its first order partial derivatives are square integrable on Ω^i, then Green's identity states that

$$W\varphi(\mathbf{x}) - V\frac{\partial\varphi}{\partial n}(\mathbf{x}) = \begin{cases} -\varphi(\mathbf{x}), & \mathbf{x}\in\Omega^i \\ -(1/2)\varphi(\mathbf{x}), & \mathbf{x}\in\Gamma^R \\ 0 & \mathbf{x}\in\Omega^e, \end{cases} \tag{2}$$

where Γ^R denotes Γ minus its corner points. (If $\mathbf{x}$ is a corner point of Γ, then a different multiple of $\varphi(\mathbf{x})$ appears on the right hand side. This detail is unimportant here.) On the other hand, if φ satisfies (1) in Ω^e, and if φ and its first order partial derivatives are square integrable in all bounded subdomains of Ω^e, and if φ satisfies also the "radiation" conditions:

$$\varphi(\mathbf{x}) = O(1/|\mathbf{x}|) \ , \quad \text{and} \quad |\nabla\varphi(\mathbf{x})| = O(1/|\mathbf{x}|^2) \ , \text{ as } |\mathbf{x}|\to\infty, \tag{3}$$

then we have

$$W\varphi(\mathbf{x}) - V\frac{\partial\varphi}{\partial n}(\mathbf{x}) = \begin{cases} \varphi(\mathbf{x}), & \mathbf{x}\in\Omega^e \\ (1/2)\varphi(\mathbf{x}), & \mathbf{x}\in\Gamma^R \\ 0 & \mathbf{x}\in\Omega^i. \end{cases} \tag{4}$$

If (1) is now to be solved in Ω^i (Ω^e), subject to mixed boundary conditions (and condition (3) in Ω^e), then the middle equation of (2) ((4)) gives a boundary integral equation to be solved for the unknown parts of φ or $\partial\varphi/\partial n$ on Γ. Once this boundary data is found the first equation of (2) (or (4)) then gives a representation for φ at any point $\mathbf{x}$ in its domain of definition. When the boundary condition is purely Neumann, the integral equation to be solved is of the second kind; a first kind equation results from Dirichlet boundary conditions. For mixed boundary conditions a 2×2 system (with essentially one equation of each type) results. We shall discuss only the first two cases here, and remark that there are still many open questions concerning practical methods for solving the full mixed problem (althlough [8] contains pioneering results on the Galerkin method for that problem.)

3. NEUMANN PROBLEMS

If (1), (3) are to be solved in Ω^e subject to $\partial\varphi/\partial n = h$ on Γ, then (4) yields the equation:

$$(I - 2W)u = -2Vh, \tag{5}$$

to be solved for $u:=\varphi|_\Gamma$. On the other hand, if (1) is to be solved in Ω^i, subject to $\partial\varphi/\partial n = h$ on Γ, then (2) yields

$$(I + 2W)u = 2Vh \tag{6}$$

to be solved for $u:=\varphi|_\Gamma$.

At the moment these equations hold only at points of Γ^R, but they can be extended to all of Γ by taking appropriate limits of the integral operators at the corners (see, for example, [10]). Introduce now the space $L^2(\Gamma)$ of square-integrable real-valued functions on Γ, with inner product denoted by $(\cdot,\cdot)$. Then, as is well-known, the operators $(I \pm 2\mathcal{W})$ are Fredholm operators of index zero on $L^2(\Gamma)$. The operator $(I - 2\mathcal{W})$ has a trivial null-space, and hence is invertible on $L^2(\Gamma)$. Thus (5) has a unique solution $u \in L^2(\Gamma)$, when $\mathcal{V}h \in L^2(\Gamma)$.

However $(I + 2\mathcal{W})$ has null-space span$\{1\}$, where 1 is the constant unit function on Γ. Thus if (6) has a solution at all, it is unique only up to an additive constant. Moreover for (6) to have a solution, $\mathcal{V}h$ must be orthogonal to the null-space of the adjoint operator $(I + 2\mathcal{W}^*)$. Classical arguments for smooth Γ (e.g. [15]) may be generalised easily to the polygonal case, showing that the null-space in question is span$\{v_0\}$, where $v_0 \in L^2(\Gamma)$ and $\omega \in \mathbf{R}$ solve the system

$$\mathcal{V}v_0 = \omega$$

(7)

$$(v_0, 1) = 1$$

This problem has a unique solution (for example [14]). Now the condition needed for the existence of a solution to (6) becomes $(\mathcal{V}h, v_0) = 0$, or equivalently, since $\mathcal{V}$ is self-adjoint, $(h, \mathcal{V}v_0) = 0$. By (7) this is equivalent to

$$(h, 1)\omega = 0 \tag{8}$$

However, recall that h is the Neumann data of a solution of (1) in Ω^i, so by the Divergence Theorem $(h, 1) = 0$. So (8) is satisfied, and (6) always has a (non-unique) solution.

To remove the non-uniqueness, suppose L is a linear functional on $L^2(\Gamma)$ which does not have 1 in its null-space. Then consider, instead of (6), the equation

$$(I + 2\mathcal{W})u + Lu = 2\mathcal{V}h. \tag{9}$$

We shall see that (9) has a unique solution in $L^2(\Gamma)$, which is also a solution of the original equation (6). To see this, first observe that the operator on the left hand side of (9) is still Fredholm of index zero, and is invertible if it is injective. So suppose v solves the homogeneous equation

$$(I + 2\mathcal{W})v + Lv = 0. \tag{10}$$

Then

$$(v_0, (I + 2\mathcal{W})v + Lv) = 0,$$

and so

$$((I + 2\mathcal{W}^*)v_0, v) + (v_0, Lv) = 0.$$

By properties of v_0, this implies

$$0 = (v_0, Lv) = (v_0, 1)Lv = Lv. \tag{11}$$

Substituting (11) into (10), we see that v is in the null-space of $(I + 2\mathcal{W})$, and so v is constant. Since L is linear and 1 is not in its null-space, it follows from (11) that $v = 0$. Hence we have shown that the operator on the left hand side of (9) is injective, and so (9) has a unique solution u. Now taking the inner product of each side of (9) with v_0 and using the properties of v_0 and $\mathcal{V}$, we obtain

$$(v_0, Lu) = (v_0, 2\mathcal{V}h)$$
$$= 2(\mathcal{V}v_0, h)$$
$$= 2(1, h)\omega = 0.$$

Hence

$$Lu = (v_0, Lu) = 0,$$

and the unique solution of (9) is indeed a solution of the unaugmented equation (6), as claimed. The above results can be extended to the space $C(\Gamma)$ of continuous real-valued functions on Γ. If h is sufficiently smooth then (5) has a unique solution in $C(\Gamma)$ which has singularities in its first and higher derivatives at the corners of Γ. If, in addition, L is a linear functional on $C(\Gamma)$ which does not have 1 in its null-space then (9) also has a unique solution in $C(\Gamma)$, with the same smoothness properties as the solution of (5).

A related technique for removing the non-uniqueness in (6) is given in Atkinson [1]. However Atkinson's reformulation of (6) requires, in principle, the prior calculation of v_0, so it will be more costly than (9).

The Nystrom method established in [10] can now be used to solve (5) and (9). Since (5) is well documented, we discuss (9) here. For concreteness (and, it turns out, convenience), we assume $Lu = u(\mathbf{x}^*)$, where $\mathbf{x}^*$ is any particular corner of Γ. Then rearrange (9) as

$$u + 2\sum_{l=1}^{r}\mathcal{W}(u^l - u(\mathbf{x}_{2l})1^l) + 2\sum_{l=1}^{r}u(\mathbf{x}_{2l})\mathcal{W}1^l + u(\mathbf{x}^*) = 2\mathcal{V}h. \qquad (12)$$

A numerical solution u_h ($h = 1/n$) to (12) is then defined as follows. First, for any $v \in C(\Gamma)$, define an approximation $\mathcal{W}_h v$ to $\mathcal{W}v$ by applying a fixed interpolatory quadrature rule (e.g. midpoint, trapezoidal, etc.) to each segment of the mesh on Γ. Then replace the $\mathcal{W}$ appearing in the first sum on the left hand side of (12) by $\mathcal{W}_h$. The integrals $\mathcal{W}1^l$ appearing in the second sum are known analytically, and collocation of the approximate equation at the quadrature points and the corner points then yields a linear system for u_h at those points. This method can be thought of as a discrete collocation method, although that observation is of no practical significance. The results of [10] can then be extended to show that, provided the mesh is graded appropriately near each of the corners of Γ, and provided the method is stable, then it will converge (in the uniform norm) with rate $O(N^{-R})$, where R is the order of the underlying quadrature rule. (This estimate assumes that the right hand side $2\mathcal{V}h$ is calculated analytically, or at least is approximated sufficiently accurately by quadrature.)

However, proving stability for this method turns out to be tricky, and in [10] was established in general only for a slightly modified method. This requires the contributions from i^* subintervals on each side of each corner be removed from $\mathcal{W}_h$, and [10] proved (for (5)) that a fixed i^* exists which guarantees stability, and which preserves the rate of convergence of the approximate solution. The same result will hold for (9).

Altlhough the experiments in [7] showed that a stability theory of unmodified methods is not possible in general, it is well observed that most practical methods are stable without modification. The modification is still useful in practice, however, as a means of accelerating the convergence of multigrid solvers for (12) (In effect, by enhancing the smoothing property of the numerical integral operator near the corners of Γ ([3]).

It turns out that a similar modification is useful in the stability theory of numerical methods for first kind integral equations, and also in the acceleration of (preconditioned) multigrid methods for the resultant linear systems. We now turn our attention to this case.

4. DIRICHLET PROBLEMS

In this section we consider boundary integral equations of the form

$$2\mathcal{V}u = g \tag{13}$$

If (1), (3) are solved in Ω^e subject to φ given on Γ, then (4) yields (13) with $g = (-I + 2\mathcal{W})\varphi|_\Gamma$. On the other hand, if (1) is solved in Ω^i, subject to φ given on Γ, then by (2) we have (13) with $g = (I + 2\mathcal{W})\varphi|_\Gamma$. In each case the unknown is $u = \partial\varphi/\partial n|_\Gamma$.

We shall assume for convenience that the transfinite diameter of Γ is not equal to 1, so that (13) has a unique solution in $L^2(\Gamma)$ for all $g \in L^2(\Gamma)$ such that the tangential derivative of g is also in $L^2(\Gamma)$. As we saw in §3, collocation-type methods of arbitrarily high order are known to converge for second kind equations. Much less is known about such methods for first kind equations (when Γ is polygonal), and here we restrict ourselves to discussing the numerical solution of (13) by the following very simple collocation scheme.

(a) The piecewise constant collocation method

Recall the mesh introduced in §1. Then approximate u by a function u_h (where $h = 1/n$), which is constant on each segment of the mesh, and which satifies

$$2\mathcal{V}u_h(\mathbf{x}^*) = g(\mathbf{x}^*) , \tag{14}$$

where the collocation points $\mathbf{x}^*$ range over all mid-points of the mesh segments.

This method is quite old (having been originally proposed by Symm [19]), but it's analysis is still something of an open question. Progress was recently made by Arnold, Saranen and Wendland [4], [17], whose wide-ranging results include an analysis of method (14) when Γ is smooth. However their methods, based on Fourier analysis, do not extend

126

naturally to polygonal Γ.

A new approach for polygonal Γ was given in [22], and is extended in [11]. We now give an overview of the results in [11].

The theory in [11] assumes that the mesh introduced in §1 is uniform on each side of the polygon Γ (i.e. each Γ_j is subdivided into $m_j N$ *equal* pieces). However, we shall also make some remarks later on the performance of graded meshes. Now choose a parametrisation $\gamma:[-\pi,\pi]\to\Gamma$ with the property that $|\gamma'|$ is constant on each Γ_j, and γ maps the uniform mesh with n subintervals on $[-\pi,\pi]$ to the mesh on Γ described above. Any $v:\Gamma\to\mathbf{R}$ is then identified with $v(s)=v(\gamma(s)),s\in[\pi,\pi]$. Then (13) is equivalent to

$$-\frac{1}{\pi}\int_{-\pi}^{\pi} \log|\gamma(s)-\gamma(\sigma)|\,|\gamma'(\sigma)|\,u(\sigma)\,d\sigma = g(s), \quad s\in[-\pi,\pi]$$

which we abbreviate as

$$Kw = g \quad, \tag{15}$$

where $w = |\gamma'|\,u$. Let V_h denote the space of real valued functions on $[-\pi,\pi]$ which are piecewise constant with respect to the uniform mesh of n subintervals on $[-\pi,\pi]$. Let Q_h denote the projection operator onto V_h, which takes any continuous function on $[-\pi,\pi]$ to its interpolant at the mid-points of subintervals. Then (14) is equivalent to seeking $w_h:=|\gamma'|u_h$ to satisfy

$$Q_h K w_h = Q_h g. \tag{16}$$

Let A denote the operator

$$Av(s) = -\frac{1}{\pi}\int_{-\pi}^{\pi}\log|2\sin(s-\sigma)/2|\,v(\sigma)d\sigma + \frac{1}{2\pi}\int_{-\pi}^{\pi} v(\sigma)\,d\sigma \quad.$$

Then A is invertible (on appropriate function spaces), and (15) may be recast as

$$(I + A^{-1}(K - A))\,w = f := A^{-1}g \quad. \tag{17}$$

Thus w satisfies a (non standard) second kind equation. The principal idea in [22] was to observe that (16) can similarly be converted into a (non-standard) projection method for (17), and then to invoke well-known arguments for the analysis of such methods.

The first step in this procedure is to study the operator $Q_h A_h : V_h \to V_h$. (Here A_h denotes the restriction of A to V_h.) Choose the usual basis of V_h consisting of the characteristic functions of the subintervals, and let $\mathcal{A}$ denote the matrix of $Q_h A_h$ with respect to that basis. Then it turns out that $\mathcal{A}$ is a *circulant* matrix. That is, if $\alpha_{j,k}$ denotes its $(j,k)th$ element, then, for $j=1,\cdots,n-1$,

$$\alpha_{j+1,k+1} = \alpha_{j,k}, \qquad k = 1,\cdots,n-1,$$

and

$$\alpha_{j+1,1} = \alpha_{j,n}.$$

Circulants have very nice properties (e.g. [9]), and using these the eigenvalues of $\mathcal{A}$ can be calculated explicitly. They all turn out to be positive, and so $\mathcal{A}$ is invertible, and $Q_h A_h$ is invertible on V_h. Moreover the inverse of $\mathcal{A}$ can be explicitly computed from the first row of $\mathcal{A}$ by (essentially) two n-dimensional discrete Fourier tranforms. Thus $Q_h A_h$ can be inverted in $O(n \log n) = O(N \log N)$ multiplications.

Now write the collocation method (16) as

$$Q_h A (I + A^{-1}(K - A)) w_h = Q_h A f,$$

or, equivalently, premultiplying by $(Q_h A_h)^{-1}$,

$$(I + B_h A^{-1}(K - A)) w_h = B_h f \tag{18}$$

where $B_h := (Q_h A_h)^{-1} Q_h A$.

From now on we use L^2 to denote the space of real-valued square integrable functions on $[-\pi, \pi]$. To compare (18) with (17), the crucial questions are the properties of B_h and $A^{-1}(K - A)$. Let $|\cdot|$ denote the norm in L^2, and use the same notation for the operator norm in L^2. The norm of B_h is calculated in the following lemma, which is proved in [22].

LEMMA 1. $|B_h| \leq B$, *for all* $N \in \mathbf{N}$, *where* $1.34 < B < 1.35$.

The following decomposition of $A^{-1}(K - A)$ is proved in [11].

LEMMA 2. $A^{-1}(K - A) = -HR + E$. *Here E is compact on* L^2, *and H is the Hilbert transform which is an isometry on* L^2. *The operator R is a non-compact convolution operator with the property that, for any* $\varepsilon > 0$,

$$|R| < \max\{ (1 - \cos \frac{\chi_l \pi}{2}) : l = 1, \cdots, r \} + \varepsilon, \tag{19}$$

provided m_{2j}/m_{2j+1} *is close enough to* $|\Gamma_{2j}|/|\Gamma_{2j+1}|$, *for each* $j = 1, \cdots, r$. (*Here* $\Gamma_{2r+1} = \Gamma_1, m_{2r+1} = m_1$.)

REMARK. The splitting of the operator K into $A + (K - A)$ has been extensively used in the literature (for example, see [13]). Related remarks on the operator $A^{-1}(K - A)$ are given in [5], where the integral equation (17) is also considered, but from a different point of view.

We henceforth assume that the $\{m_j\}$ are chosen so that $|R| < 1$. (This is clearly possible by Lemma 2). By Lemma 2, (17),(18) then become

$$(I - (HR - E))w = f,$$

$$(I - B_h(HR - E))w_h = B_h f.$$

Since HR is a contraction and E is compact, the operator $(I - (HR - E))$ is Fredholm of index zero on L^2. Known uniqueness results for (13) then imply that $(I - (HR - E))$ is injective and hence invertible. Evidently B_h reduces to the identity on V_h. Since V_h is dense in L^2, as $N \to \infty$, we have $B_h \to I$ pointwise on L^2. Using this, and the fact that B_h is uniformly bounded on L^2, as $N \to \infty$, standard methods for second kind operators show that $(I - B_h(HR - E))$ is invertible on L^2, and its inverse is uniformly bounded as $N \to \infty$, provided $|B_h(HR)|$ is bounded below 1, as $N \to \infty$. However the crude bound

$$|B_h HR| \le |B_h| |R|$$

leads, via Lemma 2, to the requirement that

$$\max\{ (1 - \cos \frac{\chi_l \pi}{2}) : l = 1, \cdots, r \} < B^{-1} < (1.34)^{-1},$$

which in turn led in [22] to unnatural restrictions on the angles at the corners of the polygon. These are similar to the angle restrictions which occurred in [2] for the equations in §3. To avoid such restrictions we consider, as in [7], a slightly different method.

(b) The modified collocation method

On each side of each of the corners of Γ, choose the nearest i^* segments of the mesh to the corner. Denote the union of all these chosen segments by Γ^h. Let $I^h \subseteq [-\pi, \pi]$ denote the set $\gamma^{-1}(\Gamma^h)$. Then define the truncation operator T^h by

$$T^h v(s) = 0, \ s \in I^h, \quad \text{and} \quad T^h v(s) = v(s), \ s \in [-\pi, \pi] \setminus I^h.$$

Using T^h, we then define the modified operator K^h by

$$K^h = A + (K - A) T^h . \tag{20}$$

When $i^* = 0$, $I^h = \varnothing$, and $K^h = K$. When $i^* > 0$, K^h differs from K in that the kernel $\log |\gamma(\sigma) - \gamma(\sigma)|$ of $Kv(s)$ is replaced by $\log |2e^{-\frac{1}{2}} \sin (s - \sigma)/2|$ for $\sigma \in I^h$, and all $s \in [-\pi, \pi]$. We then define the modified collocation solution w_h to (15) by

$$Q_h K^h w_h = Q_h g . \tag{21}$$

The following technical result is proved in [11].

LEMMA 3. *Let P_h denote the orthogonal projection of L^2 onto V_h. Then, for each $\varepsilon > 0$, there exists i^* independent of N, such that*

$$|(I - P_h) R T^h| < \varepsilon ,$$

for all N sufficiently large.

The proof requires a careful treatment of the convolution operator R, and uses methods developed in [7], [10] for dealing with the second kind equations described in §3. Lemma 3 leads directly to the following stability theorem for method (21). From now on C will denote a generic constant independent of N.

THEOREM 4. *There exists a fixed $i* \geq 0$ such that, for all N sufficiently large, (21) defines $w_h \in V_h$ uniquely, and*

$$|w - w_h| \leq C \max \left\{ |w - B_h w| \, , \, |w - T^h w| \right\}.$$

PROOF. Using Lemma 2 and (20), equation (21) may be written

$$(I - B_h(HR - E) \, T^h) \, w_h = B_h \, f \ . \tag{22}$$

Recall that, $(I - (HR - E))$ is invertible on L^2, and hence so is $I + (I - HR)^{-1}E$. Recall also that for any $L : L^2 \to L^2$, the identities

$$(I - L \, T^h)^{-1} = I + L(I - T^h L)^{-1} \, T^h \tag{23}$$

$$(I - T^h L)^{-1} = I + T^h(I - L \, T^h)^{-1} L \tag{24}$$

are valid (when the inverses involved exist).

Now a further technical result proved in [11] is that $|B_h H_h| \leq 1$, where H_h denotes the operator H restricted to the space V_h. Hence, using this result and Lemma 1, we have

$$|B_h H R T^h| \leq |B_h H P_h R T^h| + |B_h H(I - P_h)R T^h|$$

$$\leq |R| + B|(I - P_h)R T^h|.$$

Since $|R| < 1$, Lemma 3 shows that there exists fixed $i*$ such that $|B_h H R T^h|$ is bounded below 1 as $N \to \infty$, and so

$$|(I - B_h H R T^h)^{-1}| \leq C \, , \quad \text{as } N \to \infty \, .$$

By (24), then

$$|(I - T^h B_h HR)^{-1}| \leq C \, , \quad \text{as } N \to \infty \, . \tag{25}$$

Thus we have the bound

$$|(I - T^h B_h HR)^{-1} \, T^h B_h E - (I - HR)^{-1} \, E|$$

$$\leq |(I - T^h B_h HR)^{-1}(T^h B_h - I)E|$$

$$+|(I - T^h B_h HR)^{-1}(T^h B_h - I)HR(I - HR)^{-1} \, E|$$

and each of these terms approaches zero as $N \to \infty$, since E is compact, and $T^h B_h \to I$ pointwise on L^2. The operator $I + (I - T^h B_h HR)^{-1} T^h B_h E$ therefore tends in norm to the invertible operator $I + (I - HR)^{-1}E$, and thus has an inverse which is uniformly bounded, as $N \to \infty$. By (25), then

$$|(I - T^h B_h (HR - E))^{-1}| \le C , \quad \text{as } N \to \infty ,$$

and so, by (23),

$$|(I - B_h (HR - E) T^h)^{-1}| \le C , \quad \text{as } N \to \infty . \tag{26}$$

Hence for sufficiently large N, (22) is uniquely solvable, and

$$w - w_h = (I - B_h (HR - E) T^h)^{-1} (w - B_h ((HR - E) T^h w + f))$$

$$= (I - B_h (HR - E) T^h)^{-1} \left[(w - B_h w) + B_h (HR - E)(w - T^h w) \right]$$

Taking $| \cdot |$ and using the triangle inequality with Lemma1 and (26) yields the theorem.

The rate of convergence of w_h can easily be deduced from Theorem 4. It is known that, when $\gamma(s_l) = x_{2l}$ then, for s near s_l, we have, in general, $w(s) = O(|s - s_l|^{\beta_l - 1})$, where $\beta_l = (1 + |\chi_l|)^{-1}$. Since $(I - B_h)$ annililates V_h, we have, by Lemma 1 that $|w - B_h w|$ is of the order of the error in the best approximation to w from V_h. It follows (see, e.g. [22]) that $|w - B_h w| = O(N^{1/2 - \beta})$, where $\beta = \max\{ \beta_l : l = 1, \cdots, r \} \in (1/2, 1)$. Since i^* is fixed, $|w - T^h w|$ is of the same order. So, overall,

$$|w - w_h| = O(N^{1/2 - \beta}).$$

A better rate of convergence will be obtained with graded meshes. Unfortunately the stability theory does not yet extend to this case. Nevertheless, as a numerical experiment we solved (13), with $g(x) = \dfrac{1}{\pi}(x_1^2 + x_2^2)$, and with Γ the boundary of the L-shaped domain with vertices (in anticlockwise order): (0,0), (1,0), (1,1/2), (1/2,1/2), (1/2,1), and (0,1). The sides of length 1 are subdivided into N segments (N even), while the other sides (all of length 1/2) are subdivided into $N/2$ segments. Then $n = 4N$, and $h = 1/(4N)$. The mesh is graded near each corner in the usual way (e.g. [6], [7], [10], [20]). For example, for given $q > 1$, the mesh on the side joining (0,0) to (1,0) consists of the points

$$((1/2)(2i/N)^q , 0), \quad i = 1, \cdots, N/2,$$

plus their reflections in the line $x_1 = 1/2$. The mesh is similarly graded near all the other corners, using the same grading exponent q. The modification was not needed to ensure the stability of the method (i.e. the method used was simply (14)).

Rather than calculating the rate of convergence of u_h, we look at the induced potential

$$\varphi_h(x) := \mathcal{V} u_h(x)$$

at points $x = (0.25, 0.25)$, $(-0.25, -0.25)$. The integrals required for implementation of the collocation method are computed analytically. Those required in the calculation of the potential are computed by Simpson's rule. The results, given in the tables below, show an estimated order of convergence (EOC) which climbs to a maximum of about 3 as the grading is increased. Since $O(N^{-3})$ is the rate of convergence which would be expected for potentials when Γ is smooth [16], the results with the graded meshes are very encouraging. However at present the theory does not explain properly the observed success of the method.

q	N	$\pi \times \varphi_h(0.25, 0.25)$	EOC	$\pi \times \varphi_h(-0.25, -0.25)$	EOC
1	16	0.25651		-0.34490	
	32	0.25621		-0.34604	
	64	0.25611	1.59	-0.34649	1.33
	128	0.25608	1.43	-0.34668	1.33

q	N	$\pi \times \varphi_h(0.25, 0.25)$	EOC	$\pi \times \varphi_h(-0.25, -0.25)$	EOC
2	16	0.26113		-0.33704	
	32	0.25710		-0.34457	
	64	0.25622	2.19	-0.34642	2.56
	128	0.25608	2.66	-0.34673	2.56

q	N	$\pi \times \varphi_h(0.25, 0.25)$	EOC	$\pi \times \varphi_h(-0.25, -0.25)$	EOC
3	16	0.25749		-0.34491	
	32	0.25632		-0.34652	
	64	0.25609	2.37	-0.34676	2.78
	128	0.25606	2.86	-0.34679	2.91

(c) An iterative solver

The theory of the previous section has an interesting application to the acceleration of two grid methods for solving the collocation equations. The idea is to solve (16) for $h = 2\pi/n$, $n = MN$, and N large iteratively by a process involving direct inversion of (16) only for some larger h, say $\tilde{h} = 2\pi/\tilde{n}$, $\tilde{n} = M\tilde{N}$, with $\tilde{N}$ small. For convenience we assume

$$N = k\tilde{N} , \tag{27}$$

for some $k \in \mathbf{N}$, although the algorithm does not depend strongly on this assumption. Note that (27) implies

$$V_{\tilde{h}} \subseteq V_h .$$

Then our (preconditioned) two grid algorithm for (16) is given as follows. Choose $w_h^{(0)}$. Then, for $v \geq 0$, until $|w_h^{(v+1)} - w_h^{(v)}|$ is sufficiently small, perform the following steps.

Input $\qquad\qquad\qquad\qquad w_h^{(v)}$

Compute residual $\qquad\qquad d_h^{(v)} := Q_h(g - Kw_h^{(v)})$

Precondition $\qquad\qquad d_h^{(v)} := (Q_hA_h)^{-1}d_h^{(v)}$

Smoothing step $\qquad\qquad \delta_h^{(v)} := Q_h(K - A)d_h^{(v)}$

Precondition $\qquad\qquad \delta_h^{(v)} := (Q_hA_h)^{-1}\delta_h^{(v)}$

Coarse grid correction $\qquad \delta_h^{(v)} := -(Q_{\tilde{h}}K_{\tilde{h}})^{-1}(Q_{\tilde{h}}A)\delta_h^{(v)}$

Output $\qquad\qquad\qquad w_h^{(v+1)} = w_h^{(v)} + d_h^{(v)} + \delta_h^{(v)},$

$$\tag{28}$$

where A_h, K_h denote the restrictions of A, K to V_h.

Some elementary algebra shows this algorithm is equivalent to the following algorithm for the non-standard second kind equation (18).

Input $\qquad\qquad\qquad\qquad w_h^{(v)}$

Compute residual $\qquad\qquad d_h^{(v)} := B_h f - (I + L_h)w_h^{(v)}$

Smoothing step $\qquad\qquad \delta_h^{(v)} := L_h d_h^{(v)}$

Coarse grid correction $\qquad \delta_h^{(v)} := -(I + L_{\tilde{h}})^{-1}B_{\tilde{h}}\delta_h^{(v)}$

Output $\qquad\qquad\qquad w_h^{(v+1)} := w_h^{(v)} + d_h^{(v)} + \delta_h^{(v)},$

$$\tag{29}$$

where $L_h := B_h A^{-1}(K - A)$.

Standard arguments (e.g. [12]) applied to this second kind algorithm then yield the error estimate

$$|w_h - w_h^{(v+1)}| \le C_{\tilde{N},k} |w_h - w_h^{(v)}| \ ,$$

where

$$C_{\tilde{N},k} = C_{\tilde{N},k}^1 \ C_{\tilde{N},k}^2 \ ,$$

and

$$C_{\tilde{N},k}^1 = |(I + L_{\tilde{h}})^{-1}|$$

$$C_{\tilde{N},k}^2 = |(B_{\tilde{h}} - I) L_h + (B_{\tilde{h}} L_h - L_{\tilde{h}}) L_h| \ .$$

Observe now that the operators $B_{\tilde{h}} B_h - B_h$ and $B_{\tilde{h}} B_h - B_{\tilde{h}}$ both annihilate $V_{\tilde{h}}$, and hence, as $\tilde{N} \to \infty$,

$$B_{\tilde{h}} B_h - B_h \to 0, \quad \text{and} \quad B_{\tilde{h}} B_h - B_{\tilde{h}} \to 0 \quad \text{pointwise on } L^2.$$

Thus, when Γ is smooth, $A^{-1}(K - A)$ is compact, and $C_{\tilde{N},k}^2$ may be made arbitrarily small and $C_{\tilde{N},k}^1$ is uniformly bounded as $\tilde{N} \to \infty$. So for $\tilde{N}$ sufficiently large $C_{\tilde{N},k} < 1$, and the two grid algorithm for (16) converges linearly. Then $w_h^{(v)}$ converges to within $0(1/N) = 0(h)$ tolerance of w_h in $0(\log N)$ iterations. Each iteration costs $0(N^2)$ multiplications, provided the coarse grid correction is considered negligible (recall that each preconditioning step costs $0(N \log N)$ multiplications). Convergence is thus obtained in $0(N^2 \log N)$ multiplications. (See [18] for related observations about a multigrid Galerkin-type method for (13).)

When Γ is a polygon, $A^{-1}(K - A) = -HR + E$, and the above convergence proof fails. However the modified collocation scheme (21) may also be solved iteratively by applying algorithm (28) with K replaced by K^h. This is again equivalent to (29), but this time with

$$L_h = B_h(-HR + E) T^h \ ,$$

and

$$L_{\tilde{h}} = B_{\tilde{h}}(-HR + E) T^h \ ,$$

In [11] we prove the following result.

THEOREM 5. *Choose* $i^* = \tilde{N}$. *Then, for sufficiently large k, (28) with K replaced by K^h converges linearly for all $\tilde{N}$ sufficiently large.*

NOTE 1. For second kind equations on domains with very sharp corners, the two grid algorithm actually fails to converge, and modification is necessary to restore convergence

134

[3]. This theorem demonstrates that the modification will have a similar effect on the convergence of (28) (equivalently (29)), should the non smoothness of $A^{-1}(K - A)$ mean it fails to converge. Further numerical experiments will be necessry to determine if and when the modification is needed.

NOTE 2. If i^* is allowed to depend on $\tilde{N}$, as in Theorem 5, the limit function of w_h of the sequence $w_h^{(v)}$ defined by (28) does not converge as $\tilde{N} \to \infty$. This is because the component $|w - T^h w|$ in the error given by Theorem 4 does not converge as $\tilde{N} \to \infty$. However for k chosen sufficiently large initially this error could be made small compared with the other component $|w - B_h w|$ thus making w_h respectably close to w. In the second kind case, experiments suggest that even in the case of extremely sharp angles relatively small i^* is needed to achieve convergence of the iteration for moderately large $\tilde{N}$.

ACKNOWLEDGEMENTS. I. G. Graham was supported by a Royal Society Travel Grant, and by the Australian Research Council's program grant "Numerical analysis for integrals, integral equations and boundary value problems". Thanks are also due to Ken Atkinson, Graeme Chandler and Ian Sloan for useful discussions, and to Jean-Paul Berrut for referring us to relevant papers from the conformal mapping literature.

REFERENCES

[1] K.E. Atkinson, "The solution of non-unique linear integral equations", *Numer. Math.* 10 (1967), 117-124.

[2] K. E. Atkinson and F. R. de Hoog, "The numerical solution of Laplace's equation on a wedge", *IMA J. Numer. Anal.* 4 (1984), 19-41.

[3] K. E. Atkinson and I. G. Graham, "An iterative variant of the Nystrom method for boundary integral equations on nonsmooth boundaries", in *The Mathematics of Finite Elements and Applications* (ed. J. R. Whiteman), (Academic Press, London, 1988).

[4] D. N. Arnold and W. L. Wendland, "The convergence of spline collocation for strongly elliptic equations on curves", *Numer. Math.* 47 (1985), 317-341.

[5] J.-P. Berrut, "A Fredholm integral equation of the second kind for conformal mapping", *J. Comput. Appl. Math.* 14 (1986), 99-110.

[6] G. A. Chandler, "Galerkin's method for boundary integral equations on polygonal domains", *J. Austral. Math. Soc. (Series B)* 26 (1984), 1,13.

[7] G. A. Chandler and I. G. Graham, "Product integration - collocation methods for non-compact integral operator equations", *Math. Comp.* 50 (1988), 125-138.

[8] M. Costabel and E. P. Stephan, "Boundary integral equations for mixed boundary value problems in polygonal domains and Galerkin approximation", in *Mathematical Models and Methods in Mechanics*, (Banach Centre Publications 15, PWN, Warsaw, 1985), 175-251.

[9] P. J. Davis, *Circulant Matrices* (Wiley, New York, 1979).

[10] I. G. Graham and G. A. Chandler, "High order methods for linear functionals of solutions of second kind integral equations", *SIAM J. Numer. Anal.* 25 (1988), 1118-1137.

[11] I. G. Graham and Y. Yan, "Piecewise constant collocation for first kind boundary integral equations", Research Report, School of Mathematics, University of N.S.W., Sydney, 1989. (Submitted for publication.)

[12] W. Hackbusch, *Multigrid Methods and Applications* (Springer-Verlag, Berlin, 1985).

[13] P. Henrici, "Fast Fourier methods in computational complex analysis", *SIAM Rev.* 21 (1979), 481-527.

[14] W. McLean, "Boundary integral methods for the Laplace equation", Ph.D. Thesis, Australian National University, Canberra, 1985.

[15] I. G. Petrovskii, *Partial Differential Equations* (Iliffe, London, 1967).

[16] J Saranen, "The convergence of even degree spline collocation solution for potential problems in smooth domains of the plane", *Numer. Math.* 52 (1988), 499-512.

[17] J. Saranen and W. L. Wendland, "On the asymptotic convergence of collocation methods with spline functions of even degree", *Math. Comp.* 45 (1985), 91-108.

[18] H. Schippers, "Multigrid methods for boundary integral equations", *Numer. Math. 46 (1985), 351-363*.

[19] G. T. Symm, "Integral equation methods in potential theory II", *Proc. Roy. Soc.* (A) 275 (1963), 33-46.

[20] Y. Yan and I. H. Sloan, "Mesh grading for integral equations of the first kind with logarithmic kernel", *SIAM J. Numer. Anal.* 26 (1989), 574-587.

[21] Y. Yan and I. H. Sloan, "On integral equations of the first kind with logarithmic kernels", Preprint AM88/2, School of Mathematics, University of NSW, *J. Integral Equations and Applications*, to appear.

[22] Y. Yan, "The collocation method for first kind boundary integral equations on polygonal regions", Preprint AM88/21, School of Mathematics, University of NSW, *Math. Comp.*, to appear.

I.G. Graham, School of Mathematical Sciences, University of Bath, Claverton Down, Bath BA2 7AY, United Kingdom.

Y. Yan, School of Mathematics, University of N.S.W., P.O. Box 1, Sydney, N.S.W. 2033, Australia.

N.J. HIGHAM
How Accurate is Gaussian Elimination?

1   Introduction

Consider the linear system $Ax = b$, where A is the 7×7 Vandermonde matrix with $a_{ij} = j^{i-1}$ and $b_i = i$. Using Matlab we solved this system in single precision (unit roundoff $\approx 10^{-7}$) via both Gaussian elimination (GE) and Gaussian elimination with partial pivoting (GEPP), obtaining computed solutions $\hat{x}_{\text{GE}}$ and $\hat{x}_{\text{GEPP}}$, respectively. We found that

$$\frac{\|x - \hat{x}_{\text{GE}}\|_\infty}{\|x\|_\infty} \approx 4 \times 10^{-8},$$

$$\frac{\|x - \hat{x}_{\text{GEPP}}\|_\infty}{\|x\|_\infty} \approx 6 \times 10^{-3}.$$

An alternative measure of the quality of each of these computed solutions is whether it is the exact solution of a perturbed system $(A + \Delta A)x = b$ for some small ΔA (clearly, ΔA is not unique). Using results to be described in section 2 we found that the minimum value of $\|\Delta A\|_\infty / \|A\|_\infty$ is approximately 3×10^{-9} for both $\hat{x}_{\text{GE}}$ and $\hat{x}_{\text{GEPP}}$. However, the minimum value of $\max_{i,j} |\Delta a_{ij}/a_{ij}|$, which measures the perturbation component-wise relative to A, is 3×10^{-8} for $\hat{x}_{\text{GE}}$ and 2×10^{-6} for $\hat{x}_{\text{GEPP}}$. If we do one step of iterative refinement starting from $\hat{x}_{\text{GEPP}}$, *entirely in single precision*, we obtain an updated solution $\bar{x}$ for which the componentwise measure of the size of ΔA is 5×10^{-8} and $\|x - \bar{x}\|_\infty / \|x\|_\infty \approx 4 \times 10^{-5}$.

This example opposes the conventional wisdom that GE with partial pivoting is preferable to GE without pivoting. It also shows that iterative refinement in single precision can be beneficial. It is natural to ask: Can this behaviour be explained by a priori analysis of the problem, and can the example be generalized? More generally, what can one say about the sizes of $x - \hat{x}$ and ΔA for arbitrary A and b?

Answers to these questions are contained in this survey of the accuracy of Gaussian elimination in finite precision arithmetic. Work on this subject began in the 1940s at around the time of the first electronic computers. It reached maturity in the 1960s, largely due to Wilkinson's contributions. Research done since the mid 1970s has provided further understanding of the subject.

We begin in section 2 by discussing in detail the notion of backward error and showing how a wide class of backward errors can be computed. In section 3 we survey rounding error analysis for Gaussian elimination. To show that error analysis need not be difficult we give a derivation of one of the most useful backward error bounds. Implications of the analysis are discussed in section 4, while section 5 covers practical computation of backward error bounds. Iterative refinement, which has attracted renewed interest recently, is the subject of section 6. In section 7 we turn our attention to estimation of

137

the forward error. We offer some final thoughts in section 8, where we show how the results presented here help to explain the numerical example above.

Our notation is as follows. Throughout, A is a real $n \times n$ matrix. Computed quantities are denoted with a hat. Thus $\hat{x}$ is a computed solution to $Ax = b$ and $\hat{L}$, $\hat{U}$ are computed LU factors of A.

Much of this paper is concerned with backward error analysis. This is because there is a lot to say about the topic, not because we believe backward error analysis is of vital importance to all users of GE software. We suspect that many users are interested mainly in the size of $x - \hat{x}$. For them the most important material is in section 7, and familiarity with sections 2 and 6 is recommended.

2 Backward Error

In analysing the behaviour of GE in finite precision arithmetic two types of error are of interest. The *forward error* is any suitable norm of $x - \hat{x}$, and is easy to appreciate. *Backward error* is a more subtle concept, as we now explain.

In GE backward error is a measure of the smallest perturbations such that either

$$\hat{L}\hat{U} = A + \Delta A \tag{2.1}$$

if we are concerned with the LU factorization alone, or

$$(A + \Delta A)\hat{x} = b + \Delta b \tag{2.2}$$

for the overall process. In (2.1) ΔA is unique and the question is how to measure its size. In (2.2) there are many possible ΔA and Δb, which makes determination of the backward error nontrivial.

The usual motivation for considering backward error is that there may already be uncertainty in the data A and b (arising from rounding errors in storing or computing the data, for example). If the backward error and the uncertainty are of similar magnitude then it can be argued that the computed solution $\hat{x}$ is beyond reasonable criticism. Another attractive feature of backward error analysis is that it enables one to invoke existing perturbation theory to produce bounds for the forward error (see section 7).

We consider two ways to measure the size of the perturbations ΔA and Δb.

(a) Normwise Backward Error

Here we measure the perturbations using norms. We will use an arbitrary vector norm and the corresponding subordinate matrix norm. For the LU factorization the backward error is simply $\beta_N = \|A - \hat{L}\hat{U}\|/\|A\|$. For $\hat{x}$ it is

$$\beta_N = \min\{\omega : (A + \Delta A)\hat{x} = b + \Delta b, \ \|\Delta A\| \leq \omega\|A\|, \ \|\Delta b\| \leq \omega\|b\|,$$
$$\Delta A \in \mathbb{R}^{n \times n}, \ \Delta b \in \mathbb{R}^n\}.$$

The following result of Rigal and Gaches [34] gives a convenient expression for β_N.

138

Theorem 2.1

$$\beta_N = \frac{\|r\|}{\|A\|\,\|\widehat{x}\| + \|b\|}, \tag{2.3}$$

and perturbations which achieve the minimum in the definition of β_N are

$$\Delta A_{\min} = \frac{\|A\|\,\|\widehat{x}\|}{\|A\|\,\|\widehat{x}\| + \|b\|} r z^T,$$

$$\Delta b_{\min} = \frac{\|b\|}{\|A\|\,\|\widehat{x}\| + \|b\|} r,$$

where $r = b - A\widehat{x}$ and z is a vector dual to $\widehat{x}$, that is,

$$z^T \widehat{x} = \|z\|_D \|\widehat{x}\| = 1 \qquad \text{where} \qquad \|z\|_D = \max_{y \neq 0} \frac{|z^T y|}{\|y\|}.$$

Proof. See [34, Theorem 1]. ∎

If we set $\Delta b = 0$ in the definition of β_N then the result simplifies to $\beta_N = \|r\|/(\|A\|\,\|\widehat{x}\|)$ and $\Delta A_{\min} = r z^T$. In the case of the 2-norm, $z = \widehat{x}/\|\widehat{x}\|_2^2$ and the result with $\Delta b \equiv 0$ is well-known.

If A has some special property, then in the definition of β_N we might wish to prescribe that $A + \Delta A$ has the same property. In the case of symmetry, this more restrictive backward error measure (with $\Delta b \equiv 0$) has been investigated by Bunch, Demmel and Van Loan [4]. They show the pleasing result that the symmetry constraint does not change the backward error for the 2-norm, and it increases it by at most a factor $\sqrt{2}$ for the Frobenius norm.

(b) Componentwise Backward Error

This definition involves a matrix $E \geq 0$ and a vector $f \geq 0$ whose elements provide relative tolerances against which the components of ΔA and Δb are measured. For $\widehat{x}$, the componentwise backward error is defined as

$$\beta_C(E, f) = \min\{\omega : (A + \Delta A)\widehat{x} = b + \Delta b, \ |\Delta A| \leq \omega E, \ |\Delta b| \leq \omega f, $$
$$\Delta A \in \mathbb{R}^{n \times n}, \ \Delta b \in \mathbb{R}^n\}.$$

The matrix absolute value and matrix inequality are interpreted componentwise: thus $|X| \leq Y$ means that $|x_{ij}| \leq y_{ij}$ for all i, j. For the LU factorization the definition is simply $\beta_C(E) = \max\{|\Delta a_{ij}|/e_{ij}, \ 1 \leq i, j \leq n\}$. Two extreme choices of E and f are as follows.

1. $E = |A|$, $f = |b|$. In this case we measure the size of the perturbation in each element of A or b relative to the element itself. This is the most stringent backward error measure of general interest. Note that the constraints in the definition of $\beta_C(|A|, |b|)$ force $A + \Delta A$ and $b + \Delta b$ to have the same sparsity patterns as A and b respectively. Following [1] we call $\beta_C(|A|, |b|)$ the *componentwise relative backward error*.

2. $E = \|A\|_\infty ee^T$, $f = \|b\|_\infty e$, where $e = (1,1,\ldots,1)^T$. For this choice we are measuring perturbations in an absolute sense (in a similar, but not identical, way to the norm case (a)).

An expression for β_C is given by Oettli and Prager [29], and we present the short proof since it aids in understanding the result.

Theorem 2.2

$$\beta_C(E,f) = \max_i \frac{|b - A\widehat{x}|_i}{(E|\widehat{x}| + f)_i}, \qquad (2.4)$$

where $0/0$ is interpreted as zero, and $\xi/0$ ($\xi \neq 0$) as infinity, the latter case meaning that no finite ω exists in the definition of $\beta_C(E,f)$.

Proof. For any candidate perturbations ΔA and Δb in the definition of $\beta_C(E,f)$ we have

$$|r| = |b - A\widehat{x}| = |\Delta A\widehat{x} - \Delta b| \leq \omega E|\widehat{x}| + \omega f,$$

which implies that

$$\beta_C(E,f) \geq \max_i \frac{|r_i|}{(E|\widehat{x}| + f)_i} \equiv \theta.$$

To show that this lower bound is attained note that

$$r = D(E|\widehat{x}| + f) \qquad (2.5)$$

for a diagonal D with $|D| \leq \theta I$. Defining $\Delta A = DE \operatorname{diag}(\operatorname{sign}(\widehat{x}))$ and $\Delta b = -Df$ we have $|\Delta A| \leq \theta E$, $|\Delta b| \leq \theta f$, and, using (2.5),

$$(A + \Delta A)\widehat{x} - (b + \Delta b) = A\widehat{x} + DE|\widehat{x}| - b + Df = 0,$$

as required. ∎

This result shows that β_C can be computed using two matrix-vector multiplies. For the E and f in case (2) above the formula reduces to $\beta_C = \|r\|_\infty/(\|A\|_\infty\|\widehat{x}\|_1 + \|b\|_\infty)$, which is very similar to β_N for the ∞-norm.

3 Error Analysis of Gaussian Elimination

In this section we give a brief survey of rounding error analysis for Gaussian elimination. For further perspective on this topic we recommend the papers [31, 51, 52, 53] of Wilkinson.

Unless otherwise stated, results quoted are for floating point arithmetic. We will ignore permutations in stating backward error bounds; thus A actually denotes the original matrix after all row or column interchanges necessary for the pivoting strategy have been performed.

In the 1940s there were three major papers giving error analyses of GE. Hotelling [27] presented a short forward error analysis of the LU factorization stage of GE. Under the assumptions that $|a_{ij}| \leq 1$ and $|b_i| \leq 1$ for all i and j, and that the pivots are all of modulus unity, Hotelling derives a bound containing a factor 4^{n-1} for the error in the elements of the reduced upper triangular system. This result led to pessimism about the practical effectiveness of GE for solving large systems of equations. Three papers later in the same decade helped to restore confidence in GE.

Von Neumann and Goldstine [46] gave a long and difficult fixed-point error analysis for the inversion of a symmetric positive definite matrix A via GE. They obtained a bound proportional to $\kappa_2(A)$ for the residual $\|A\widehat{X} - I\|_2$ of the computed inverse $\widehat{X}$. Wilkinson [51] gives an interesting critique of this paper and points out that the residual bound could hardly be improved using modern error analysis techniques.

Turing [44] analysed GEPP for general matrices and obtained a bound for $\|x - \widehat{x}\|_\infty$ that contains a term proportional to $\|A^{-1}\|_\infty^2$. (By making a trivial change in the analysis Turing's bound can be made proportional only to $\|A^{-1}\|_\infty$.) Turing also showed that the factor 4^{n-1} in Hotelling's bound can be improved to 2^{n-1} and that still the bound is attained only in exceptional cases.

Fox, Huskey and Wilkinson [18] presented empirical evidence in support of GE, commenting that "in our practical experience on matrices of orders up to the twentieth, some of them very ill-conditioned, the errors were in fact quite small".

A major breakthrough in the error analysis of GE came with Wilkinson's pioneering backward error analysis [48, 49]. Wilkinson showed that with partial or complete pivoting the computed solution $\widehat{x}$ satisfies

$$(A + E)\widehat{x} = b, \tag{3.1}$$

where

$$\|E\|_\infty \leq \rho_n p(n)u\|A\|_\infty. \tag{3.2}$$

Here, p is a cubic polynomial and the growth factor ρ_n is defined by

$$\rho_n = \rho_n(A) = \frac{\max_{i,j,k} |a_{ij}^{(k)}|}{\max_{i,j} |a_{ij}|},$$

where the $a_{ij}^{(k)}$, $k = 0, 1, \ldots, n-1$, are the elements that occur during the elimination. Apart from its simplicity and elegance, the main feature that distinguishes Wilkinson's analysis from the earlier error analyses of GE is that it bounds the normwise backward error rather than the forward error or the residual.

Three of the first textbooks to incorporate Wilkinson's analysis were those of Fox [17, pp. 161–174], Wendroff [47] and Forsythe and Moler [16, Ch. 21]. Fox gives a simplified analysis for fixed-point arithmetic under the assumption that the growth factor is of order 1. Forsythe and Moler give a particularly readable backward error analysis which has been widely quoted.

Wilkinson's 1961 result is essentially the best that can be obtained by a normwise analysis. Subsequent work in error analysis for GE has mainly been concerned with bounding the backward error componentwise, that is, obtaining n^2 individual bounds for the elements of the backward error matrix E, rather than a single bound for a norm of E.

Chartres and Geuder [6] analyse the Doolittle "compact" version of GE. They derive the componentwise backward error result

$$(A + E)\hat{x} = b,$$

$$|e_{ij}| \le 2(j+3)w_{ij}c_1(n,u) + \begin{cases} 2|\hat{u}_{ij}|c_2(n,u), & i < j, \\ 3|\hat{u}_{jj}|c_2(n,u), & i = j, \\ 2|\hat{l}_{ij}\hat{u}_{jj}|c_2(n,u), & i > j, \end{cases} \tag{3.3}$$

where $c_1(n,u) \approx u$, $c_2(n,u) \approx (n+1)u$, and

$$w_{ij} = \sum_{k=1}^{m-1} |\hat{l}_{ik}\hat{u}_{kj}| = (|\hat{L}||\hat{U}|)_{ij} - |\hat{l}_{im}\hat{u}_{mj}|, \qquad m = \min(i,j).$$

We note that Wilkinson could have given a componentwise bound in place of (3.2), since most of his analysis is at the element level.

Reid [32] shows that the assumption in Wilkinson's analysis that partial pivoting or complete pivoting is used is unnecessary. Without making any assumptions on the pivoting strategy he derives the result for the LU factorization

$$\hat{L}\hat{U} = A + E,$$

$$|e_{ij}| \le 3.01n \, u \max_k |a_{ij}^{(k)}|.$$

Again, this is a componentwise bound. Note that the backward error analyses discussed so far display three different "styles" of error bound, indicating the considerable freedom one has in deciding how to develop and phrase a backward error analysis.

de Boor and Pinkus [9] give the result

$$(A + E)\hat{x} = b, \tag{3.4}$$

$$|E| \le \gamma_n(2 + \gamma_n)|\hat{L}||\hat{U}|, \tag{3.5}$$

where

$$\gamma_n = \frac{nu}{1 - nu}.$$

They refer to the original 1972 German edition of [40] for a proof of the result, and explain several advantages to be gained by working with the matrix-level bound (3.5) (see section 4.2).

We briefly mention some other relevant work:

- Demmel [10] shows how existing backward error analyses for GE can be modified to take into account the possibility of underflow (the analyses we have described assume that underflow does not occur).

- Backward error analysis has been used to investigate how the accuracy and stability of GE are affected by scaling, and by use of row pivoting instead of column pivoting. van der Sluis [45] and Stewart [39] employ norm analysis, while Skeel [35, 37] uses a componentwise approach.

- Specialized backward error analyses have been done for the Cholesky factorization of positive (semi-) definite matrices; see [25] and the references therein.

- Forward error analyses have been done for GE. The analyses are more complicated and more difficult to interpret than the backward error analyses. See [30] and [41, 42].

The "$|\hat{L}||\hat{U}|$" componentwise-wise style of backward error analysis is now well-known, as evidenced by its presence in several textbooks [8, 19, 40]. To emphasize the simplicity of the analysis we give a short proof of (3.5). This proof is modelled on one in [40]. See also [7] for a similar presentation.

We will use the standard model of floating point arithmetic:

$$fl(x \text{ op } y) = (x \text{ op } y)(1 + \delta), \qquad |\delta| \leq u, \quad \text{op} = +, -, *, /, \tag{3.6}$$

where u is the unit roundoff. The low-level technical details can be confined to two lemmas.

Lemma 3.1 *If $|\delta_i| \leq u$ and $p_i = \pm 1$ for $1 \leq i \leq n$, and $nu < 1$, then*

$$\prod_{i=1}^{n}(1 + \delta_i)^{p_i} = 1 + \theta_n,$$

where $|\theta_n| \leq \gamma_n \equiv nu/(1 - nu)$.

Proof. A straightforward induction. ∎

Lemma 3.2 *If the expression $s = (c - \sum_{i=1}^{k-1} a_i b_i)/b_k$ is evaluated in floating point arithmetic, in whatever order, then the computed value $\hat{s}$ satisfies*

$$\left| c - \sum_{i=1}^{k-1} a_i b_i - \hat{s} b_k \right| \leq \gamma_k \left(|\hat{s} b_k| + \sum_{i=1}^{k-1} |a_i||b_i| \right). \tag{3.7}$$

Proof. Straightforward manipulation. If we assume that the expression is evaluated in the natural left-right order then an intermediate result in the proof (and a backward error result in its own right) is

$$\hat{s} b_k (1 + \theta_k) = c - \sum_{i=1}^{k-1} a_i b_i (1 + \theta_i) \tag{3.8}$$

where the θ_i satisfy $|\theta_i| \leq \gamma_i$. It is not hard to see that this inequality holds whatever ordering is used when evaluating s if we replace each θ_i on the right-hand side by θ_k (different in each instance). ∎

Recall that the GE algorithm comprises three nested loops, and that there are six ways of ordering the loops. Each version sustains precisely the same rounding errors under the model (3.6) because each version does the same floating point operations with the same arguments—only the order in which the operations are done differs. We find it convenient to analyse the Doolittle, or "jik" variant, which computes L a column at a time and U a row at a time according to

$$u_{ij} = a_{ij} - \sum_{k=1}^{i-1} l_{ik} u_{kj}, \qquad j \geq i$$

$$l_{ij} = (a_{ij} - \sum_{k=1}^{j-1} l_{ik} u_{kj})/u_{jj}, \qquad i > j$$

Applying Lemma 3.2 to these equations we obtain

$$|a_{ij} - \sum_{k=1}^{i} \hat{l}_{ik} \hat{u}_{kj}| \leq \gamma_i \sum_{k=1}^{i} |\hat{l}_{ik} \hat{u}_{kj}|, \qquad j \geq i,$$

$$|a_{ij} - \sum_{k=1}^{j} \hat{l}_{ik} \hat{u}_{kj}| \leq \gamma_j \sum_{k=1}^{j} |\hat{l}_{ik} \hat{u}_{kj}|, \qquad i > j,$$

where we have defined $\hat{l}_{ii} = l_{ii} \equiv 1$. These inequalities may be written in matrix form as

$$\hat{L}\hat{U} = A + E, \qquad |E| \leq \gamma_n |\hat{L}||\hat{U}|. \tag{3.9}$$

Finally, we have to consider the forward and back substitutions, $Ly = b$, $Ux = y$. From (3.8) we immediately obtain

$$(\hat{L} + \Delta\hat{L})\hat{y} = b, \qquad \Delta\hat{L} \leq \gamma_n |\hat{L}|, \tag{3.10}$$

$$(\hat{U} + \Delta\hat{U})\hat{x} = \hat{y}, \qquad \Delta\hat{U} \leq \gamma_n |\hat{U}|. \tag{3.11}$$

Thus $(\hat{L} + \Delta\hat{L})(\hat{U} + \Delta\hat{U})\hat{x} = b$, and combining (3.9)–(3.11) we arrive at (3.5). (Actually, we obtain (3.5) with the '2' replaced by '3'—a slightly more refined analysis of the substitution stages produces the '2'.)

Note that if A has bandwidth k ($a_{ij} = 0$ for $|i - j| > k$) then, since L and U have the same bandwidth, we can replace γ_n by γ_{k+1} in the above analysis.

4   Interpreting the Error Analysis

In this section we interpret the backward error analyses summarized in section 3 and look at their implications. It is instructive to make comparisons with the "ideal" bounds

$$\|E\|_\infty \leq u\|A\|_\infty \qquad \text{(small normwise backward error)}, \tag{4.1}$$

$$|E| \leq u|A| \qquad \text{(small componentwise relative backward error)}, \tag{4.2}$$

which hold if, for example, $A + E$ is the rounded version of A. (Note that (4.2) implies (4.1)).

4.1 Normwise Analysis

Wilkinson's backward error result is usually explained as follows. The bound (3.2) differs from the ideal bound (4.1) by having the extra factor $\rho_n p(n)$. The $p(n)$ term is fixed, and hence is beyond our control. Also, it is "pessimistic", since it arises from repeated use of triangle inequalities and taking of matrix norms. Wilkinson [49, pp. 62, 108] comments that the bound is usually not sharp even if we replace $p(n)$ by its square root. Therefore our attention is focussed on the size of the growth factor.

Let ρ_n, ρ_n^p, ρ_n^c denote the growth factors for GE with, respectively, no pivoting, partial pivoting and complete pivoting. It is easy to see that $\rho_n(A)$ can be arbitrarily large, and so GE without pivoting is unstable in general. For partial pivoting, ρ_n^p is almost invariably small in practice ($\rho_n^p < 10$, say) but a parametrized family of matrices is known for which it achieves its maximum of 2^{n-1} [26]. The situation is similar for ρ_n^c, except that a much smaller upper bound is known, and it has been conjectured that $\rho_n^c(A) \leq n$ for real A. Recent work has shed more light on the behaviour of ρ_n^p and ρ_n^c. Higham and Higham [26] present several families of real matrices from practical applications for which $\rho_n^p(A)$ and $\rho_n^c(A)$ are about $n/2$. These examples show that moderately large growth factors can be achieved on non-contrived matrices. Trefethen and Schreiber [43] develop a statistical model of the average growth factor for partial pivoting and complete pivoting. Their model supports their empirical findings that for various distributions of random matrices the average growth factor (normalized by the standard deviation of the initial matrix elements) is close to $n^{2/3}$ for partial pivoting and $n^{1/2}$ for complete pivoting.

For certain classes of matrix special bounds are known for the growth factor (see [48], [50, pp. 218–220] and [38, p. 158]):

- If A is diagonally dominant by columns ($|a_{jj}| > \sum_{i \neq j} |a_{ij}|$ for all j) then $\rho_n(A) = \rho_n^p(A) \leq 2$ (and no row interchanges are performed with partial pivoting).

- If A is tridiagonal then $\rho_n^p(A) \leq 2$ and if A is upper Hessenberg then $\rho_n^p(A) \leq n$.

- If A is symmetric positive definite then $\rho_n(A) \leq 1$.

- If the LU factors of A have nonnegative elements then $\rho_n(A) \leq 1$. $L \geq 0$ and $U \geq 0$ is guaranteed if A is *totally nonnegative*, that is, if the determinant of every submatrix of A is nonnegative.

4.2 Componentwise Analysis

The componentwise bound (3.5) is weaker than the ideal bound (4.2) in general, but it matches it (up to a factor n) if $|\hat{L}||\hat{U}| \leq c|A|$ with c a small constant, which is true in several cases as we now explain. We assume the use of GE *without* pivoting.

First, we mention the important case of triangular systems: here the componentwise relative backward error is always small, as shown by (3.10)–(3.11). The implications of this fact are explored in detail in [24].

Next, following [9], consider totally nonnegative matrices A. For such A, the exact LU factors have nonnegative elements, and the same is true of the computed ones if the unit roundoff is sufficiently small. In this case we have, using (3.9),

$$|\hat{L}||\hat{U}| = |\hat{L}\hat{U}| = |A + E| \leq |A| + \gamma_n|\hat{L}||\hat{U}|,$$

that is,

$$|\hat{L}||\hat{U}| \leq \frac{1}{1 - \gamma_n}|A|.$$

Hence the backward error matrix E in (3.5) satisfies

$$|E| \leq \frac{\gamma_n(2 + \gamma_n)}{1 - \gamma_n}|A|,$$

that is, the componentwise relative backward error is pleasantly small. The same is true for some important classes of tridiagonal matrix. Higham [22] shows that if A is tridiagonal and either

- symmetric positive definite,

- an M-matrix ($a_{ij} \leq 0$ for all $i \neq j$ and $A^{-1} \geq 0$), or

- diagonally dominant by columns or by rows,

then

$$(A + E)\hat{x} = b, \qquad |E| \leq f(u)|A|,$$

where $f(u) = cu + O(u^2)$, c being a constant of order unity.

Note that in the above examples row interchanges in the LU factorization destroy the required matrix properties and so we lose the favourable backward error bounds. In these cases it is advantageous *not* to pivot!

5   Computing Backward Error Bounds

In cases where it is not known a priori that the backward error is sufficiently small it is desirable to obtain an a posteriori bound for it. Several authors have considered how to compute or estimate the bounds of section 3.

Consider first Wilkinson's bound (3.2), in which the only "nontrivial" term is the growth factor ρ_n. The growth factor can be computed by monitoring the size of elements during the elimination, at a cost of $O(n^3)$ comparisons. This has been regarded as rather expensive, and more efficient ways to estimate ρ_n have been sought.

Businger [5] describes a way to obtain an upper bound for ρ_n in $O(n^2)$ operations. This approach is generalized by Erisman and Reid [14] who apply the Holder inequality to the equation

$$a_{ij}^{(k)} = a_{ij} - \sum_{r=1}^{k} l_{ir}u_{rj}, \qquad i, j > k,$$

to obtain the bound

$$|a_{ij}^{(k)}| \leq |a_{ij}| + \|(l_{i1}, \ldots, l_{ik})\|_p \|(u_{1j}, \ldots, u_{kj})\|_q,$$
$$\leq \max_{i,j} |a_{ij}| + \max_i \|(l_{i1}, \ldots, l_{i,i-1})\|_p \max_j \|(u_{1j}, \ldots, u_{j-1,j})\|_q, \qquad (5.1)$$

where $p^{-1} + q^{-1} = 1$. In practice, $p = 1, 2, \infty$ are the values of interest. Barlow [2] notes that application of the Holder inequality instead to

$$a_{ij}^{(k)} = \sum_{r=k+1}^{\min(i,j)} l_{ir} u_{rj}$$

yields the sometimes sharper bound

$$|a_{ij}^{(k)}| \leq \max_i \|L(i, :)\|_p \max_j \|U(:, j)\|_q,$$

where $L(i, :)$ is the ith row of L and $U(:, j)$ is the jth column of U.

It is interesting to note that in the light of experience with the bound (5.1) Reid [33] recommends computing the growth factor explicitly in the context of sparse matrices, arguing that the expense is justified because (5.1) can be a very weak bound. See [13] for some empirical results on the quality of the bound.

Chartres and Geuder [6] propose computing the n^2 bounds in (3.3) explicitly. They note that since the terms $\hat{l}_{ik} \hat{u}_{kj}$ which make up W are formed anyway during the LU factorization, W can be computed in parallel with the factorization at a cost of $n^3/3$ additions.

Chu and George [7] observe that the ∞-norm of the matrix $|\hat{L}||\hat{U}|$ can be computed in $O(n^2)$ operations without forming the matrix explicitly, since

$$\| |\hat{L}||\hat{U}| \|_\infty = \| |\hat{L}||\hat{U}|e \|_\infty = \| |\hat{L}|(|\hat{U}|e) \|_\infty.$$

Thus one can cheaply compute a bound on $\|E\|_\infty$ from a componentwise backward error bound such as (3.5).

All the methods discussed above make use of an a priori error analysis to compute bounds on the backward error. Because the bounds do not take into account the statistical distribution of rounding errors, and because they have somewhat pessimistic constant terms, they cannot be expected to be very sharp. Thus it is important not to forget that, as shown is section 2, it is straightforward to compute the backward error itself! To obtain the backward error in the LU factorization we have to compute $\hat{L}\hat{U}$, which costs $O(n^3)$ operations. If the normwise backward error is wanted then one could instead *estimate* $\|A - \hat{L}\hat{U}\|_1$ in $O(n^2)$ operations using the matrix norm estimator described in section 7. The backward error for $\hat{x}$ can be computed in $O(n^2)$ operations via just one or two matrix-vector products, from the formulas in section 2. We discuss the effect of rounding error on this computation in section 7.

6 Iterative Refinement

What can we do if the computed solution $\hat{x}$ to $Ax = b$ does not have a small enough backward error? The traditional answer to a slightly different question—what to do if $\hat{x}$ does not have a small enough forward error—is to use iterative refinement, and it is usually stressed that residuals must be computed in higher precision (see, e.g., [16]). Work by Jankowski and Woźniakowski [28] and Skeel [36] shows that iterative refinement using *single precision* residuals is usually sufficient to yield a small backward error, although it will not necessarily produce a small forward error. Jankowski and Woźniakowski deal with normwise backward error in their analysis and cater for arbitrary linear equation solvers. Skeel specializes to GEPP and uses the stronger componentwise relative backward error. Skeel's analysis and results are intricate, but the gist may be stated simply:

> If the product of $\text{cond}(A^{-1}) \equiv \| \, |A||A^{-1}| \, \|_\infty$ and $\sigma(A, x) \equiv \max_i(|A||x|)_i / \min_i(|A||x|)_i$ is less than $(f(A, b)u)^{-1}$, where $f(A, b)$ is typically $O(n)$, then after GEPP with one step of single precision iterative refinement $\beta_C(|A|, |b|) \leq (n + 1)u$.

Thus after just one step ·of iterative refinement in single precision GEPP—which is already stable in the sense of having small normwise backward error—is much more strongly stable: it has a small componentwise relative backward error, provided that the problem is not too ill-conditioned ($\text{cond}(A^{-1})$ is not too large) or too badly scaled ($\sigma(A, x)$ is not too large).

Arioli, Demmel and Duff [1] consider in detail the practical use of iterative refinement in single precision with GEPP, with an emphasis on sparse matrices. They note that $\sigma(A, x)$ can be very large for systems in which both A and b are sparse, in which case Skeel's result is not applicable. To circumvent this problem they change the backward error measure from $\beta_C(|A|, |b|)$ to $\beta_C(|A|, f)$, where f is chosen in an a posteriori way in which f_i is permitted to exceed $|b_i|$. Thus in the backward error definition the sparsity of A is preserved while that of b may be sacrificed in order to make the backward error small after one or more steps of iterative refinement. See [1] for further details, and a comprehensive suite of numerical tests. For a thorough survey of iterative refinement in linear equations and other contexts see [3].

7 Estimating and Bounding the Forward Error

The usual way to explore the forward error $\|x - \hat{x}\|$ is by applying perturbation theory to a backward error analysis. Corresponding to the normwise backward error result

$$(A + \Delta A)\hat{x} = b + \Delta b, \qquad \|\Delta A\| \leq \epsilon\|A\|, \qquad \|\Delta b\| \leq \epsilon\|b\|,$$

we have the well-known bound

$$\frac{\|x - \hat{x}\|}{\|x\|} \leq \frac{2\epsilon\kappa(A)}{1 - \epsilon\kappa(A)} \qquad (\epsilon\kappa(A) < 1), \tag{7.1}$$

where the condition number $\kappa(A) = \|A\|\|A^{-1}\|$. For the componentwise analysis

$$(A + \Delta A)\hat{x} = b + \Delta b, \qquad |\Delta A| \leq \omega E, \qquad |\Delta b| \leq \omega f,$$

a straightforward generalization of a result in [35, Theorem 2.1] yields

$$\frac{\|x - \hat{x}\|_\infty}{\|x\|_\infty} \leq \frac{\omega \kappa_{E,f}(A, b)}{1 - \omega \kappa_E(A)} \qquad (\omega \kappa_E(A) < 1), \qquad (7.2)$$

where, in the notation of [1],

$$\kappa_{E,f}(A, b) \equiv \frac{\||A^{-1}|E|x| + |A^{-1}|f\|_\infty}{\|x\|_\infty},$$

$$\kappa_E(A) \equiv \||A^{-1}|E\|_\infty.$$

Of particular interest is the condition number for the componentwise relative backward error, $\kappa_{|A|,|b|}(A, b)$. This is easily seen to differ by no more than a factor 2 from, using Skeel's notation [35],

$$\text{cond}(A, x) \equiv \frac{\||A^{-1}||A||x|\|_\infty}{\|x\|_\infty}.$$

The maximum value of $\text{cond}(A, x)$ occurs for $x = e$ and is

$$\text{cond}(A) \equiv \||A^{-1}||A|\|_\infty.$$

It is an important fact that $\text{cond}(A)$ is never bigger than $\kappa_\infty(A)$, and it can be much smaller. This is because $\text{cond}(A, x)$ is independent of the row scaling of A while $\kappa_\infty(A)$ is not. One implication of this row scaling independence is that $\text{cond}(A)$ may differ greatly from $\text{cond}(A^T)$. Thus in the sense of componentwise perturbations $Ax = b$ can be much more or less ill-conditioned than $A^T y = c$ (see [24] for some specific examples). Another implication is that (7.2) can sometimes provide a much smaller upper bound than (7.1). Note also that the forward error depends on the right-hand side b, and that (7.2) displays this dependency but (7.1) does not.

We note that under assumptions on the statistical distribution of the perturbations ΔA and Δb an expression can be derived for the expected 2-norm of the forward error [15]; this may be preferable to (7.1) and (7.2) when statistical information about ΔA and Δb is available.

Both bounds (7.1) and (7.2) contain a condition number involving A^{-1}. In most cases exact computation of the condition number would necessitate forming A^{-1}. To avoid this expense it is standard practice to compute an inexpensive estimate of the condition number using a condition estimator. Various condition estimation techniques have been developed—see [21] for a survey. The most well-known estimator is the one used in LINPACK, which provides a lower bound for $\kappa_1(A)$. The method underlying this estimator does not generalize to the estimation of $\kappa_{E,f}(A, b)$. A more versatile estimator with this capability is one developed by Hager [20] and Higham [23]. This estimator treats the general problem of estimating $\|B\|_1$, where B is not known explicitly. The estimator assumes that B is described by a "black box" that can evaluate Bx or $B^T x$ given x. Typically, 4 or 5 such matrix-vector products are required to produce a lower bound for $\|B\|_1$, and the bound is almost invariably within a factor 3 of $\|B\|_1$. To estimate $\kappa_1(A)$ we take $B = A^{-1}$, and the estimator requires the solution of linear systems with A and A^T as coefficient matrices, which is inexpensive if an LU factorization of A is available. Arioli, Demmel and Duff [1] show how to apply the estimator to $\kappa_{E,f}(A, b)$. The problem is basically that of estimating $\| |A^{-1}|g \|_\infty$, where $g \geq 0$. With $G = \text{diag}(g_1, g_2, \ldots, g_n)$ the equalities

$$\| |A^{-1}|g \|_\infty = \| |A^{-1}|Ge \|_\infty = \| |A^{-1}G|e \|_\infty = \| |A^{-1}G| \|_\infty = \|A^{-1}G\|_\infty$$

show that the problem reduces to estimating $\|B\|_1$ where $B = (A^{-1}G)^T$ and where Bx and $B^T y$ can be formed by solving linear systems involving A^T and A respectively.

There are certain circumstances in which one can compute $\kappa_{E,f}(A, b)$ *exactly* with the same order of work as solving a linear system given an LU factorization of A. Higham [22] shows that if the nonsingular matrix A is tridiagonal and nonsingular and has an LU factorization with $|L||U| = |A|$ then

$$|U^{-1}||L^{-1}| = |A^{-1}|. \tag{7.3}$$

Recall from section 4.2 that the condition $|L||U| = |A|$ implies a small componentwise relative backward error for GE; it holds when the tridiagonal matrix is symmetric positive definite, totally nonnegative, or an M-matrix [22]. To see the significance of (7.3) consider a 3×3 bidiagonal matrix and its inverse:

$$U = \begin{bmatrix} d_1 & e_2 & \\ & d_2 & e_3 \\ & & d_3 \end{bmatrix}, \qquad U^{-1} = \begin{bmatrix} d_1^{-1} & -e_2 d_1^{-1} d_2^{-1} & e_2 e_3 d_1^{-1} d_2^{-1} d_3^{-1} \\ & d_2^{-1} & -e_3 d_2^{-1} d_3^{-1} \\ & & d_3^{-1} \end{bmatrix}.$$

Clearly we have

$$|U^{-1}| = M(U)^{-1} \qquad \text{where} \quad M(U) = \begin{bmatrix} |d_1| & -|e_2| & \\ & |d_2| & -|e_3| \\ & & |d_3| \end{bmatrix}$$

($M(U)$ is called the comparison matrix for U). The relation $|B^{-1}| = M(B)^{-1}$ is true for all bidiagonal B, so

$$|A^{-1}|g = |U^{-1}||L^{-1}|g = M(U)^{-1} M(L)^{-1} g,$$

which may be computed by solving two bidiagonal systems. Thus for the three classes of tridiagonal matrix mentioned above it costs only $O(n)$ operations to compute $\kappa_{E,f}(A, b)$ given an LU factorization of A. The same technique can be used for M-matrices in general, because $A^{-1} \geq 0$ implies $|A^{-1}|g = A^{-1}g$.

Next we describe further details of estimating the forward error in practice. To evaluate bounds (7.1) and (7.2) we need the backward error, or a bound for it. We could use the backward error bounds from section 3, but as explained in section 5 it is better to compute the backward error directly using the formulas from section 2.

A more direct approach is to use the following bound, which is so straightforward that it is easily overlooked: from $x - \hat{x} = A^{-1}r$, where $r = b - A\hat{x}$,

$$\|x - \hat{x}\|_\infty \leq \| |A^{-1}||r| \|_\infty.$$

Note that the inequality arises solely from ignoring signs of terms in the matrix-vector product. Assuming r is computed in single precision, the rounding errors in its formation are accounted for by (using a variation of Lemma 3.2)

$$\hat{r} = r + \Delta r, \qquad |\Delta r| \leq \gamma_{n+1}(|b| + |A||\hat{x}|).$$

Thus, our final practical bound, which we estimate using the norm estimator described above, is

$$\frac{\|x - \hat{x}\|_\infty}{\|\hat{x}\|_\infty} \leq \frac{\| |A^{-1}|(|\hat{r}| + \gamma_{n+1}(|b| + |A||\hat{x}|)) \|_\infty}{\|\hat{x}\|_\infty}. \tag{7.4}$$

Note that the $|b| + |A||\hat{x}|$ term needs to be computed anyway if we are evaluating the componentwise relative backward error from (2.4).

Arioli, Demmel and Duff [1] found in their experiments that after iterative refinement (7.4) gave similar sized bounds to (7.2) with $E = |A|$, $f = |b|$, and with ω computed a posteriori using (2.4).

We conclude this section by noting the need for care in interpreting the forward error. Experiments in [24] show that simply changing the order of evaluation of an inner product in the substitution algorithm for solution of a triangular system can change the forward error in the computed solution by orders of magnitude. This means, for example, that it is dangerous to compare different codes or algorithms solely in terms of observed forward errors.

8 Concluding Remarks

The work we have described falls broadly into two areas:

- normwise error analysis and iterative refinement in double precision, and

- componentwise analysis and iterative refinement in single precision.

The first area might be described as being "traditional" or "in the style of Wilkinson", while the second has been the subject of most of the recent research in this subject.

An indication of the success of the recent work is that LAPACK [11], the successor to LINPACK and EISPACK that is currently under development, will incorporate iterative refinement in single precision with computation of the componentwise relative backward error and estimation of $\kappa_{|A|,|b|}(A, b)$ (see [12]).

Finally, we return to the numerical example in section 1. The Vandermonde matrix A in this example is totally nonnegative, so we know from section 4.2 that the componentwise relative backward error must be small for GE. For GEPP there is no such guarantee since the permuted matrix "PA" is not totally nonnegative—as we saw, only the normwise backward error is small in this example. We have $\kappa_\infty(A) \approx 4 \times 10^7$, and $\mathrm{cond}(A, x) \approx 2 \times 10^4$ ($\mathrm{cond}(A) \approx 9 \times 10^4$). The forward errors for the two computed solutions are consistent with the "forward error $\leq$ condition number $\times$ backward error" results (7.1) and (7.2). Concerning iterative refinement in single precision, the product $\mathrm{cond}(A^{-1})\sigma(A, x) \approx 10^3 u$, so Skeel's result quoted in section 6 is not applicable. Nevertheless, the first iterate has a small componentwise relative backward error and its forward error is consistent with the forward error bounds.

Acknowledgements

I thank Des Higham and Steve Vavasis for carefully reading the manuscript and suggesting numerous improvements.

REFERENCES

[1] M. Arioli, J.W. Demmel and I.S. Duff, Solving sparse linear systems with sparse backward error, SIAM J. Matrix Anal. Appl., 10 (1989), pp. 165–190.

[2] J.L. Barlow, A note on monitoring the stability of triangular decomposition of sparse matrices, SIAM J. Sci. Stat. Comput., 7 (1986), pp. 166–168.

[3] Å. Björck, Iterative refinement and reliable computing, in *Reliable Numerical Computation*, M.G. Cox and S.J. Hammarling, eds., Oxford University Press, 1989.

[4] J.R. Bunch, J.W. Demmel and C.F. Van Loan, The strong stability of algorithms for solving symmetric linear systems; to appear in SIAM J. Matrix Anal. Appl.

[5] P.A. Businger, Monitoring the numerical stability of Gaussian elimination, Numer. Math., 16 (1971), pp. 360–361.

[6] B.A. Chartres and J.C. Geuder, Computable error bounds for direct solution of linear equations, J. Assoc. Comput. Mach., 14 (1967), pp. 63–71.

[7] E. Chu and A. George, A note on estimating the error in Gaussian elimination without pivoting, ACM SIGNUM Newsletter, 20 (1985), pp. 2–7.

[8] S.D. Conte and C. de Boor, *Elementary Numerical Analysis*, Third Edition, McGraw-Hill, Tokyo, 1980.

[9] C. de Boor and A. Pinkus, Backward error analysis for totally positive linear systems, Numer. Math., 27 (1977), pp. 485–490.

[10] J.W. Demmel, Underflow and the reliability of numerical software, SIAM J. Sci. Stat. Comput., 5 (1984), pp. 887–919.

[11] J.W. Demmel, J.J. Dongarra, J.J. Du Croz, A. Greenbaum, S.J. Hammarling and D.C. Sorensen, Prospectus for the development of a linear algebra library for high-performance computers, Technical Memorandum No. 97, Mathematics and Computer Science Division, Argonne National Laboratory, Illinois, 1987.

[12] J.W. Demmel, J.J. Du Croz, S.J. Hammarling and D.C. Sorensen, Guidelines for the design of symmetric eigenroutines, SVD, and iterative refinement and condition estimation for linear systems, LAPACK Working Note #4, Technical Memorandum 111, Mathematics and Computer Science Division, Argonne National Laboratory, 1988.

[13] A.M. Erisman, R.G. Grimes, J.G. Lewis, W.G. Poole and H.D. Simon, Evaluation of orderings for unsymmetric sparse matrices, SIAM J. Sci. Stat. Comput., 8 (1987), pp. 600–624.

[14] A.M. Erisman and J.K. Reid, Monitoring the stability of the triangular factorization of a sparse matrix, Numer. Math., 22 (1974), pp. 183–186.

[15] R. Fletcher, Expected conditioning, IMA Journal of Numerical Analysis, 5 (1985), pp. 247–273.

[16] G.E. Forsythe and C.B. Moler, *Computer Solution of Linear Algebraic Systems*, Prentice-Hall, Englewood Cliffs, New Jersey, 1967.

[17] L. Fox, *An Introduction to Numerical Linear Algebra*, Oxford University Press, 1964.

[18] L. Fox, H.D. Huskey and J.H. Wilkinson, Notes on the solution of algebraic linear simultaneous equations, Quart. J. Mech. and Applied Math., 1 (1948), pp. 149–173.

[19] G.H. Golub and C.F. Van Loan, *Matrix Computations*, Johns Hopkins University Press, Baltimore, Maryland, 1983.

[20] W.W. Hager, Condition estimates, SIAM J. Sci. Stat. Comput., 5 (1984), pp. 311–316.

[21] N.J. Higham, A survey of condition number estimation for triangular matrices, SIAM Review, 29 (1987), pp. 575–596.

[22] N.J. Higham, Bounding the error in Gaussian elimination for tridiagonal systems, Technical Report 88-953, Department of Computer Science, Cornell University, 1988.

[23] N.J. Higham, FORTRAN codes for estimating the one-norm of a real or complex matrix, with applications to condition estimation, ACM Trans. Math. Soft., 14 (1988), pp. 381–396.

[24] N.J. Higham, The accuracy of solutions to triangular systems, Numerical Analysis Report No. 158, University of Manchester, England, 1988; to appear in SIAM J. Numer. Anal.

[25] N.J. Higham, Analysis of the Cholesky decomposition of a semi-definite matrix, in *Reliable Numerical Computation*, M.G. Cox and S.J. Hammarling, eds., Oxford University Press, 1989.

[26] N.J. Higham and D.J. Higham, Large growth factors in Gaussian elimination with pivoting, SIAM J. Matrix Anal. Appl., 10 (1989), pp. 155–164.

[27] H. Hotelling, Some new methods in matrix calculation, Ann. Math. Statist., 14 (1943), pp. 1–34.

[28] M. Jankowski and H. Woźniakowski, Iterative refinement implies numerical stability, BIT, 17 (1977), pp. 303–311.

[29] W. Oettli and W. Prager, Compatibility of approximate solution of linear equations with given error bounds for coefficients and right-hand sides, Numer. Math., 6 (1964), pp. 405–409.

[30] F.W.J. Olver and J.H. Wilkinson, *A posteriori* error bounds for Gaussian elimination, IMA Journal of Numerical Analysis, 2 (1982), pp. 377–406.

[31] G. Peters and J.H. Wilkinson, On the stability of Gauss-Jordan elimination with pivoting, Comm. ACM, 18 (1975), pp. 20–24.

[32] J.K. Reid, A note on the stability of Gaussian elimination, J. Inst. Maths. Applics., 8 (1971), pp. 374–375.

[33] J.K. Reid, Sparse matrices, in *The State of the Art in Numerical Analysis*, A. Iserles and M.J.D. Powell, eds., Oxford University Press, 1987, pp. 59–85.

[34] J.L. Rigal and J. Gaches, On the compatibility of a given solution with the data of a linear system, J. Assoc. Comput. Mach., 14 (1967), pp. 543–548.

[35] R.D. Skeel, Scaling for numerical stability in Gaussian elimination, J. Assoc. Comput. Mach., 26 (1979), pp. 494–526.

[36] R.D. Skeel, Iterative refinement implies numerical stability for Gaussian elimination, Math. Comp., 35 (1980), pp. 817–832.

[37] R.D. Skeel, Effect of equilibration on residual size for partial pivoting, SIAM J. Numer. Anal., 18 (1981), pp. 449–454.

[38] G.W. Stewart, *Introduction to Matrix Computations*, Academic Press, New York, 1973.

[39] G.W. Stewart, Research, development, and LINPACK, in *Mathematical Software III*, J.R. Rice, ed., Academic Press, New York, 1977, pp. 1–14.

[40] J. Stoer and R. Bulirsch, *Introduction to Numerical Analysis*, Springer-Verlag, New York, 1980.

[41] F. Stummel, Forward error analysis of Gaussian elimination, Part I: Error and residual estimates, Numer. Math., 46 (1985), pp. 365–395.

[42] F. Stummel, Forward error analysis of Gaussian elimination, Part II: Stability theorems, Numer. Math., 46 (1985), pp. 397–415.

[43] L.N. Trefethen and R.S. Schreiber, Average-case stability of Gaussian elimination, Numerical Analysis Report 88-3, Department of Mathematics, M.I.T., 1988; to appear in SIAM J. Matrix Anal. Appl.

[44] A.M. Turing, Rounding-off errors in matrix processes, Quart. J. Mech. and Applied Math., 1 (1948), pp. 287–308.

[45] A. van der Sluis, Condition, equilibration and pivoting in linear algebraic systems, Numer. Math., 15 (1970), pp. 74–86.

[46] J. von Neumann and H.H. Goldstine, Numerical inverting of matrices of high order, Bull. Amer. Math. Soc., 53 (1947), pp. 1021–1099.

[47] B. Wendroff, *Theoretical Numerical Analysis*, Academic Press, New York, 1966.

[48] J.H. Wilkinson, Error analysis of direct methods of matrix inversion, J. Assoc. Comput. Mach., 8 (1961), pp. 281–330.

[49] J.H. Wilkinson, *Rounding Errors in Algebraic Processes*, Notes on Applied Science No. 32, Her Majesty's Stationery Office, London, 1963.

[50] J.H. Wilkinson, *The Algebraic Eigenvalue Problem*, Oxford University Press, 1965.

[51] J.H. Wilkinson, Modern error analysis, SIAM Review, 13 (1971), pp. 548–568.

[52] J.H. Wilkinson, The state of the art in error analysis, NAG Newsletter 2/85, Numerical Algorithms Group, Oxford, 1985.

[53] J.H. Wilkinson, Error analysis revisited, IMA Bulletin, 22 (1987), pp. 192–200.

Nicholas J. Higham
Department of Computer Science
Cornell University
Ithaca
New York 14853
U.S.A.
(On leave from the University of Manchester.)

C. JOHNSON

The Streamline Diffusion Finite Element Method for Compressible and Incompressible Fluid Flow

1. INTRODUCTION.

A basic problem in computational fluid mechanics is to design finite element or finite difference methods which combine high accuracy with good stability properties in cases where convection dominates diffusion and the mesh size h is not fine enough to resolve all details of the flow (more precisely if $h > \epsilon/a$, where ϵ is the (small) diffusion coefficient and a is the modulus of the convection velocity). In particular, this problem comes up in the vanishing viscosity limit $(\epsilon = 0)$ where the mathematical models for compressible flow take the form of hyperbolic systems of conservation laws and solutions may present discontinuities (shocks) which cannot be fully resolved on any mesh of finite mesh size. The basic design goals in the limit $\epsilon = 0$ may be stated as follows:

convergence to a physical (entropy) solution,	(1.1a)
sharp oscillation-free representation of discontinuities (shocks),	(1.1b)
high accuracy in domains where the solution is smooth.	(1.1c)

With (small) diffusion $\epsilon > 0$ discontinuities are replaced by layers with rapid but continuous variation of the flow variables and we may then either seek to resolve the layers by choosing h sufficiently small, or essentially stay in the vanishing viscosity limit without accurately resolving layers. In any case the ultimate design goal including (1.1) in particular, may be summarized as follows:

efficient and reliable adaptive quantitative control of the discretization (1.2)
error in some norm (e.g. the L_1 or L_2-norm).

We remark that adaptive methods in computational fluid mechanics are now starting to appear and promise to open fascinating new possibilities of efficiently modeling complicated flows with features of vastly different scales using locally refined meshes. An ultimate challenge is here to design an efficient adaptive method e.g. for the Navier-Stokes equations with reliable error control making it possible to guarantee that the discretization error is below a certain tolerance level.

Failure to satisfy (1.1a) or (1.1b) may be viewed as resulting from poor stability and thus (1.1a–c) may be summarized as requiring both good stability and high accuracy. We recall that classical methods fail to satisfy one of these requirements: On the one hand we have centered finite difference schemes and standard Galerkin finite element schemes which are formally higher order accurate, but lack in stability and may produce approximate solutions with spurious oscillations which do not converge to physically correct entropy solutions. On the other hand we have Godunov schemes based on solving local Riemann problems with piecewise constant initial data, and classical artificial diffusion schemes like the Lax–Friedrichs' scheme (or simple upwind schemes) obtained by adding second order artifical viscosity with viscosity coefficient proportional to the mesh size h. These methods may satisfy (1.1a) but are only first order accurate (also in regions where the exact solution is smooth) and smear discontinuities (in particular contact discontinuities) over many mesh points.

For finite difference methods, the basic approaches in recent years to construct methods which combine accuracy and stability have followed two lines: One in the direction of higher order Godunov methods with discontinuous piecewise linear or parabolic approximations, and one in the direction of "hybrid" schemes with "switches" or "limiters" through which dynamically during the computational process different schemes are chosen in different parts of the computational region; typically a hybrid scheme is constructed starting from a first order monotone scheme with heavy artificial viscosity in the entire region which is modifed so as to decrease the amount of artificial viscosity in regions of smoothness. The simplest switch changes from a first order monotone scheme to a higher order scheme if $|DU| < A$ where DU is some difference quotient of the current approximate solution U and A is a constant to be determined by the user. A scheme with this switch may be analyzed as a small perturbation of the basic first order monotone scheme but unfortunately also performs accordingly with e.g. poor resolution of contact discontinuities. On the other hand, methods with more complicated switches may be difficult to analyze and may also require problem dependent "tuning" of parameters.

For finite element methods attempts to improve the stability of Galerkin methods for hyperbolic problems were made in the late seventies by perturbing the test functions in various ways giving so called Petrov–Galerkin methods. For one-dimensional problems successful such Petrov–Galerkin methods were developed based on test functions weighted in the "upwind direction". However, accurate extensions to more than one dimension of such "upwind methods" were not easy to accomplish, and instead another technique for perturbing the test functions was proposed by Hughes and Brooks

156

1979 ([7]) for linear stationary convection-diffusion problems under the acronym SUPG (Streamline Upwind Petrov Galerkin).This method was analyzed and extended to time-dependent problems by Johnson and Nävert ([5], [16], [25]) and was referred to as the streamline diffusion method (cf. below). The method has subsequently been developed into a general finite element method for hyperbolic type problems, including hyperbolic systems of conservation laws such as the compressible Euler equations and also the incompressible Euler and Navier-Stokes equations (see [8]-[14], [17]-[23], [28]-[30], [26]). The purpose of this paper is to give an overview of some recent theoretical and computational results for the streamline diffusion method (SD-method for short below).

The SD-method is obtained from a standard Galerkin method through two basic modifications: First, the test functions are modified by adding a multiple of a linearized form of the hyperbolic operator involved which results in a particular Petrov-Galerkin method with a certain least squares control of the residual of the finite element solution built in, where the residual is obtained by inserting the finite element solution into the given continuous equation. The term "streamline diffusion" refers to this modification with terminology suggested by a natural interpretation in the case of a scalar convection problem. Secondly, artificial viscosity is added with viscosity coefficient proportional to the local absolute value of the residual of the finite element solution and a certain power of the local mesh size. The streamline diffusion modification, which was introduced first, eliminates most of the spurious oscillations in the standard Galerkin method, while the second modification referred to as "shock-capturing" further improves the quality of the numerical solution at discontinuities giving almost monotone discrete shock profiles. A further feature of the SD-method is the use of space-time finite elements for time-dependent problems with the basis functions being continuous in space and discontinuous in time, which leads to implicit time stepping methods. Using isoparametric space-time elements approximately oriented along characteristics makes it also possible to give the SD-method the advantages of methods of characteristics (e.g. large time steps) while avoiding the disadvantages of such methods since in the SD-method only crude mesh alignment independently over each time step is sufficient.

The dramatically improved performance of the SD-method may be viewed as resulting from the increased stability introduced through the streamline diffusion and the shock capturing modifications. Both modifications naturally fit into a finite element framework and do not seem to have known finite difference counterparts, although the resulting methods of course may be interpreted (on regular grids) as new finite difference methods. The shock capturing artificial viscosity depending on the local absolute value of the residual may be viewed as a certain "switch", adding automatically considerable viscosity at shocks where the residual will be large and little viscosity in regions of smoothness where the residual may be small. However, the construction of this switch is different from those used for finite difference methods for which the concept of residual of the discrete solution is not naturally available or at least has not been used. Further, the "upwind concept" so important in the derivation of finite difference methods is not used in the SD-method (except in the time variable for time-dependent problems). This is the motivation for using the term "streamline diffusion" which emphasizes that stability is the basic concept being obviously related to diffusion, instead of SUPG originally used by Hughes and Brooks indicating that "upwind" is the bearing idea. More precisely, "upwind" seems to be fruitful as design concept only together with discontinuous finite element approximations and accordingly appears (as a "jump term") in the SD-method only for the discretization in time. This does not mean of course that SD-solutions with continuous basis functions lack upwind features concerning e.g. the propagation of effects, only that these features appear as a result of the presence of diffusion-related stability (as in a convection-diffusion problem where the sign of the diffusion term and boundary data determine the upwind direction of the convection), rather than being explicitly imposed.

The basic theoretical results for the SD-method may be briefly summarized as follows:

L_2-error estimates of order $\mathcal{O}(h^{k+\frac{1}{2}})$, where k is   (1.3a)
the of the piecewise polynomials, for general linear
problems assuming that derivatives of order $k+1$ of
the exact solution belong to L_2 ([16], [25]).

Localization results e.g. for linear scalar problems   (1.3b)
stating that the presence of (1.3b) discontinuities or
layers in the exact solution does not degrade the accuracy
in regions where the exact solution is smooth ([16], [22]).

Convergence to the unique entropy solution for   (1.3c)
general scalar conservation laws in several dimensions
also with boundary conditions in bounded domains, with
polynomials of arbitrary degree ([18]-[20], [22], [28-30]).

For systems of conservation laws it is proved that   (1.3d)
limits of finite element solutions are weak solutions
of the system andsatisfy any associated entropy inequality
(general form of the Lax-Wendroff theorem (see [22], [30]).

For the incompressible Euler and Navier-Stokes equa-   (1.3e)
tions in the case of high Reynolds number, error
estimates of order $\mathcal{O}(h^{k+\frac{1}{2}})$ using stream function-
vorticity or velocity-pressure (with k = 1) formulations
([17], [6]).

Some of these results may be proved with only the streamline modification
present (1.3a,b,e), while the convergence results for conservation laws in general require
also the shock-capturing term. Recently, also some results for a linear model problem on
an adaptive version of the SD-method connecting with the design goal (1.2) have been
obtained ([23]). The adaptive method is based on an a posteriori error bound involving
the residual of the computed finite element solution and it is proved that the corresponding adaptive algorithm is reliable and efficient in a certain sense. In particular, it is
proved that the a posteriori bound will tend to zero with decreasing mesh size also in
cases where the exact solution is discontinuous. To be able to prove this result, which is
not valid for instance for the standard Galerkin method, we use the strong control of the
residual built in the SD-method through both the streamline and shock capturing modification. Thus, the further demand (1.2) of adaptive error control seems to naturally fit
with the design of the SD-method. As far as we know, the result in [23] is the first to
prove that reliable and efficient quantitative error control is possible for a hyperbolic
model problem; previous adaptive techniques for such problems have been qualitative in
nature only using some criterion for local mesh refinement, but without attempting to
control the total error quantitatively in an efficient way by balancing the error
contribution from different sources. A common problem here seems to be the use of
refinement criteria which e.g. tend to put too many mesh points at shocks and too few at
contact discontinuities or in more regular parts of the flow.

To sum up, the theoretical and numerical results for the SD-method obtained so
far seem to indicate that the method satisfies the criteria (1.1a-c). Also, the recent
result on an adaptive version gives evidence that (1.2) may be satisfied to some extent,
and sheds some further light on the role of the two modifications involved. We again
emphasize that both modifications have full finite element flavour and do not seem to
have any known finite difference analogues, except of course the general notion of

artificial viscosity which is pervading the whole subject. We may further note that the results (1.3) including in particular (1.3c,d) concerning conservation laws are more general than those obtained earlier for finite difference methods, cf. Section 3 below.

An outline of the remainder of this note is as follows. In Section 2 we introduce the SD-method for a linear scalar model problem and also state the basic results for an adaptive version of the method in this case. In Section 3 we consider the case of nonlinear conservation laws including the Euler equations for compressible flow, in Section 4 we report on the results available for the incompressible Euler and Navier-Stokes equations and finally in Section 5 we present some numerical results. By C we denote a positive constant, not necessarily the same at each occurence.

2. THE SD-METHOD FOR A LINEAR SCALAR MODEL PROBLEM

As a model problem we consider the following linear convection-diffusion problem:

$$\begin{cases} u_\beta + u - \mathrm{div}(\epsilon \nabla u) = f & \text{in } \Omega, \\ u = g & \text{on } \Gamma, \end{cases} \tag{2.1}$$

where $u_\beta = \beta \cdot \nabla u$, Ω is a bounded convex domain in $\mathbb{R}^2$ with polygonal boundary Γ, β is a given smooth velocity field, f and g an given data, and ϵ is a given positive diffusion coefficient. In the vanishing viscosity limit with $\epsilon = 0$ the problem takes the form

$$\begin{cases} u_\beta + u = f & \text{in } \Omega, \\ u = g & \text{on } \Gamma_-, \end{cases} \tag{2.2}$$

where $\Gamma_- = \{x \in \Gamma : \beta(x) \cdot n(x) < 0\}$, where $n(x)$ is the outward normal to Γ at $x \in \Gamma$. We recall that the solution of (2.1) typically will have a boundary layer of thickness $\mathcal{O}(\epsilon)$ at the outflow boundary $\Gamma_+ = \{x \in \Gamma : \beta(x) \cdot n(x) > 0\}$ and internal layers of width $\mathcal{O}(\sqrt{\epsilon})$ along streamlines given by β if e.g. the inflow data g is discontinuous.

To define the SD-method for (2.1-2) let $T_h = \{K\}$ be a finite element triangulation of Ω into triangles K of diameter h_K and define for a given $k \geq 1$,

$$V_h = \{v \in C(\Omega) : v|_K \in P_k(K) \quad \forall K \in T_h\},$$

where $P_k(K)$ denotes the set of polynomials of degree at most k on K, and $C(\Omega)$ denotes the set of continuous functions on $\bar{\Omega} = \Omega \cup \Gamma$. We note that V_h consists of continuous piecewise polynomials of degree k on Ω. The SD-method for (2.2) can now be formulated as follows: Find $U \in V_h$ such that

$$B(U,v) \equiv \int_\Omega U_\beta(v+\delta v_\beta)dx - \int_{\Gamma_-} Uv\beta \cdot nds + \int_\Omega \hat{\epsilon}\nabla U \cdot \nabla v dx \qquad (2.3)$$
$$= \int_\Omega f(v+\delta v_\beta)dx - \int_{\Gamma_-} gv\beta \cdot nds \equiv L(v) \qquad \forall v \in V_h,$$

where

$$\delta = C \frac{h}{|\beta|},$$
$$\hat{\epsilon} = \bar{C}h^\alpha |r(U)|,$$

where $C, \bar{C}$ and α are positive constants with $\frac{3}{2} \le \alpha < 2$, and $r(U) = U_\beta - f$ is the residual of the finite element solution. Further, $h: \bar{\Omega} \to \mathbb{R}$ is a continuous function measuring the local mesh size satisfying

$$ch_K \le h(x) \le Ch_K, \quad x \in K, \quad \forall K \in T_h,$$

with c and C positive constants. The δ-term is the streamline diffusion modification and the $\hat{\epsilon}$-term is the shock-capturing modification with viscosity coefficient $\hat{\epsilon} = \bar{C}h^\alpha |r(U)|$ involving the residual $r(U)$ of the finite element solution. The method (2.3) can be directly extended to the full problem (2.1) as follows for simplicity with $g = 0$ and $k = 1$: Find $U \in \overset{\circ}{V}_h \equiv \{v \in V_h: v = 0 \text{ on } \Gamma\}$ such that

$$B_\epsilon(U,v) = L(v) \qquad \forall v \in \overset{\circ}{V}_h, \qquad (2.4)$$

where B_ϵ is obtained from B by setting $\delta(x) = 0$ if $h(x)|\beta(x)| < C\epsilon$ and replacing ϵ by $\max(\bar{C}h^\alpha |r(U)|,\epsilon)$. We note that here the boundary condition is strongly imposed while in (2.3) only weakly through the variational formulation. Further, we see that the streamline diffusion modification is shut down if h is sufficiently small and that no shock-capturing artificial viscosity is added if $\bar{C}h^\alpha |r(U)| \le \epsilon$.

The basic stability result for the SD-method (2.3), which is valid e.g. if $- \text{div } \beta$ is bounded below by a positive constant, cf [16], and clearly displays the stabilizing effect

161

of the two modifications, reads as follows:

$$\|U\| + \|\delta^{\frac{1}{2}}(U_\beta - f)\| + \|\hat{\epsilon}^{\frac{1}{2}}\nabla U\| + (\int_\Gamma U^2 |\beta \cdot n| ds)^{\frac{1}{2}} \tag{2.5}$$
$$\leq C(\|f\| + (\int_{\Gamma_-} g^2 |\beta \cdot n| ds)^{\frac{1}{2}}),$$

where $\|\cdot\|$ denotes the $L_2(\Omega)$-norm. Using this stability result it is possible to prove the following a priori estimate for the error $e = u-U$ in the case $\epsilon = 0$:

$$\|e\| + \|\delta^{\frac{1}{2}}e_\beta\| \leq C\|h^{k+\frac{1}{2}}D^{k+1}u\|, \tag{2.6}$$

where $D^{k+1}u = (\sum_{|\alpha|=k+1} |D^\alpha u|^2)^{\frac{1}{2}}$ with usual multi-index notation and the constant C depends on k and characteristics of T_h as usual, and also the maximum norm of the gradient of u (see [16]). Note that $e_\beta = u_\beta - U_\beta = f - U_\beta = -r(U)$ and thus (2.4) in particular gives an optimal order $\mathcal{O}(h^k)$ - estimate of the residual $r(U)$. It is also possible to prove localized versions of (2.4) showing e.g. that if the exact solution u is discontinuous across a streamline given by β, then the error outside a numerical layer typically of width $\mathcal{O}(h^{3/4})$ will be of order $\mathcal{O}(h^{k+\frac{1}{2}})$. For the problem (2.1) with discrete analogue (2.4) an outflow boundary layer will be captured in a numerical layer of width of order $\mathcal{O}(h)$. The error estimate (2.6) may be naturally extended to the case with positive ϵ and there is then a continuous transition to the standard estimate $\|e\| \leq C\|h^{k+1}D^{k+1}u\|$ valid if $\epsilon = \mathcal{O}(1)$, (see [16], [21]).

For use below we note that we expect that in an internal numerical layer of width $O(\hat{\epsilon}^{1/2})$ we would have $r(U)=O(h^{-1/2})$ if $\alpha = 2$, since by (2.6) $r \sim h/\hat{\epsilon} \sim h^{-1}/r$ in the layer. Further, in an outflow layer we expect $r(U)=O(h^{-1})$. Thus, we expext that $\hat{\epsilon}=O(h^{3/2})$ in an internal layer and $\hat{\epsilon}=O(h)$ in an outflow layer if $\alpha=2$. Also, we expect in the same case that in smooth regions of the flow $\hat{\epsilon}=O(h^3)$.

We now briefly discuss the adaptive version of the SD-method (2.4) given in [23]. The object is to design an efficient algorithm for finding a mesh such that the error in the corresponding SD-solution is below a certain tolerance in the L_2-norm. The error control is based on decomposing the total error $e=u-U$ into $e=e_1+e_2=(\hat{u}-U)+(u-\hat{u})$, where u is the solution of a continuous perturbed problem of the form (2.1) with $\hat{\epsilon}$ replaced by ϵ. We now first present an a posteriori estimate for the discretization error

162

e_1 , then comment on the perturbation error e_2 and finally discuss some features of the adaptive algorithm.

The a posteriori estimate for e_1 given in [23] reads essentially as follows,

$$\|e_1\| \leq C\|\min(1,\hat{\epsilon}^{-1}h^2)R(U)\| \equiv E_1(U,f,h),\tag{2.7}$$

where $R(U)$ is the total residual defined by

$$R(U)= |U_\beta\text{-}f| + |\text{div}_h(\hat{\epsilon}\nabla U)|,$$

where $\text{div}_h(\hat{\epsilon}\nabla\cdot)$ is a discrete analogue of $\text{div}(\hat{\epsilon}\nabla\cdot)$, cf [23], [24]. The estimate holds under certain smoothness assumptions on ϵ ($|\nabla\hat{\epsilon}|\leq C\hat{\epsilon}^{\frac{1}{2}}$) which may require smoothing of $\hat{\epsilon}$ (and is proved in [23] in the special case with Ω a square). In the adaptive algorithm we now seek in order to control $\|e_1\|$ a mesh (by successive local refinement/unrefinement starting with a coarse mesh) such that

$$E_1(U,f,h) \cong TOL,\tag{2.8}$$

where TOL is the given tolerance. For the remaining perturbation error e_2 it is also possible, for the model problem, to prove an a posteriori estimate and seek to control the corresponding quantity as in (2.8). However, the proof is somewhat technical, while it is easy to prove a a priori estimate for e_2 of the form

$$\|e_2\| \leq C\min(\|(\hat{\epsilon}\text{-}\epsilon)^{1/2}\nabla u\|,\|\hat{\epsilon}d\Delta u\|),$$

where d is the distance to the outflow boundary From this estimate one may conclude that the contribution to $\|e_2\|$ will be of order $O(\hat{\epsilon}\text{-}\epsilon)$, $O((\hat{\epsilon}\text{-}\epsilon)^{1/4})$ and $O((\hat{\epsilon}\text{-}\epsilon)^{1/2})$ from regions where the solution is smooth, from an internal layer and from an outflow layer, respectively. Now, from the discussion following (2.6) we expect that in "smooth regions" $\hat{\epsilon}=O(h^3)$ and $R(U)=O(h)$, in internal layers of width $\hat{\epsilon}^{1/2}$ we have $\hat{\epsilon}=O(h^{3/2})$ and $R(U)=O(h^{-1/2})$, and in outflow layers of width $O(h)$ we have $\hat{\epsilon}=O(h)$ and $R(U)=O(h^{-1})$. We thus expect that (2.8) will give the local mesh size $h=O(TOL)$, $h=O(TOL^{8/3})$ and $h=O(TOL^2)$ in a smooth region, in an internal layer and in an outflow layer, respectively. It follows that the above contributions to $\|e_2\|$ would be of order $O(TOL^3)$, $O(TOL)$ and $O(TOL)$, respectively. Thus, the contribution to $\|e_2\|$ from smooth regions would normally be neglible, whereas the contribution from layers would be of the order of the tolerance. To sum up, it appears that $\|e_2\|$ will be dominated by $\|e_1\|$ and thus that we may base the adaptive method on (2.8).

Alternatively, we may continue the adaptive refinement until $\hat{\epsilon}=\epsilon$, in which case of course $e_2=0$. It apperas that if $\hat{\epsilon}=\epsilon$, then the mesh is fine enough to resolve all details of the flow. Now, if in a given problem we choose TOL sufficiently small, then we expect (2.8) to enforce $\hat{\epsilon}=\epsilon$ and resolution of all details of the flow. However, if TOL is not chosen small enough, then L_2-control may be possible to realize without resolving (e.g. outflow) layers. Continuing the refinement until $\hat{\epsilon}=\epsilon$ in such a case then requires a certain "over-refinement". We remark that even if $\hat{\epsilon}=\epsilon$ and no shock capturing is effectively added at the end of the adaptive process, there is a need for the shock capturing modification in earlier stages of the process when the mesh does not resolve all details.

Summing up, it appears that we may base the adaptive algorithm on (2.8) only, or add also the requirement that $\hat{\epsilon}=\epsilon$ to guarantee resolution of all details. An adaptive algorithm based on (2.8) only, seems reasonably efficient and appears to indicate the correct refinement in various parts of the flow (cf. Section 5 below). In particular, one can prove that $E_1(U,f,h)$ will tend to zero with decreasing h also in the presence of discontinuities (if $\epsilon=0$). We note that the a posteriori estimate for the standard Galerkin method would read as (2.7) with $\hat{\epsilon}$ replaced by ϵ, that is in the case $\epsilon=0$ as follows:

$$\|u-U\| \leq C\|r(U)\|. \tag{2.10}$$

In this case, however, the right hand side will not tend to zero with decreasing h in the presence of a discontinuity (unless the mesh is aligned with the discontinuity) and thus an adaptive method based on (2.10) would not work. Further, we note that even in a case when the solution u of (2.2) is continuous and h so small that $|r(U)| < h^{1-\alpha/2}$, in which case the residual terms in (2.7) and (2.10) would be the same, it is advantageous to use the streamline diffusion modification since by (2.6) in this case $\|r(U)\| = \mathcal{O}(h^k)$, while for the standard Galerkin method $\|r(U)\| = \mathcal{O}(h^{k-1})$ in general.

The presented results only concern a model problem, but it seems to be possible to generalize (2.7) to more general problems (e.g certain non-linear problems). The basic difficulty is there the stability of a continuous linearized dual problem reflecting the error propagation properties of the SD-method.

3. THE SD-METHOD FOR CONSERVATION LAWS

Until recently the computational scene for hyperbolic systems of conservation laws was dominated by finite difference methods and the design and theoretical analysis of these methods by a tradition to seek approximate solutions for which a bound on the total variation could be demonstrated. From a uniform bound on the total variation for a sequence of approximate solutions with decreasing mesh size, follows by compactness that a subsequence converges to a limit. With schemes on so called conservation form one can then conclude that the limit is a weak solution of the system of conservation laws and a step towards a full convergence proof has been taken. However, to be able to guarantee that such a limit is a physically correct weak solution, one has to also verify that it satisfies relevent entropy inequalities and to this end bounded variation in itself is not sufficient. For scalar conservation laws convergence has been proved by the indicated bounded variation estimates (BV-estimates) for first order monotone schemes in multi-dimensions and for certain higher order hybrid schemes in one space dimension, but no complete results for higher order schemes in more than one dimension seem to be available (cf. [3], [27], [32]).

When work to extend the SD-method to non-linear conservation laws started in the mid eighties chances to obtain results at first seemed slim since there was little hope of being able to prove BV-estimates on general meshes. Fortunately, some new mathematical techniques using the concepts of compensated compactness and measure valued solutions became available at that time through the work of DiPerna, Murat and Tartar ([31]) which opened new possibilities of proving convergence without BV-estimates. The new techniques were developed to obtain new existence results and were applied to prove convergence of approximate solutions of viscous regularizations of the continuous problem, but turned out to be very naturally applicable to prove also convergence of the approximate solutions given by the SD-method. Our work in this direction together with A. Szepessy started with a convergence proof for the SD-method without shock capturing for Burgers' equation in one space dimension using compensated compactness assuming that the (piecewise linear) finite element solutions were bounded in the maximum norm ([18]). This result was then completed in [22] for the full SD-method with shock capturing by proving the maxnorm bound. In the thesis [30] by Szepessy this result was generalized to general scalar conservation laws in several dimensions, also including boundary conditions on bounded domains, with piecewise polynomials of arbitrary degree by using a technique of DiPerna based on a uniqueness result for measure valued solutions. To prove convergence of a sequence of approximate solutions to the unique physically correct entropy solution it is in this approach sufficient to establish

a bound on the maxnorm of the approximate solutions,   (3.1a)

consistency with all entropy inequalities,   (3.1b)

consistency with the initial condition.   (3.1c)

By consistency with all entropy inequalities we mean that a weak limit (which may be a measure-valued function) of the sequence of approximate solutions satisfies in a weak sense all entropy inequalities associated with the conservation law. Further, (3.1c) requires such a limit to satisfy the given initial condition in a certain sense. We emphasize that BV-estimates are not needed in this approach.

For systems of conservation laws the SD-method was first applied using so called entropy variables (cf. below) in [9] and it was proved in [22] that a limit of finite element solutions is a weak solution of the system and satisfies the entropy inequality associated with the choice of entropy variables. In [30] the SD-method was formulated also in the usual conservation variables and a similar result to that just stated was proved in this case, now for any entropy inequality associated with the system. Several computational results for systems (the compressible Euler equations) are now available using entropy variables in two and three space dimensions (see [12], [26], [22], [8]). The quality of these results is comparable to that of good finite difference methods but the implementation is somewhat heavy due to the use of entropy variables and the impliciteness of the time-stepping. The implementation in conservation variables promises to be cheaper, cf [30], where some numerical results for one-dimensional compressible flow using conservation variables are given. In Figure 5.3 below we give a result for a two-dimensional shock reflection problem using entropy variables.

We now formulate the SD-method for a system of $m \geq 1$ conservation laws using conservation and entropy variables. We consider the following intial value problem for simplicity in one space dimension: Find $u = (u_1,..., u_m): \mathbb{R} \times \mathbb{R}_+ \to \mathbb{R}^m$ such that

$$L(u) \equiv u_t + f(u)_x = 0 \quad x \in \mathbb{R}, \ t \in \mathbb{R}_+, \quad\quad (3.2a)$$
$$u(x,0) \equiv u_0(x) \quad\quad x \in \mathbb{R}, \quad\quad (3.2b)$$

where $f: \mathbb{R}^m \to \mathbb{R}^m$ is a given smooth flux, $u_0 = \mathbb{R} \to \mathbb{R}^m$ is a given bounded initial value with compact support and $\mathbb{R}_+$ denotes the set of positive real numbers. We assume that there exists a convex entropy $\eta: \mathbb{R}^m \to \mathbb{R}_+$ with corresponding entropy flux $q: \mathbb{R}^m \to \mathbb{R}$

166

satisfying the compatibility relation

$$\nabla \eta f' = \nabla q, \tag{3.3}$$

where f' is the Jacobian of f. The corresponding entropy inequality for (3.2) is then given by

$$\eta(u)_t + q(u)_x \leq 0, \tag{3.4}$$

with interpretation in the distribution sense, which implies entropy stability in the form

$$\int_{\mathbb{R}} \eta(u(x,t))dx \leq \int_{\mathbb{R}} \eta(u_0(x))dx, \quad t \in \mathbb{R}_+. \tag{3.5}$$

We recall that (3.4) is obtained by multiplying (viscous regularizations of) (3.2a) by $\nabla \eta(u)$.

To define the SD-method for (3.2) we introduce for $h > 0$ a sequence of discrete time levels $0 = t_0 < t_1 < t_2 < \dots$ with time steps $t_{n+1} - t_n$ of order h, and for $n = 0,1,2,\dots$, we let V_1^n be the space of continuous piecewise linear functions on a triangulation $T_h^n = \{K\}$ of the space-time strip $S_n = \Omega \times I_n$, $I_n = (t_n, t_{n+1})$, of mesh size of order h. We also introduce the space

$$V_2^n = \{v: \nabla \eta(v) \in V_1^n\}, \quad n = 0,1,2,\dots .$$

We will now for $i = 1$ and 2 seek a discrete solution U such that $U|_{S_n} \in V_i^n$ for $n = 0,1,\dots$, where the usual conservation variables U will be piecewise linear if $i = 1$, while if $i = 2$ instead the so called conservation variables $\nabla \eta(U)$ will be piecewise linear. We note that in either case U may be discontinuous across the discrete time levels t_n and to account for this fact we introduce the notation

$$U_\pm^n = \lim_{s \to 0\pm} U(\cdot, t_n + s).$$

The SD-method can now be formulated as follows for $i = 1$ (conservation variables piecewise linear) or $i = 2$ (entropy variables piecewise linear): For $n = 0,1,2,\dots$, find $U = U|_{S_n} \in V_i^n$ such that

$$\int\limits_{S_n} L(U) \cdot (v + \delta(v_t + f'(U)^T v_x)) dxdt + \int\limits_{\mathbb{R}} (U_+^n - U_-^n) v_+^n dx$$

(3.6)

$$+ \int\limits_{S_n} \epsilon_1(U)(U_x v_x + U_t \cdot v_t) dxdt + \int\limits_{S_n} \epsilon_2(U) U_x \cdot v_x dxdt = 0 \quad \forall v \in V_1^n,$$

where $U_-^0 = u_0$, $f'(U)^T$ is the transpose of $f'(U)$,

$$\epsilon_1(U)|_K = \bar{\delta} \int\limits_K |L(U)| dxdt / \int\limits_K dxdt, \quad K \in T_h^n,$$

$$\epsilon_2(U)|_K = \bar{\delta} \int\limits_{K_n} |U_+^n - U_-^n| dx / \int\limits_{K_n} dx, \quad K_n = K \cap \{x,t\}: t = t_n\}, \quad K \in T_h^n,$$

$$\delta = hC_{0,i}(U)/(1 + |f'(U)|),$$

$$\bar{\delta} = C_1 h^{\alpha_1}, \quad \bar{\bar{\delta}} = C_2 h^{\alpha_2},$$

$$\tfrac{3}{2} \le \alpha_1 < 2, \quad \tfrac{1}{2} \le \alpha_2 < 1.$$

Here C_1 and C_2 are positive constants and $C_{0,i}(U)$ is a certain positive definite
m×m matrix. It is natural to choose $C_{0,1} = C_{0,2} \eta''$ where η'' is the Hessian of η. For
relevant choices of $C_{0,1}$ we refer to [12], [22]. The simplest reasonable choice is (up to a
poitive constant) $C_{0,2} = I$ corresponding to $C_{0,1} = \eta''$.

We note the presence of the "streamline diffusion" modification related to the
δ-term giving some control of the residual choosing $v = \nabla \eta(U)$ (cf. below) and noting
that by differentiating (3.3), we have $f'^T \eta'' = (f'^T \eta'')^T = \eta'' f'$ so that

$$(\nabla \eta(U))_t + f'(U)^T (\nabla \eta(U))_x = \eta''(U) U_t + f'(U)^T \eta''(U) U_x$$
$$= \eta''(U)(U_t + f'(U) U_x) = \eta''(U) L(U).$$

Further, we note the term with the jump $U_+^n - U_-^n$, present since U may be
discontinuous across the discrete time levels, and the shock-capturing terms with
artificial viscosities $\epsilon_i(U)$ involving the two parts of the total residual, namely $L(U)$
and the jump $U_+^n - U_-^n$. The method can be naturally extended to several dimensions

168

and higher order polynomials, see [30].

We note that since limits of finite element solutions given by (3.6) are proved to be weak solutions of the system (3.2), we may view (3.6) as effectively having "conservation form", although this is not strictly true with the definition of conservation form normally used in finite difference methodology, which requires a very specific form of the scheme. This form seems unnecessarily restricted and is not the only one for which one can verify that a limit of discrete solutions is a weak solution. As indicated, the convergence proof for scalar problems is based on verifying the condition (3.1a-c) and the entropy consistency is here at the heart of the matter (including as a special case consistency as a weak solution just discussed). To be able to prove entropy consistency for a certain entropy η we want basically to choose $v = \nabla\eta(U)$ in the finite element equations. Using entropy variables related to η this is possible, since $\nabla\eta(U)$ is then piecewise polynomial. However, in general e.g. to be able to verify (3.4) simultaneously for many entropies, we have to choose v equal to a piecewise polynomial interpolant $\pi\nabla\eta(U)$ of $\nabla\eta(U)$ and account for the error $E(U) = \nabla\eta(U) - \pi\nabla\eta(U)$. The design of the shock-capturing modification has now been made with the purpose of being able to control $E(U)$. Thus, although the shock capturing acts as a certain "switch", it has not been designed as such in some more or less ad hoc fashion as is often done, but rather by intrinsic features of the problem. As a result there are in the SD-method in principle no parameters to be determined by the user; the constants C_1 and C_2 for a given α only depend on basic features of discretization process such as degree of the polynomials and smallest angle of the finite elements involved, and do not depend on the specific problem (e.g. the flux f) or the exact solution u and neither on the computed solution U. Similarly the matrix $C_{0,i}(U)$ has a given form. To say that the SD-method is "parameter free" does not mean that improvements in the quality of the finite element solution are not possible in specific problems by somewhat varying the constants C_i and α, but doing so is not necessary to guarantee basic convergence properties of the scheme.

Again we emphasize that the basic design criteria for the streamline diffusion modification is residual control without loss of accuracy, and for the shock-capturing modification the need to control the entropy related quantity $E(U)$ discussed above. As a result of these modifications the SD-method will have the character of an "upwind scheme" and will include a "switch", but these features have not been used as basic design principles. We tend to view this as reflecting that the SD-method is "close" to the continuous equation and has a very similar basic structure, whereas in other methods the similarity on the basic level may be less obvious and instead concentrated on achieving similarity concerning secondary features such as "upwind character" which results from the basic structure. As a result of the basic similarity of the SD-method and the underlying conservation law, it has been possible to prove for this method most of the convergence results available for continuous viscous regularizations of the given problem, which may be a convenient way of summarizing the theoretical results obtained for the SD-method applied to conservation laws.

4. THE INCOMPRESSIBLE EULER AND NAVIER-STOKES EQUATIONS

Early applications of streamline diffusion modified Galerkin methods to the incompressible Euler and Navier-Stokes equations where made by Hughes and Brooks [8] using velocity-pressure formulations. However, the formulations used were not fully consistent from a theoretical point of view and a satisfactory error analysis was lacking. In [17] Johnson and Saranen introduced a fully consistent SD-method in two space dimension using a stream function-vorticity-pressure formulation of the incompressible Euler/Navier-Stokes equations with small viscosity ϵ, and proved L_2 error estimates uniformly of order $\mathcal{O}(h^{k+\frac{1}{2}})$ in the range $0 \leq \epsilon \leq h$. This method gave stable accurate numerical solutions also for difficult problems with small viscosity or high Reynolds numbers. However, extension to three dimensions seemed complicated and for this case it was desirable to have a method using velocity-pressure variables and low order polynomials. Such a method was introduced and analyzed by Hansbo and Szepessy [6] using continuous piecewise linear approximation of both velocity and pressure and L_2-estimates for the velocities of order $\mathcal{O}(h^{3/2})$ were proved (in two and three space dimensions) again uniform in the range $0 \leq \epsilon \leq h$. We now briefly present this method together with some numerical results. As far as we know, this is the first higher order accurate (better than first order) method for the small or zero viscosity incompressible Navier-Stokes equations using continuous piecewise polynomial equal order approximation of velocity and pressure. We remark that the usual error estimates for finite element methods for incompressible Navier-Stokes requires the viscosity ϵ to be bounded away from zero, cf. [2].

We first recall the incompressible Navier-Stokes equations: Find (u,p) such that

$$u_t + u \cdot \nabla u - \epsilon \Delta u + \nabla p = f \quad \text{in } \Omega \times \mathbb{R}_+, \tag{4.1a}$$
$$\text{div } u = 0 \qquad \text{in } \Omega \times \mathbb{R}_+, \tag{4.1b}$$
$$u = 0 \qquad \text{on } \Gamma \times \mathbb{R}_+, \tag{4.1c}$$
$$u = u_0 \qquad \text{in } \Omega \text{ for } t = 0, \tag{4.1d}$$

where Ω is a bounded domain in $\mathbb{R}^d$, $d = 2,3$, with boundary Γ, $u(x,t) \in \mathbb{R}^d$ is the velocity and $p(x,t) \in \mathbb{R}$ the pressure, f and u_0 are given data, and ϵ is the viscosity.

To define the SD-method for (4.1) let for $h > 0$, $0 = t_0 < t_1 < t_2 \ldots$ be a sequence of time steps with $t_{n+1} - t_n \sim h$, let for each $n=1,2,..$, $T_h^n = \{\tau\}$ be a finite element triangulation of Ω, let $K_h^n = \{\kappa\}$ be the corresponding subdivision of the "slab" $S_n = \Omega \times I_n$, $I_n = (t_n, t_{n+1})$, into elements $\kappa = \tau \times I_n$ of diameter $\sim h$ (see Fig. 5.1), and introduce the spaces

$$V_h^n = \{v \in C(S_n)^d : v|_\kappa \in P_1(\tau) \times P_1(I_n)$$
$$\forall \kappa = \tau \times I_n \in K_h^n, \; v = 0 \; \text{on} \; \Gamma \times I_n\},$$
$$Q_h^n = \{q \in C(S_n) : q|_\kappa \in P_1(\tau) \times P_1(I_n), \qquad \forall \kappa = \tau \times I_n \in K_h^n\}.$$

In other words, we use continuous bilinear space-time elements on each slab S_n.

The SD-method for (4.1) can now be formulated as follows in the case $\epsilon < Ch$:
For $n = 0,1,2,\ldots$ find $U \equiv U|_{S_n}$ and $P \equiv P|_{S_n}$ such that

$$(U_t + U \cdot \nabla U + \nabla P, v + \delta_1(v_t + U \cdot \nabla v + \nabla q))_{S_n}$$
$$(r(\text{div } U)U, v)_{S_n} + (\text{div } U, q + \delta_2(\text{div } U)\text{div } U)_{S_n} + \epsilon(\nabla U, \nabla v)_{S_n} \qquad (4.2)$$
$$+ (U_+^n - U_-^n, v_+^n)_\Omega = (f, v + \delta_1(v_t + U \cdot \nabla v + \nabla q))_{S_n} \qquad \forall (v,q) \in V_h^n \times Q_h^n,$$

where $(\cdot, \cdot n)_\omega$ denotes the scalar product in $[L_2(\omega)]^m$, $m = 1, d$,

$$w_\pm^n = \lim_{s \to 0^\pm} w(\cdot, t_n + s),$$
$$r(y) = \begin{cases} 0 & \text{if } y \leq K, \\ y/2 & \text{if } y > K, \end{cases}$$
$$\delta_2(y) = \begin{cases} C_1 h & \text{if } y \leq K, \\ C_1 & \text{if } y > K, \end{cases}$$
$$\delta_1 = C_2 h,$$

where C_1, C_2 and K are certain constants independent of h and ϵ. We note that for simplicity no shock capturing modification is present in (4.2). We further note that (4.2) may be viewed as a natural extension of the basic method (2.4) with certain additional non-standard nonlinear terms including the factors $r(\text{div } U)$ and $\delta_2(\text{div } U)$. These terms are introduced for stability reasons related to the convection term. In particular

171

we note that by Green's formula
$$(U \cdot \nabla U, U) + (r(\text{div } U)U, U)$$
$$= ((r(\text{div } U) - \frac{1}{2} \text{div } U)U, U) \geq - K(U,U),$$
which leads to basic L_2 control through a Gronwall inequality. L_2-error estimates of order $\mathcal{O}(h^{3/2})$ for the velocity and of order $\mathcal{O}(h^{\frac{1}{2}})$ for the gradients of the pressure are proved in [6]. An essential step in the analysis is to prove that div U may be larger than K only in a small part of the domain and as a result the presence of a non-zero factor r and a factor $\delta_2 = C_1$ in this part does not destroy overall L_2-accuracy. Again, we note that the difficulty in the error analysis is the fact that ϵ may be arbitrarily small and thus the diffusion term cannot be used to control convection terms. If $\epsilon = \mathcal{O}(1)$, then choosing $\delta_1 = Ch^2$, one can prove e.g. H^1-estimates of order $\mathcal{O}(h)$ for the velocities (see [6], and also [1] and [13] for a related method for the Stokes equations).

In Section 5 we present some numerical results for (4.2). The quality of these results indicate good accuracy and stability and gives some hope that the method may be a candidate for a simple yet accurate method for complicated flows in three dimensions.

5. NUMERICAL EXAMPLES

In this section we briefly present the results of some numerical experiments with the SD-method. We first consider the adaptive method (2.8) for the linear convection problem (2.2). In this example displayed in Fig. 5.2 the velocity β is discontinuous and as indicated in the figure the exact solution has a "contact discontinuity" across parallel characteristics and a "shock" with crossing characteristics. The meshes generated by the adaptive algorithm during the adaptive process together with the level curves of the approximate solution on the final mesh are displayed. We also give the actual L_2-error and the estimated L_2-error through the a posteriori estimate (2.7) on the different meshes.

In Fig. 5.3 we give the result of applying the method (3.6) with i = 2, i.e. in entropy variables, to a (stationary) shock reflection problem for the compressible Euler equations for a perfect gas.

Finally, in Fig 5.4 and 5.5 we give some results with the method (4.2) for the incompressible Navier-Stokes equation in the case of two-dimensional flow around a cylinder at Reynolds number 100 and a time dependent cavity problem at Reynolds number 400.

172

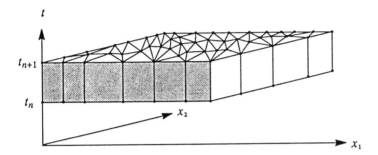

Fig. 5.1   Space–time discretization

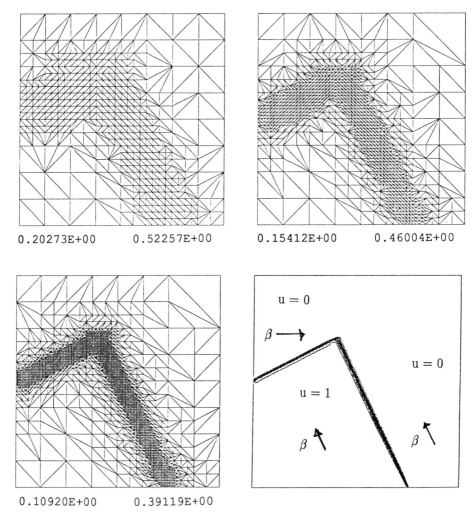

0.20273E+00   0.52257E+00   0.15412E+00   0.46004E+00

0.10920E+00   0.39119E+00

Fig. 5.2

Adaptive refinement for stationary convection problem with "contact" and "shock" discontinuity together with level curves of approximate solution on final mesh. Note that the flow direction is discontinuous as indicated. The actual (left) and estimated (right) L_2-errors on each mesh are given.

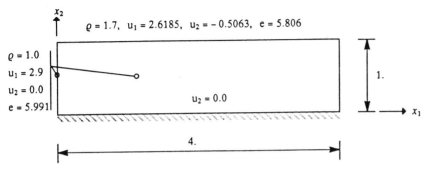

The computational domain with boundary and initial conditions

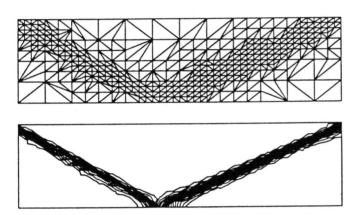

Mesh and contour lines for density

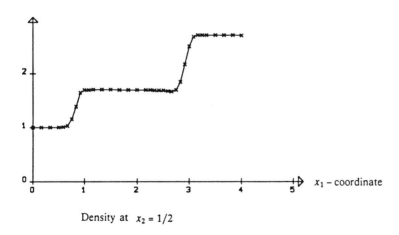

Density at $x_2 = 1/2$

Fig. 5.3   Shock reflection problem for the compressible Euler equations.

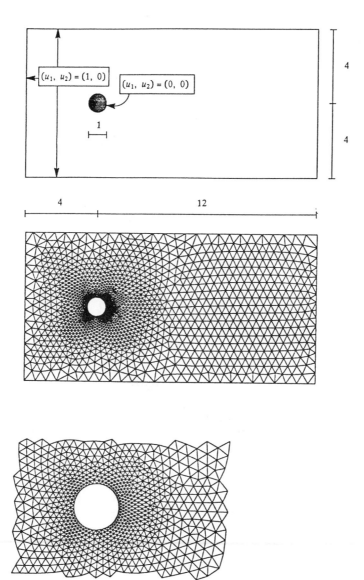

Fig. 5.4

Domain and computational mesh for flow around a cylinder for the incompressible Navier–Stokes equations at Reynolds number 100.

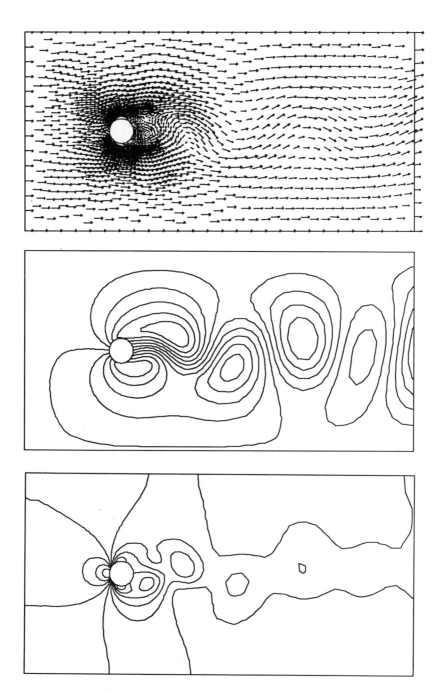

Fig. 5.4 continued:

Velocities, stationary streamlines and pressure of computed solution.

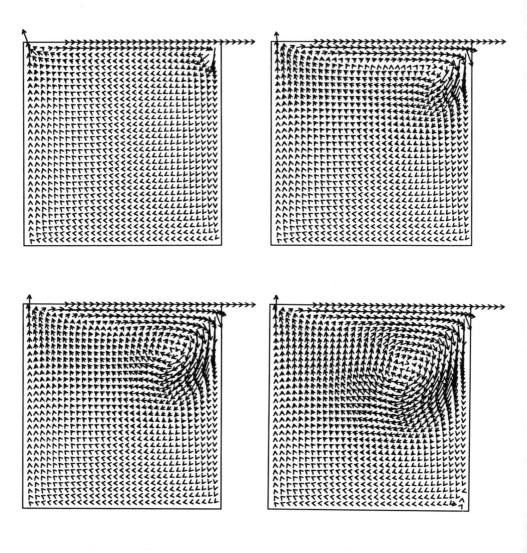

Fig. 5.5

Time dependent cavity problem for the incompressible Navier-Stokes
equations at Reynolds number 400 (velocities at time t = 0.1, t = 1, t = 2,
and t = 4).

References

[1] Brezzi, F. and Douglas, J., *Stabilized mixed methods for the Stokes problem*, Numer. Math., Vol 53 (1988), 225-235.

[2] Girault, V. and Raviart, P.-A., *Finite Element Approximation of the Navier-Stokes Equations*, Lecture Notes in Math. vol. 749, Springer, Berlin, 1979.

[3] Goodman, J.B. and Le Veque, R.J., *On the accuracy of stable schemes for 2D scalar conservation laws*, Math. Comp. 45 (1965), 697-715.

[4] Hansbo, P., *Finite Element Procedures for Conduction and Convection Problems*, Publication 86:7, Dept. of Structural Mechanics, Chalmers Univ. of Technology.

[5] Hansbo, P., *Adaptivity and Streamline Diffusion Procedures in the Finite Element Method*, Thesis, Publication 89:2, Dept. of Structural Mechanics, Chalmers Univ. of Technology.

[6] Hansbo, P. and Szepessy, A., *A velocity pressure streamline diffusion finite element method for the incompressible Navier-Stokes equations*, to appear in Computer Methods in Applied Mechanics and Engineering.

[7] Hughes, T.J.R. and Brooks, A., *A multidimensional upwind scheme with no crosswind diffusion*, AMD-vol 34, Finite Element Methods for Convection Dominated Flows, e.d. T.J.R. Hughes, ASME New York 1979.

[8] Hughes, T.J.R. and Brooks, A., *Streamline Upwind/Petrov-Galerkin Formulations for Convection Dominated Flows with Particular Emphasis on the Incompressible Navier-Stokes Equations,*

[9] Hughes, T.J.R., Franca, L.P. and Mallet, M., *A new finite element formulation for computational fluid dynamics: I Symmetric forms of the compressible Euler and Navier-Stokes equations and the second law of thermodynamics,* Computer methods in applied mechanics and engineering 54 (1986) 223-234.

[10] Hughes, T.J.R., Mallet, M. and Mizukami, A., *A New Finite Element formulation for Computational Fluid Dynamics: II. Beyond SUPG*, Comput. Meths. Appl. Mech. Engrg. Vol. 54 (1986), 341-355.

[11] Hughes, T.J.R. and Mallet, M., *A New Finite Element Formulation for Computational Fluid Dynamics: III The General Streamline Operator for Multidimensional Advective-Diffusive Systems*, Comput. Meths. Appl. Mech. Engrg. Vol. 58 (1986), 329-336.

[12] Hughes, T.J.R. and Mallet, M., *A New Finite Element Formulation for Computational Fluid Dynamics: IV. A Discontinuity-Capturing Operator for Multidimensional Advective-Diffusive Systems*, Comput. Meth. Appl. Mech. Engrg., Vol. 58 (1986), 329-336.

[13] Hughes, T.J.R., Franca, L.P., and Balestra, M., *A new finite element formulation for computational fluid dynamics: V. Circumventing the Babuska-Brezzi Condition: A Stable Petrov-Galerkin Formulation of the Stokes-Problem Accomodating Equal-order Interpolations*, Comput. Meths. Appl. Mech. Engrg., Vol. 59 (1986), 85-99.

[14] Hughes, T.J.R., Franca, L.P., Hulbert, G.M., Johan, Z. and Shakib, F., *The Galerkin/lest-squares method for advective-diffusive equations*, Recent Developments in Computational Fluid Dynamics, (ed. T.E. Tezduyar and T.J.R. Hughes) AMD-Vol. 95, ASME, Chicago, 1988.

[15] Johnson, C. and Nävert, U., *An analysis of some finite element methods for advection-diffusion problems*, in Axelsson, Frank, van der Sluis, eds., Analytical and Numerical Approaches to Asymptotic Problems in Analysis (North-Holland, Amsterdam, (1981).

[16] Johnson, C., Nävert and Pitkäranta, *Finite Element Methods for Linear Hyperbolic Problems*, Comput. Meths. Appl. Mech. Engrg., Vol. 45 (1984), 285-312.

[17] Johnson, C., and Saranen, J., *Streamline Diffusion Methods for the incompressible Euler and Navier-Stokes equations*, Math. Comp., Vol. 47 (1986), 1-18.

[18] Johnson, C. and Szepessy, A., *On the convergence of a finite element method for a nonlinear hyperbolic conservation law*, Math. Comp., vol. 49, N. 180 Oct 1987, p. 427-444.

[19] Johnson, C., and Szepessy, A., *On the convergence of streamline diffusion finite element methods for hyperbolic conservation laws*, Numerical Methods for Compressible Flow - Finite Difference, Finite Element and Volume Techniques - AMD-vol. 78 (1986), eds. T.E. Tezduyar and T.J.R. Hughes.

[20] Johnson, C., and Szepessy, A., *Shock-capturing streamline diffusion finite element methods for nonlinear conservation laws*, Recent Developments in Computational Fluid Mechanics - AMD-vol. 95 (1988), eds. T.E. Tezduyar and T.J.R. Hughes.

[21] Johnson, C., Schatz, A.H. and Wahlbin, L.B., *Crosswind smear and pointwise errors in streamline diffusion finite element methods.* Math. Comp., 47, July 1987, 25-38.

[22] Johnson, C., Szepessy, A. and Hansbo, P., *On the convergence of shock-capturing streamline diffusion finite element methods for hyperbolic conservation laws*, to appear in Math. Comp.

[23] K. Eriksson and Johnson, C., *An adaptive streamline diffusion finite element method for convection-diffusion problemes*, to appear.

[24] K. Eriksson and Johnson, C., *Adaptive finite element methods for parabolic problmes I: A linear model problem*, Preprint 88:31, Dept. of Mathematics, Chalmers Univ. of Technology.

[25] Nävert, U., *A finite element method for convection-diffusion problems.* Thesis, Dept. of Computer Science, Chalmers Univ. of Technology.

[26] Shakib, F., *Finite element analysis of the compressible Euler and Navier-Stokes equations, Thesis*, Standford University, Standford, California, 1988.

[27] Shu, C. and Osher, S. *Efficient implementation of essentially non-oscillatory shock-capturing schemes*, J. Comput. Phys. 77 (1988), 4390471.

[28] Szepessy, A., *Convergence of a shock-capturing streamline diffusion finite element method for a scalar conservation law in two space dimensions*, to appear in Math. Comp. (Oct. 1989).

[29] Szepessy, A., *Measure valued solutions of scalar conservation laws with boundary conditions*, to appear in Arch. Rat. Mech. Anal.

[30] Szepessy, A., *Convergence of the Streamline Diffusion Finite Element Method for Conservation Laws*, Thesis, Dept. of Mathematics, Chalmers Univ. of Technology, 1989.

[31] Tartar, L, *Compensated compactness and Applications to Partial Differential Equations, in Nonlinear Analysis and Mechanics,* Heriot-Watt Symposium, IV, p. 136-192, Research Notes in Mathematics, Pitman.

[32] Vila, J.P., C.R. Acad. Sc. Paris, t.299, série I, no. 5, 1984, 157-160.

T. LYCHE
Condition Numbers for B-Splines

1. INTRODUCTION

In univariate spline theory there are still some unsolved problems. One problem is to determine how large B-spline coefficients a spline of unit norm can have. For such a condition number as it is customary called, we would like to know exact values for low orders and asymptotic estimates for high ordes [5].

It can be shown that the spline with largest B-spline coefficient behaves like a Chebyshev polynomial in that it alternates n times between $+1$ and -1, where n is the dimension of the spline space. Such equioscillating splines or Chebyshev splines is the topic of Section 3.

The conditon number also play a role in spline interpolation. Choosing the interpolation points at the extrema of the Chebyshev polynomial gives a polynomial interpolant which is good in some sense. For splines, Mørken [16] has shown that interpolating at the extrema of the Chebyshev spline gives an interpolant with minimal supremum norm condition number of the B-spline collocation matrix. In Section 4 we present some numerical experiments from [7] comparing spline interpolation at knot averages and interpolation at the extrema of the Chebyshev spline.

In the final section we consider condition numbers with respect to knot insertion. We give precise estimates on how much the size of the B-spline coefficients can be reduced by inserting one knot. This amounts to a study of condition numbers for discrete splines.

2. A CONDITION NUMBER

We define the *condition number* of a basis (ϕ_j) of a normed linear space S as the number

$$\kappa = \sup_{\mathbf{c}} \frac{\|\sum_j \phi_j c_j\|}{\|\mathbf{c}\|} \sup_{\mathbf{c}} \frac{\|\mathbf{c}\|}{\|\sum_j \phi_j c_j\|}. \tag{1}$$

We consider the case

$$S = S_{k,\mathbf{t}} = span(B_j)_{j=1}^n,$$

of B-splines $B_j = B_{j,k}(x) = B(x \mid t_j, \ldots, t_{j+k})$ of order k (degree $k - 1$) on a nondecreasing knot sequence $\mathbf{t} = (t_j)_{j=1}^{n+k}$. We assume that $n \geq k$, $t_k < t_{k+1}$, $t_n < t_{n+1}$, and $t_{j+k} > t_j$ for $j = 1, 2, \ldots, n$. Also, we assume that the B-splines are normalized to sum to one. Since B-splines form a nonnegative partition of unity, the first supremum in (1) is equal to 1 and we obtain

$$\kappa_{k,\mathbf{t}} = \sup_{\mathbf{c}} \frac{\|\mathbf{c}\|}{\|\sum_j B_j c_j\|}, \tag{2}$$

where in (2) we shall only consider the infinity norms

$$\|f\| = \sup_{t_1 < \leq x \leq t_{n+k}} |f(x)|, \qquad \|\mathbf{c}\| = \max_{1 \leq j \leq n} |c_j|.$$

We are mainly interested in the knot independent number

$$\kappa_k = \sup_{\mathbf{t}} \kappa_{k,\mathbf{t}}.$$

In 1973 de Boor [1] showed that $\kappa_k < \infty$. By some effort ([3]) he gave an upper bound

$$\kappa_k \leq 2k9^{k-1}, \tag{3}$$

that only grows exponentially in k.

To show that κ_k must indeed grow exponentially consider the Chebyshev polynomial T_{k-1} of degree $k - 1$. On the knot sequence

$$\mathbf{t} = (\overbrace{-1, -1, \ldots, -1}^{k \text{ times}}, \overbrace{1, 1, \ldots, 1}^{k; \text{times}}) = (t_j)_{j=1}^{2k}$$

the B-spline representation of T_{k-1} is

$$T_{k-1} = \sum_{j=1}^{k} (-1)^{k-j} B_{j,k} a_j, \tag{4}$$

where

$$a_1 = 1, \qquad a_j = \frac{(2k-3)(2k-5)\cdots(2k-2j+1)}{1 \cdot 3 \cdots (2j-3)}, \qquad j > 1,$$

and

$$B_{j,k}(x) = 2^{1-k} \binom{k-1}{j-1} (1-x)^{k-j} (1+x)^{j-1}.$$

Equation (4) follows from an explicit expression for the Jacobi polynomial

$$P_n^{(\alpha,\beta)} = 2^{-n} \sum_{m=0}^{n} \binom{n+\alpha}{m} \binom{n+\beta}{n-m} (x-1)^{n-m} (x+1)^m$$

of degree n and the fact that $T_n = \binom{n-1/2}{n}^{-1} P^{(-\frac{1}{2},-\frac{1}{2})}$. The largest coefficient d_k in (4) is the middle one. Hence, ([11])

$$\frac{k-1}{k}2^{k-3/2} \le d_k \le \frac{k}{k-1}2^{k-3/2}.$$

Since $\kappa_k \ge d_k$ we see that κ_k grows exponentially in k.

Exact values of κ_k are easily determined for $k = 2,3$. Indeed, since $\|f\| = \|c\|$ for any spline of order 2 we obviously have $\kappa_2 = 1$. For the B-spline coefficients c_j of a spline f of order 3 we can use the formula ([4,p 127])

$$c_j = -\frac{1}{2}f(t_{j+1}) + 2f(\frac{t_{j+1} + t_{j+2}}{2}) - \frac{1}{2}f(t_{j+2}).$$

From this it follows that $\kappa_3 \le 3$. Since $d_3 = 3$ we therefore have $\kappa_3 = 3$. For the cubic case it was recently shown in [5] that $\kappa_4 > d_4$. For estimates and analysis in the cubic case see [14]. In [7] κ_k has been estimated by numerical methods for $k \le 11$. These estimates indicate that $\kappa_k = O(2^k)$ so that κ_k might not be much bigger than d_k.

3. THE CHEBYSHEV SPLINE

In this section we will define spline analogs of the Chebyshev polynomials. Recall that if p^* is the best uniform approximation to the monomial x^n on $[-1,1]$ by polynomials of degree less than n, then

$$T_n(x) = \frac{x^n - p^*(x)}{\|x^n - p^*(x)\|}$$

is the Chebyshev polynomial of degree n. A similar construction can be used to define Chebyshev splines, i.e. we consider approximating B_n by the first $n-1$ B-splines $(B_j)_{j=1}^{n-1}$. Recall (see [8] for a simple proof) that B-splines are totally positive, i.e. for any $x_1 < x_2 < \cdots < x_n$, the n by n matrix with element $B_j(x_i)$ in row i and column j has only nonnegative minors. This implies that $(B_j)_{j=1}^{n-1}$ is a weak Chebyshev system. By results in [10], there exists a spline f in span$(B_j)_{j=1}^{n-1}$ so that $p_n = B_n - f$ alternates n times between $+\|p_n\|$ and $-\|p_n\|$. We call

$$C_t = \frac{p_n}{\|p_n\|}$$

the Chebyshev spline of order k on $\mathbf{t}$. This function has the following properties

(i) $C_t \in S_{k,t}$,

(ii) $\|C_t\| \leq 1$,

(iii) $C_t(\xi_j) = (-1)^{n-j}$ for n points $\xi_1, \ldots, \xi_n$, with $\xi_1 < \xi_2 < \cdots < \xi_n$.

In [16] it is shown that there is a unique function in $S_{k,t}$ which satisfies (i),(ii), and (iii). Some other properties of

$$C_t = \sum_{j=1}^{n} B_j c_j^*$$

shown in [16,17] are

(iv) $(-1)^{n-j} B_j(\xi_j) c_j^* > 0$, $\qquad j = 1, \ldots, n$,

(v) $\xi_j \in [t_{j+1}, t_{j+k-1}]$ for $k \geq 2$,

(vi) $|C_t(x)| < 1$ if $x \neq \xi_j$, $\qquad j = 1, \ldots, n$,

(vii) $\kappa_{k,t} = \|\mathbf{c}^*\|$.

Property (iv) follows from Property (v) and the strong version of the variation diminishing property of B-splines [18, p. 179]. Property (vii) shows that C_t is the worst conditioned spline in $S_{k,t}$. To see this, we note that for each j the spline

$$p_j = C_t/c_j^* = B_j + \sum_{i \neq j} B_i c_i^*/c_j^*$$

is well defined by (iv) and alternates n times. By [10] p_j is the error in best approximation of B_j by elements of $S_j = \text{span}(B_i)_{i \neq j}$. It follows that

$$\|p_j\| = 1/|c_j^*| = \text{dist}(B_j, S_j).$$

But (2) can be written

$$\kappa_{k,t} = \max_j \frac{1}{\text{dist}(B_j, S_j)}.$$

Hence (vii) follows.

To find the Chebyshev spline corresponding to a knot sequence t is in general a nontrivial task. A Matlab program is presented in [5]. The extrema have to be found iteratively in some way. A Remes type method has been implemented in [7]. To describe one iteration suppose an estimate for the extrema of C_t is available. To compute a (hopefully) better approximation one can proceed in a standard way. First, compute a spline in $S_{k,t}$ by interpolating to $+1$ and -1 at the current extrema. Then one finds the extrema of this spline. To start the iteration one could choose the knot averages as an initial guess.

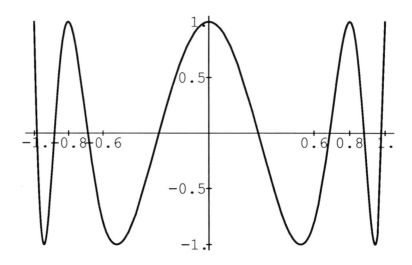

Figure 1

Figure 1 shows a quartic ($k = 5$) Chebyshev spline with knots

$$(-1, -1, -1, -1, -1, -0.8, -0.6, 0.6, 0.8, 1, 1, 1, 1, 1)$$

. The computed B-spline coefficients are (1.00000000, -3.23382456, 3.34285092, -7.70966426, 9.54118953, -7.70966426, 3.34285092, -3.23382456, 1.00000000).

4. GOOD INTERPOLATION POINTS

Let a function $f \in C[a, b]$ and a spline space $S_{k,t}$ with $a \le t_k$ and $b \ge t_{n+1}$ be given. We want to construct a good approximation to f from $S_{k,t}$ using interpolation. For given knots t the question is how to choose the interpolation points $\rho = (\rho_1, \rho_2, \ldots, \rho_n)$. A measure of goodness is given by the size of the norm of the interpolation projector

$$P_\rho : C[a, b] \rightarrow S_{k,t}, \qquad (P_\rho f)(\rho_i) = f(\rho_i), \quad i = 1, 2, \ldots, n.$$

We would like a ρ so that

$$\|P_\rho\| = \sup \frac{\|P_\rho f\|}{\|f\|},$$

is small and preferably bounded by a constant indepentent of meshratios.

An upper bound for the norm of the interpolation projector is given by the norm of the B-spline collocation matrix $\mathbf{B}_\rho = (B_j(\rho_i))$. Indeed, since ([2])

$$\kappa_{k,t}^{-1} \|\mathbf{B}_\rho^{-1}\| \le \|P_\rho\| \le \|\mathbf{B}_\rho^{-1}\|,$$

where $\kappa_{k,t} \leq \kappa_k$ is the B-spline condition number, we see that bounding $||P_\rho||$ and $||\mathbf{B}_\rho^{-1}||$ is equivalent.

For cubic spline interpolation at knots we cannot bound P_ρ independently of mesh ratios. ([2]). In practice this means that if the knot spacing is highly nonuniform then there exist data f so that the interpolant will oscillate.

The situation is somewhat better for the choice

$$\hat{\rho}_i = t_{i,k} = (t_{i+1} + \cdots + t_{i+k-1})/(k-1), \qquad \hat{\rho} = (t_{1,k}, \ldots t_{n,k}), \tag{5}$$

i.e. interpolation at the *knot averages* $t_{i,k}$. For $k = 3$ we have $||P_{\hat{\rho}}|| = 2$, see [15], while for $k = 4$ $||\mathbf{B}_\rho^{-1}|| \leq 27$, [2]. This lead to the conjecture that for any order k, the projector $P_{\hat{\rho}}$ can be bounded independently of $\mathbf{t}$. That this conjecture is false was shown by Jia[9].

Consider now instead using the extreme points $\rho = \xi$ of the Chebyshev spline $C_\mathbf{t}$ in $S_{k,\mathbf{t}}$. It was discovered independently by Demko[6] and Mørken[16] that for this choice the interpolation projector can be bounded independently of $\mathbf{t}$. Indeed,

$$||\mathbf{B}_\xi^{-1}|| = \kappa_{k,t} \leq \kappa_k. \tag{6}$$

To prove (6) we observe that for any increasing ρ the inverse $\mathbf{B}_\rho^{-1}$ of the collocation matrix $\mathbf{B}_\rho$ is a checkerboard matrix, i.e. the elements are alternatively nonnegative and nonpositive. This follows since $\mathbf{B}_\rho$ has all minors nonnegative. From the sign structure of $\mathbf{B}_\rho^{-1}$ we have

$$||\mathbf{B}_\rho^{-1}|| = ||\mathbf{c}||,$$

where the vector $\mathbf{c} = (c_1, \ldots, c_n)^T$ is the solution of

$$\mathbf{B}_\rho \mathbf{c} = \mathbf{s}, \qquad \mathbf{s} = ((-1)^{n+1}, (-1)^n, \ldots, -1, 1)^T.$$

But if $\rho = \xi$ then this system is precisely the determining system for the B-spline coefficients $\mathbf{c}^*$ of the Chebyshev spline. Hence (6) follows.

Since $||\mathbf{B}_\rho|| = 1$, (6) implies that the *condition number with respect to inversion*

$$||\mathbf{B}_\rho|| ||\mathbf{B}_\rho^{-1}||$$

is bounded by κ_k when $\rho = \xi$. If we are looking for well conditioned matrices then interpolating at the Chebyshev extrema is as good as we can do. Indeed, Mørken [16] shows that

$$\min_\rho ||\mathbf{B}_\rho^{-1}|| = ||\mathbf{B}_\xi^{-1}||.$$

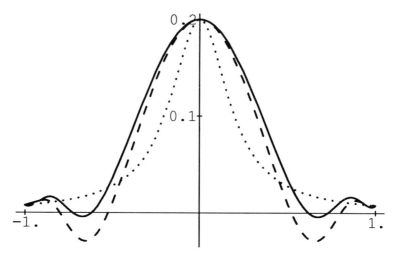

Figure 2: $k = 5$

To illustrate the above discussion we present some numerical experiments. Consider the Runge function $f(x) = 0.2/(1 + 25 x^2)$ on $[-1, 1]$. On the knot sequence

$$(\overbrace{-1, \ldots, -1}^{k}, -0.8, -0.6, 0.6, 0.8, \overbrace{1, \ldots, 1}^{k})$$

we have computed spline interpolants of order $k = 5, 9, 17$. For each k, we constructed two interpolants. One using knot averages, and the other interpolating at the extreme points of the Chebyshev spline of order k. The results are shown in Figure 2-4. Here we have $f(x)$ (dotted), knot averages (dashed) and extreme points (solid).

For the interpolant at the knot averages we observe increasing oscillations for large k. This is similar to the well known Runge phenomenon, where large oscillations occur in high degree equally spaced polynomial interpolation to f.

5. KNOT INSERTION AND DISCRETE B-SPLINES

Suppose $\tau = (\tau_j)_{j=1}^n$ and $\mathbf{t} = (t_i)_{i=1}^m$ are nondecreasing sequences with $\tau \subset \mathbf{t}$. If $p \in S_{k,\tau}$, say

$$p = \sum_{j=1}^n B_{j,k,\tau} c_j,$$

then also $p \in S_{k,\mathbf{t}}$, so that

$$p = \sum_{i=1}^m B_{i,k,\mathbf{t}} d_i.$$

188

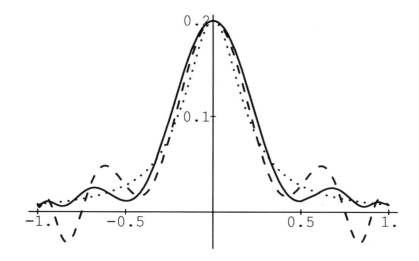

Figure 3: $k = 9$

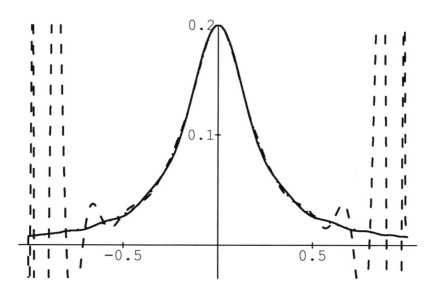

Figure 4: $k = 17$

We will assume that $\tau_{j+k} > \tau_j$ and $t_{i+k} > t_i$ for all integers i, j. We are interested in the transformation which maps the B-spline coefficients $\mathbf{c} = (c_1, \ldots, c_n)^T$ in $S_{k,\tau}$ into the B-spline coefficients $\mathbf{d} = (d_1, \ldots, d_m)^T$ in $S_{k,t}$. We call the m, n matrix $\mathbf{A}$ such that $\mathbf{d} = \mathbf{Ac}$ for the *knot insertion matrix of order k from τ to $\mathbf{t}$*. The elements of $\mathbf{A}$ have properties which makes it natural to call them *discrete B-splines*. $\mathbf{A}$ itself has the same band structure and normalization as a B-spline collocation matrix (cf. the previous section). In particular, the elements are nonnegative with row sums equal to one, i.e. $||\mathbf{A}|| = 1$. We also note that $\mathbf{A}$ has rank n since it is a basis transformation from one linear space to another. We refer to [12] for an introduction to discrete B-splines and B-spline basis transformations.

We define the discrete spline condition number as

$$\delta_{k,\tau,t} = \sup_c \frac{||\mathbf{Ac}||}{||\mathbf{c}||} \sup_c \frac{||\mathbf{c}||}{||\mathbf{Ac}||}.$$

Since $||\mathbf{A}|| = 1$, the first supremum is one and we have

$$\delta_{k,\tau,t} = \sup_c \frac{||\mathbf{c}||}{||\mathbf{Ac}||}. \tag{7}$$

With $\mathbf{d} = \mathbf{Ac}$ this number measures how much the B-spline coefficients can be reduced by knot insertion. The following result gives an upper bound for this reduction and again brings in the condition number for B-splines.

Proposition 1. *For the number $\delta_{k,\tau,t}$ in (7) we have for fixed τ*

$$\sup_{\tau \subset t} \delta_{k,\tau,t} = \kappa_{k,\tau}.$$

where $\kappa_{k,\tau}$ is given by (2).

Proof. Suppose $\tau \subset t$ and let $p \in S_{k,\tau}$ be a spline with B-spline coefficients $\mathbf{c}$. By (2) and the fact that the B-splines on t form a nonnegative partition of unity, we have

$$\kappa_{k,\tau}^{-1}||\mathbf{c}|| \leq ||p|| \leq ||\mathbf{Ac}||.$$

But then $\delta_{k,\tau,t} \leq \kappa_{k,\tau}$. Now for the ith B-spline coefficient $d_i = (\mathbf{Ac})_i$ of p on $\mathbf{t}$, we have ([4, p. 159])

$$d_i - p(t_{i,k}) = O((t_{i+k-1} - t_{i+1})^2),$$

where $t_{i,k}$ is a knot average given by (5). It follows that in some sense $\lim_{\Delta t \to 0} \mathbf{Ac} = p$, and the proposition follows.

190

With $\# u$ denoting the number of elements in a sequence u consider now

$$\delta_k^\ell = \sup_{\tau,\mathbf{t}} \{\delta_{k,\tau,\mathbf{t}} \ : \ \tau \subset \mathbf{t} \ \& \ \#\mathbf{t} - \#\tau = \ell\}.$$

This number measures how much the size of a B-spline coefficient can be reduced by inserting exactly ℓ new knots. In [13] we prove the following result in the case $\ell = 1$.

Theorem 1. The number δ_k^1 satisfies

$$\delta_k^1 = k, \qquad k \text{ odd};$$

$$k - \frac{2}{k} \le \delta_k^1 < k, \qquad k \text{ even}.$$

For k odd the extremal situation occurs when we insert one knot at 0 in the Bernstein knot sequence

$$(\overbrace{-1,\ldots,-1}^{k}, \overbrace{1,\ldots,1}^{k}).$$

On the refined knot sequence the extremal polynomial has B-spline coefficients alternating between $+1$ and -1, except for the two middle coefficients which have the same sign. Further details can be found in [13].

REFERENCES

1. de Boor, C., The quasi-interpolant as a tool in elementary polynomial spline theory, in *Approximation Theory*, G. G. Lorentz (ed.), Academic Press, New York, 1973, 269–276.
2. de Boor, C., On bounding spline interpolation, J. Approx. Th. **14** (1975), 191–203.
3. de Boor, C., On local linear functionals which vanish at all B-splines but one, in "Theory of Approximation with Application", A. G. Law and B. N. Sahney (eds.), Academic Press, New York, 1976, 120–145.
4. de Boor, C., *A practical guide to splines*, Springer Verlag, New York, 1978.
5. de Boor, C., The exact condition of the B-spline basis may be hard to determine, Center for the Mathematical Sciences Technical Summary Report #89-1, University of Madison, Wisconsin.
6. Demko, S., On the existence of interpolating projections onto spline spaces, J. Approx. Th. **43** (1985), 151–156.

7. Glærum, S., Condition numbers for the B-spline basis, Cand Scient Thesis, University of Oslo, 1989.

8. Goodman, T. N. T., Shape preserving representations, in *Mathematical Methods in Computer Aided Geometric Design*, T. Lyche and L. Schumaker (eds.), Academic Press, N. Y., 1989, , 333–351.

9. Jia, Rong-Qing, Spline interpolation at knot averages, Constr. Approx. **4** (1988), 1–7.

10. Jones, R. C. and L. A. Karlovitz, Equioscillation under nonuniqueness in the approximation of continuous functions, J. Approx. Th. **3** (1970), 138–145.

11. Lyche, T., A note on the condition number of the B-spline basis, J. Approx. Th. **22** (1978), 202–205.

12. Lyche, T., Discrete B-splines and conversion problems, preprint.

13. Lyche, T. and K. Mørken, How much can the size of the B-spline coefficients be reduced by inserting one knot? preprint.

14. Lyche, T. and K. Mørken, On the condition number of the B-spline basis with special emphasis on the cubic case, preprint.

15. Marsden, M. J., Quadratic spline interpolation, Bull. Amer. Math. Soc. **80** (1974), 903–906.

16. Mørken, K., On two topics in spline theory: discrete splines and the equioscillating spline, Cand. Real. Thesis, University of Oslo, 1984.

17. Mørken, K., Contributions to the theory and applications of splines, Ph.D. dissertation, University of Oslo, Norway, 1989.

18. Schumaker, L. L., *Spline Functions: Basic Theory,* John Wiley & Sons, New York, 1981.

Tom Lyche,
Institute of Informatics,
University of Oslo,
P O Box 1080 Blindern,
0316 Oslo 3,
Norway.

M.J.D. POWELL

The Updating of Matrices of Conjugate Directions in Optimization Algorithms

Abstract   In optimization algorithms it is usual to store approximations to second derivative matrices in factored form. We study the factorization that gives mutually conjugate directions of the second derivative approximation explicitly, which is used by Goldfarb and Idnani [6], Han [7] and Powell [10]. A rank three family of updating formulae is presented for revising the conjugate directions in order to satisfy a quasi-Newton equation, which means that we are using a change in gradients to improve our second derivative estimates. These formulae are easy to apply, and are suitable for preserving a factorization of an active constraint matrix when there are linear constraints on the variables. An advantage of the availability of conjugate directions is that, by changing their lengths, one can reduce greatly the inefficiencies that often occur in variable metric algorithms when the inverse of the second derivative approximation is nearly singular. A version of the column scaling technique is described and some numerical results are presented. They show not only that very substantial reductions occur in the amount of computation but also that the DFP method achieves an efficiency that is within a factor of two of the BFGS method.

1. Introduction

This paper is relevant to many successful algorithms for calculating the least value of a function $\{F(x) \mid x \in \mathcal{R}^n\}$, including the case when there are linear constraints on the components of the vector of variables x. Specifically, we address those methods that require the user to provide $F(x)$ and its gradient $\nabla F(x)$ for any feasible x, and that employ an $n \times n$ positive definite matrix B which can be regarded as an estimate of a second derivative matrix $\nabla^2 F$. Several algorithms of this type are described in the books by Fletcher [2] and by Gill, Murray and Wright [3], for example.

These procedures are iterative. We let x_k be the estimate of the required vector of variables at the beginning of the k-th iteration and B_k be the current B. Then the change to the variables is guided by the quadratic model

$$F(x_k+d) \approx F(x_k) + d^T \nabla F(x_k) + \tfrac{1}{2} d^T B_k d, \quad d \in \mathcal{R}^n. \tag{1.1}$$

Indeed, in the unconstrained case when $\nabla F(x_k) \neq 0$, it is usual to obtain the vector, d_k say, that minimizes expression (1.1) from the equations

$$B_k d_k = -\nabla F(x_k). \tag{1.2}$$

Then the new vector of variables has the form

$$x_{k+1} = x_k + \alpha_k d_k, \tag{1.3}$$

193

where α_k is a positive steplength that is chosen to give the reduction $F(x_{k+1}) < F(x_k)$ in the objective function and to satisfy the relation

$$(x_{k+1} - x_k)^T [\nabla F(x_{k+1}) - \nabla F(x_k)] > 0. \tag{1.4}$$

Further, B_{k+1} is set to a positive definite matrix that obeys the "quasi-Newton equation"

$$B_{k+1} (x_{k+1} - x_k) = \nabla F(x_{k+1}) - \nabla F(x_k), \tag{1.5}$$

which completes the work of the k-th iteration.

Formula (1.2) and the positive definiteness of B_k imply $d_k^T \nabla F(x_k) < 0$ when $\nabla F(x_k) \neq 0$, so the line search function $\{\phi_k(\alpha) = F(x_k + \alpha d_k) \mid \alpha \geq 0\}$ decreases initially when α is made positive. Therefore the iteration can reduce the objective function in theory. We are going to ignore the effects of computer rounding errors. Further, assuming that $\phi_k(\alpha)$ is bounded below as $\alpha \to \infty$, its derivative must become nonnegative or less negative than the initial value $\phi_k'(0)$. Therefore condition (1.4) can be obtained too, which is necessary for the positive definiteness of any matrix B_{k+1} that satisfies the quasi-Newton equation (1.5). This equation is suitable for providing B_{k+1} with some second derivative information, because, when the objective function is twice differentiable, the average second derivatives between x_k and x_{k+1} obey the relation

$$\left[\int_0^1 \nabla^2 F(x_k + \theta \delta_k) \, d\theta \right] \delta_k = \gamma_k, \tag{1.6}$$

where δ_k and γ_k are the vectors

$$\delta_k = x_{k+1} - x_k \quad \text{and} \quad \gamma_k = \nabla F(x_{k+1}) - \nabla F(x_k). \tag{1.7}$$

Condition (1.5) admits many positive definite matrices B_{k+1} when $n \geq 2$, the BFGS choice

$$B_{k+1}^{\text{BFGS}} = B_k - \frac{B_k \delta_k \delta_k^T B_k}{\delta_k^T B_k \delta_k} + \frac{\gamma_k \gamma_k^T}{\delta_k^T \gamma_k} \tag{1.8}$$

being particularly successful in practice (see [2], for instance). It is relevant to our studies, however, that it would be inconvenient to store the elements of B_k explicitly, because then the solution of the system (1.2) would require $O(n^3)$ computer operations. Instead one can employ the Cholesky factors of the matrices $\{B_k \mid k = 1, 2, 3, \ldots\}$, or the inverse matrices $\{H_k = B_k^{-1} \mid k = 1, 2, 3, \ldots\}$, or matrices $\{Z_k \mid k = 1, 2, 3, \ldots\}$ that obey the conjugacy conditions

$$Z_k^T B_k Z_k = I. \tag{1.9}$$

In all these cases the system (1.2) can be solved in only $O(n^2)$ operations, the formula

$$d_k = -Z_k Z_k^T \nabla F(x_k) \tag{1.10}$$

being relevant when working with matrices of conjugate directions, which is the preference of Goldfarb and Idnani [6], Han [7] and Powell [10]. Further, in all these cases the analogues of formula (1.8) can be implemented in only $O(n^2)$ operations, the updating of

Cholesky factorizations being studied by Goldfarb [4] for instance, the relevant updating of $H_k = B_k^{-1}$ being the well-known formula

$$H_{k+1}^{\mathrm{BFGS}} = \left(I - \frac{\delta_k \gamma_k^T}{\delta_k^T \gamma_k} \right) H_k \left(I - \frac{\gamma_k \delta_k^T}{\delta_k^T \gamma_k} \right) + \frac{\delta_k \delta_k^T}{\delta_k^T \gamma_k}, \tag{1.11}$$

and the updating of Z_k being considered by Han [7] and Powell [8]. All these procedures have the property that, if B_k and B_{k+1} were generated from their representations, then they would satisfy equation (1.8).

We will find that the use of conjugate direction matrices $\{Z_k \mid k = 1,2,3,\ldots\}$ has some advantages. Specifically, it is shown in Section 3 that, when active set methods are employed to maintain linear constraints on the variables, then there is a convenient factorization of the matrix of constraint coefficients that can be updated easily (see also [9]). Moreover, Section 4 recommends a rescaling of any columns of Z_k that become very small, and some numerical results demonstrate that the rescaling is highly successful at avoiding the inefficiencies that tend to occur when $\nabla^2 F$ is ill-conditioned. This technique depends on the use of conjugate direction matrices, because in the other two cases the vectors that we wish to scale are neither obvious nor easily accessible.

Therefore Section 2 takes a new look at the problem of calculating Z_{k+1} from Z_k, δ_k and γ_k so that equation (1.5) is satisfied. Since the relation (1.9) implies the identity

$$H_k = B_k^{-1} = Z_k Z_k^T, \tag{1.12}$$

the quasi-Newton equation takes the form

$$H_{k+1} \gamma_k = Z_{k+1} Z_{k+1}^T \gamma_k = \delta_k. \tag{1.13}$$

We identify a general technique, depending on an arbitrary orthogonal matrix, that generates a suitable Z_{k+1}. It has the property that the rank of the difference

$$H_{k+1} - H_k = Z_{k+1} Z_{k+1}^T - Z_k Z_k^T \tag{1.14}$$

can be three. Therefore some new updating formulae occur. It is possible that none of the new procedures are useful, but our approach has the strong advantage that it presents the updating of conjugate direction matrices in a simple way. Further, we will find that the BFGS and the DFP formula

$$H_{k+1}^{\mathrm{DFP}} = Z_{k+1} Z_{k+1}^T = H_k - \frac{H_k \gamma_k \gamma_k^T H_k}{\gamma_k^T H_k \gamma_k} + \frac{\delta_k \delta_k^T}{\delta_k^T \gamma_k} \tag{1.15}$$

are of special importance within the given family of updating formulae.

2. A rank three updating formula

Given any $n \times n$ matrix Z_k, the following procedure calculates a matrix Z_{k+1} that satisfies equation (1.13).

Step 1   Pick any $n \times n$ orthogonal matrix Ω_k.

Step 2   Calculate the matrix $\tilde{Z}_k = Z_k \Omega_k$ and let its columns be $\{\tilde{z}_j \mid j = 1, 2, \ldots, n\}$.

Step 3   Let Z_{k+1} be the matrix whose columns are the vectors

$$
\left.
\begin{aligned}
\hat{z}_1 &= \delta_k / (\delta_k^T \gamma_k)^{1/2} \\
\hat{z}_j &= \left(I - \frac{\delta_k \gamma_k^T}{\delta_k^T \gamma_k}\right) \tilde{z}_j, \quad j = 2, 3, \ldots, n
\end{aligned}
\right\}. \tag{2.1}
$$

One can verify that equation (1.13) is obtained by expressing the matrix $Z_{k+1} Z_{k+1}^T$ as the sum of outer products of the columns of Z_{k+1}, which gives the identity

$$
Z_{k+1} Z_{k+1}^T \gamma_k = \sum_{j=1}^{n} \hat{z}_j \hat{z}_j^T \gamma_k = \hat{z}_1 \hat{z}_1^T \gamma_k = \delta_k. \tag{2.2}
$$

Further, because we work with real numbers, the matrix $H_{k+1} = Z_{k+1} Z_{k+1}^T$ is positive definite or positive semidefinite. It is possible for Z_{k+1} to be singular when Z_k is nonsingular: for example if Ω_k is such that $\tilde{z}_2$ is a multiple of δ_k, then $\hat{z}_2$ is zero. Therefore there are certainly some poor choices of Ω_k. On the other hand, we see that the procedure is easy to apply, so we seek the range of matrices that can be generated.

Directing our attention to the matrices $H_{k+1} = Z_{k+1} Z_{k+1}^T$, because H_{k+1} is the inverse of B_{k+1}, we find that the total freedom that is given by Ω_k is confined to the vector

$$
v_k = \left(I - \frac{\delta_k \gamma_k^T}{\delta_k^T \gamma_k}\right) \hat{z}_1 = \left(I - \frac{\delta_k \gamma_k^T}{\delta_k^T \gamma_k}\right) Z_k \Omega_k e_1, \tag{2.3}
$$

where e_1 is the first coordinate vector. Specifically, we have the identity

$$
\begin{aligned}
H_{k+1} &= \sum_{j=1}^{n} \hat{z}_j \hat{z}_j^T \\
&= \left(I - \frac{\delta_k \gamma_k^T}{\delta_k^T \gamma_k}\right) \sum_{j=2}^{n} \tilde{z}_j \tilde{z}_j^T \left(I - \frac{\gamma_k \delta_k^T}{\delta_k^T \gamma_k}\right) + \frac{\delta_k \delta_k^T}{\delta_k^T \gamma_k} \\
&= \left(I - \frac{\delta_k \gamma_k^T}{\delta_k^T \gamma_k}\right) \tilde{Z}_k \tilde{Z}_k^T \left(I - \frac{\gamma_k \delta_k^T}{\delta_k^T \gamma_k}\right) + \frac{\delta_k \delta_k^T}{\delta_k^T \gamma_k} - v_k v_k^T \\
&= \left(I - \frac{\delta_k \gamma_k^T}{\delta_k^T \gamma_k}\right) H_k \left(I - \frac{\gamma_k \delta_k^T}{\delta_k^T \gamma_k}\right) + \frac{\delta_k \delta_k^T}{\delta_k^T \gamma_k} - v_k v_k^T,
\end{aligned} \tag{2.4}
$$

where the last line depends on $\tilde{Z}_k \tilde{Z}_k^T = (Z_k \Omega_k)(Z_k \Omega_k)^T = Z_k Z_k^T = H_k$. A comparison with equation (1.11) shows that we have implemented the BFGS updating formula less the term $v_k v_k^T$, which suggests the following result.

Theorem   The procedure of this section generates a positive definite or positive semidefinite matrix of the form

$$
H_{k+1} = H_{k+1}^{\mathrm{BFGS}} - v_k v_k^T, \tag{2.5}
$$

196

where H_{k+1}^{BFGS} is the right hand side of expression (1.11) and where v_k satisfies $v_k^T \gamma_k = 0$, which is necessary to preserve the quasi-Newton equation (1.13). Conversely, if v_k is any vector such that $v_k^T \gamma_k = 0$ and the matrix (2.5) has no negative eigenvalues, and if Z_k is nonsingular, then there exists a choice of Ω_k in our procedure that causes $Z_{k+1} Z_{k+1}^T$ to be the matrix (2.5).

Proof   The first half of the theorem has been proved already. In order to establish the converse result we pick any nonzero vector v that satisfies the orthogonality condition

$$v^T \gamma_k = 0, \tag{2.6}$$

and we ask first whether our procedure can generate any matrices of the form

$$H_{k+1} = H_{k+1}^{\text{BFGS}} - \sigma\, v v^T, \tag{2.7}$$

where σ is a positive scalar. The nonsingularity of Z_k allows the first column of Ω_k to be $Z_k^{-1} v / \|Z_k^{-1} v\|_2$, which gives the column

$$\tilde{z}_1 = Z_k \Omega_k e_1 = v \,/\, \|Z_k^{-1} v\|_2. \tag{2.8}$$

It follows from equation (2.6) that the vector (2.3) is a nonzero multiple of v. Therefore v_k in equation (2.5) can have any direction that is orthogonal to γ_k.

We have some control over the factor σ of the matrix (2.7) because, as the first column of Ω_k can be any vector of unit length, we can make the choice

$$\Omega_k e_1 = Z_k^{-1}(v + \beta \delta_k) \,/\, \|Z_k^{-1}(v + \beta \delta_k)\|_2, \tag{2.9}$$

where v is as before and β is a real parameter. Thus we achieve the vector

$$v_k = \left(I - \frac{\delta_k \gamma_k^T}{\delta_k^T \gamma_k}\right) Z_k \Omega_k e_1 = v \,/\, \|Z_k^{-1}(v + \beta \delta_k)\|_2. \tag{2.10}$$

Therefore σ in equation (2.7) can be any nonnegative number that satisfies the bound

$$\sigma \le \max_{\beta} \frac{1}{\|Z_k^{-1}(v + \beta \delta_k)\|_2^2} = \frac{1}{\|Z_k^{-1} v\|_2^2 - (\delta_k^T (Z_k Z_k^T)^{-1} v)^2 / \|Z_k^{-1} \delta_k\|_2^2}, \tag{2.11}$$

the value $\sigma = 0$ being obtained when $\Omega_k e_1$ is a multiple of $Z_k^{-1} \delta_k$. It remains to show that this upper bound coincides with singularity of H_{k+1}.

Because H_{k+1}^{BFGS} is positive definite, it is elementary that the matrix (2.7) is singular if and only if σ has the value

$$\sigma = 1 \,/\, v^T (H_{k+1}^{\text{BFGS}})^{-1} v = 1 \,/\, v^T B_{k+1}^{\text{BFGS}} v. \tag{2.12}$$

Therefore, remembering condition (2.6) and $B_k = (Z_k Z_k^T)^{-1}$, we deduce from equation (1.8) that the right hand sides of expressions (2.11) and (2.12) are the same. Therefore the theorem is true.   □

Equations (1.11) and (2.5) show that the rank of the change $H_{k+1} - H_k$ is at most three, and that rank three occurs when the vectors δ_k, $H_k \gamma_k$ and v_k are linearly independent.

The implication of the theorem on $B_{k+1} = H_{k+1}^{-1}$ is as follows. Equation (2.7) provides the matrix

$$B_{k+1} = B_{k+1}^{\text{BFGS}} + \frac{\sigma \, B_{k+1}^{\text{BFGS}} v \, v^T B_{k+1}^{\text{BFGS}}}{1 - \sigma \, v^T B_{k+1}^{\text{BFGS}} v} = B_{k+1}^{\text{BFGS}} + \tau \, u u^T, \tag{2.13}$$

where $u = B_{k+1}^{\text{BFGS}} v$ and $\tau = \sigma / (1 - \sigma \, v^T B_{k+1}^{\text{BFGS}} v)$. Since v can be any nonzero vector that satisfies the condition

$$v^T \gamma_k = (B_{k+1}^{\text{BFGS}} v)^T (B_{k+1}^{\text{BFGS}})^{-1} \gamma_k = (B_{k+1}^{\text{BFGS}} v)^T \delta_k = 0, \tag{2.14}$$

we see that u can be any nonzero vector that is orthogonal to δ_k. Further, since σ can be as large as expression (2.12), τ can take any nonnegative value. Therefore the range of matrices H_{k+1} that can be generated by the procedure of this section is analogous to all second derivative approximations of the form (2.13) with $\tau \geq 0$ and $u^T \delta_k = 0$.

The importance of the BFGS method in our family of updating procedures is clear, namely that the BFGS formula gives the largest H_{k+1} that can be achieved. It is mentioned at the end of Section 1 that the DFP formula is also of special significance. We explain this assertion by asking whether $\{\hat{z}_j = \tilde{z}_j \mid j = 2, 3, \ldots, n\}$ can occur in the second line of equation (2.1). In this case γ_k would be orthogonal to the last $n-1$ columns of $\tilde{Z}_k = Z_k \Omega_k$, which is equivalent to $Z_k^T \gamma_k$ being orthogonal to the last $n-1$ columns of Ω_k. Therefore, because Ω_k is orthogonal, its first column would be one of the vectors

$$\Omega_k e_1 = \pm Z_k^T \gamma_k / \|Z_k^T \gamma_k\|_2. \tag{2.15}$$

Thus, due to $H_k = Z_k Z_k^T$, the vector (2.3) would have the value

$$v_k = \pm \left(I - \frac{\delta_k \gamma_k^T}{\delta_k^T \gamma_k} \right) H_k \gamma_k / (\gamma_k^T H_k \gamma_k)^{1/2}. \tag{2.16}$$

It follows from equation (2.5) that our algorithm would generate the matrix

$$H_{k+1} = H_{k+1}^{\text{BFGS}} - \left(I - \frac{\delta_k \gamma_k^T}{\delta_k^T \gamma_k} \right) \frac{H_k \gamma_k \gamma_k^T H_k}{\gamma_k^T H_k \gamma_k} \left(I - \frac{\gamma_k \delta_k^T}{\delta_k^T \gamma_k} \right). \tag{2.17}$$

By substituting expression (1.11) into this matrix and rearranging terms, one can deduce the formula (1.15). Therefore the DFP method is characterized by the property that it sets $\{\hat{z}_j = \tilde{z}_j \mid j = 2, 3, \ldots, n\}$ in Step 3 of the algorithm of this section.

Details of the BFGS and DFP implementations of this algorithm are given in [8]. The matrix Ω_k is chosen to be a product of $n-1$ Givens rotations, and it is explained in the next section that this construction is advantageous to linearly constrained calculations. In order to achieve the required vector (2.3), these rotations are determined by an appropriate direction of the first column of Ω_k. The direction of the DFP formula, namely the vector (2.15), can be generated in $O(n^2)$ computer operations. When the BFGS formula is preferred, however, then v_k has to be zero, which occurs if and only if $Z_k \Omega_k e_1$ is a multiple of δ_k. Therefore, if $\theta \in \mathcal{R}^n$ is known such that δ_k is proportional to $Z_k \theta$, we make the first column of Ω_k a multiple of θ. Fortunately this vector is usually available. For example, when equations (1.3) and (1.10) are applied, the value $\theta = -Z_k^T \nabla F(x_k)$ is suitable.

3. Linearly constrained calculations

In active set algorithms for linearly constrained calculations, certain constraints are treated as equations when each iteration generates the search direction d_k. Specifically, if there are m_k active constraints whose gradients are the columns of the $n \times m_k$ matrix A_k, then several algorithms let d_k be the vector in $\mathcal{R}^n$ that minimizes the quadratic function (1.1) subject to the conditions

$$A_k^T d = 0. \tag{3.1}$$

It is highly convenient if the conjugate direction matrix Z_k has the property that $Z_k^T A_k$ is upper triangular. Fortunately, this condition and equation (1.12) are compatible because, if we achieve upper triangularity by premultiplying $Z_k^T A_k$ by an orthogonal matrix, $\hat{\Omega}_k^T$ say, and if we implement this multiplication by replacing Z_k by $Z_k \hat{\Omega}_k$, then equation (1.12) is preserved. Therefore we assume without loss of generality that $Z_k^T A_k$ is an upper triangular matrix, and we assume too that the rank of A_k is m_k.

We are going to calculate d_k by a null-space method (see [3], for instance). In other words, we seek the vector θ in $\mathcal{R}^n$ such that d_k has the value $d_k = Z_k \theta$, which agrees with the notation of the last paragraph of Section 2. Substituting this form in the constraints (3.1), we deduce from the triangularity and full rank of $A_k^T Z_k$ that these constraints are satisfied if and only if the first m_k components of θ are zero. Further, in view of equation (1.9), this substitution reduces the quadratic approximation (1.1) to the simple expression

$$F(x_k + Z_k \theta) \approx F(x_k) + \theta^T Z_k^T \nabla F(x_k) + \tfrac{1}{2} \|\theta\|_2^2, \quad \theta \in \mathcal{R}^n. \tag{3.2}$$

It follows that the required last $n - m_k$ components of θ are those of $-Z_k^T \nabla F(x_k)$. Hence the constraints change the search direction (1.10) to the value

$$d_k = - \sum_{i=m_k+1}^{n} z_i\, z_i^T \nabla F(x_k), \tag{3.3}$$

where the vectors $\{z_i \mid i = 1, 2, \ldots, n\}$ are the columns of Z_k. This technique is used in [6] and [10].

The main purpose of this section is to point out that it is surprisingly easy to extend the algorithm of Section 2 in order to recover the property that $Z^T A_k$ is upper triangular. The extension is derived from the following method. Letting $\hat{Z}_{k+1}$ be the matrix whose columns have the values (2.1), we add a further step to the algorithm that postmultiplies $\hat{Z}_{k+1}$ by an orthogonal matrix $\hat{\Omega}_{k+1}$ such that $(\hat{Z}_{k+1}\hat{\Omega}_{k+1})^T A_k$ is upper triangular. Then we set $Z_{k+1} = \hat{Z}_{k+1}\hat{\Omega}_{k+1}$, which gives the same matrix $H_{k+1} = Z_{k+1}Z_{k+1}^T$ as before. Here we are assuming that the current iteration makes no change to the active set, which is not restrictive because the operations of active set changes are performed independently of the revisions of second derivative approximations.

We now consider the sparsity structures of the sequence of $n \times m_k$ matrices $Z_k^T A_k$, $\tilde{Z}_k^T A_k$ and $\hat{Z}_{k+1}^T A_k$, knowing that we begin with upper triangularity. We recall that the first column of Ω_k is to be a multiple of a certain vector, w say, that is equal to θ or $Z_k^T \gamma_k$ in the BFGS or DFP updates respectively. Equivalently, we require $\Omega_k^T w$ to be a multiple of the first coordinate vector. Therefore we express Ω_k^T as a product

$$\Omega_k^T = \Omega_k^{(1,2)}\, \Omega_k^{(2,3)} \ldots \Omega_k^{(n-1,n)} \tag{3.4}$$

of $n-1$ Givens rotations, where $\Omega_k^{(p,q)}$ is allowed to differ from the $n \times n$ unit matrix only in its p-th and q-th rows and columns. The one degree of freedom in $\Omega_k^{(i-1,i)}$ for $i = n, n-1, \ldots, 2$ must satisfy the condition that the i-th component of the vector $\Omega_k^{(i-1,i)} w_i$ is zero, where $w_n = w$ and $w_i = \Omega_k^{(i,i+1)} w_{i+1}$ for $i < n$. Thus w_1 becomes a multiple of e_1 as required. Further, we see that the matrix $\check{Z}_k^T A_k = \Omega_k^T Z_k^T A_k$ becomes upper Hessenberg.

Turning to the calculation of $\hat{Z}_{k+1}$ from $\hat{Z}_k$, we assume that the change $\delta_k = x_{k+1} - x_k$ is a multiple of d_k, as in equation (1.3). Therefore the constraint (3.1) implies $\delta_k^T A_k = 0$. It follows from equation (2.1) that we have the identities

$$\left.\begin{array}{rcl} \hat{z}_1^T A_k &=& 0 \\ \hat{z}_j^T A_k &=& \tilde{z}_j^T A_k, \quad j = 2, 3, \ldots, n \end{array}\right\}. \tag{3.5}$$

Hence the last $n-1$ rows of $\hat{Z}_{k+1}^T A_k$ are the same as the last $n-1$ rows of the upper Hessenberg matrix $\check{Z}_k^T A_k$, while the first row of $\hat{Z}_{k+1}^T A_k$ is zero. Therefore the surprisingly simple form of the proposed extension to the algorithm of Section 2 is that $\hat{\Omega}_{k+1}$ is merely a permutation matrix that forms Z_{k+1} by reordering the columns of $\hat{Z}_{k+1}$, and there is a similar reordering of the rows of $\hat{Z}_{k+1}^T A_k$.

It is best to absorb the new Step 4 into the earlier operations of the algorithm. Therefore, instead of the matrix $\tilde{Z}_k = (\tilde{z}_1 \, \tilde{z}_2 \, \ldots \, \tilde{z}_n)$ that is defined in Section 2, we let $\tilde{Z}_k$ have the columns

$$(\tilde{z}_2 \, \tilde{z}_3 \, \ldots \, \tilde{z}_{m_k+1} \, \tilde{z}_1 \, \tilde{z}_{m_k+2} \, \ldots \, \tilde{z}_n), \tag{3.6}$$

where the vectors $\{\tilde{z}_j \mid j = 1, 2, \ldots, n\}$ are exactly as before. Similarly the new Z_{k+1} is the matrix

$$Z_{k+1} = (\hat{z}_2 \, \hat{z}_3 \, \ldots \hat{z}_{m_k+1} \, \hat{z}_1 \, \hat{z}_{m_k+2} \, \ldots \, \hat{z}_n) \tag{3.7}$$

with no change to the definitions (2.1). In other words we form the new Z_{k+1} by replacing the (m_k+1)-th column of the new $\tilde{Z}_k$ by $\delta_k / (\delta_k^T \gamma_k)^{1/2}$ and by applying the operator $(I - \delta_k \gamma_k^T / \delta_k^T \gamma_k)$ to the other $n-1$ columns of $\tilde{Z}_k$. Further, when $\tilde{Z}_k^T A_k$ is calculated by premultiplying $Z_k^T A_k$ by the Givens rotations (3.4), then, due to equation (3.6), the new i-th row of $\tilde{Z}_k^T A_k$ is the old $(i+1)$-th row of $\check{Z}_k^T A_k$ for $i = 1, 2, \ldots, m_k$, but we do not disturb the zero (m_k+1)-th row of $Z_k^T A_k$ when applying the Givens rotations because we know that this row of $Z_{k+1}^T A_k$ is going to be zero. It follows from equation (3.5) that the required matrix $Z_{k+1}^T A_k$ is just $\tilde{Z}_k^T A_k$.

An equivalent view of this reordering is that the new $\tilde{Z}_k$, shown in expression (3.6), is defined by the construction

$$\tilde{Z}_k^T = \Omega_k^{(1,m_k+1)} \Omega_k^{(2,m_k+1)} \ldots \Omega_k^{(m_k,m_k+1)} \Omega_k^{(m_k+1,m_k+2)} \ldots \Omega_k^{(n-1,n)} Z_k^T, \tag{3.8}$$

where $\Omega_k^{(p,q)}$ still denotes a Givens rotation that differs from the unit matrix in its p-th and q-th rows and columns. These rotations should have the property that now $\Omega_k^T w$ becomes a multiple of the (m_k+1)-th coordinate vector instead of a multiple of the first coordinate vector.

These operations take a very convenient form in BFGS updating when equation (1.3) provides $\delta_k = \alpha d_k$. We recall that in this case the direction of w is defined by the condition that $Z_k w$ is proportional to d_k. Therefore, in view of equation (3.3), the first m_k

components of w are all zero. It follows that the vector

$$w_{m_k+1} = \Omega_k^{(m_k+1,m_k+2)} \Omega_k^{(m_k+2,m_k+3)} \ldots \Omega_k^{(n-1,n)} w, \tag{3.9}$$

which is introduced soon after equation (3.4), is a multiple of the coordinate vector e_{m_k+1}. Therefore we leave out the rotations $\{\Omega_k^{(j,m_k+1)} \mid j=1,2,\ldots,m_k\}$ of the construction (3.8). Thus, not only is there a reduction in the work of calculating $\tilde{Z}_k$, but also we have the useful identity

$$Z_{k+1}^T A_k = \tilde{Z}_k^T A_k = Z_k^T A_k, \tag{3.10}$$

which saves even more work. This interesting property of the BFGS formula is mentioned by Goldfarb [5] and by Powell [9].

4. Scaling the columns of the conjugate direction matrix

This section addresses a technique, originally proposed in [8], that scales up any small columns of Z_k automatically. It reduces greatly the inefficiencies that usually occur if the inverse of the true second derivative matrix has some eigenvalues that are much larger than those of the initial approximation $H_1 = Z_1 Z_1^T \approx (\nabla^2 F)^{-1}$. These inefficiencies and the success of column scaling will be demonstrated by numerical examples.

In order to explain the need for the technique, we consider the extreme case when H_1 is singular, when every iteration makes the change of variables

$$x_{k+1} - x_k = \delta_k = -\alpha_k H_k \nabla F(x_k), \tag{4.1}$$

and when there are no computer rounding errors. In this case, if the BFGS or DFP formula is applied, it follows from equation (1.11) or (1.15) that all vectors in the null space of H_k are also in the null space of H_{k+1}. Further, because the vector v_k in equation (2.5) is a linear combination of δ_k and $Z_k \Omega_k e_1$, we deduce from $H_k = Z_k Z_k^T$ that all the updating formulae of Section 2 have this property. Thus a simple inductive argument shows that the steps $\{x_{k+1} - x_k \mid k=1,2,3,\ldots\}$ are confined to the column space of H_1. Therefore it is usually impossible for the calculation to reach a vector of variables that minimizes the objective function.

The analogous behaviour when H_k is nearly singular for some k can be interpreted conveniently by making use of our factorization $H_k = Z_k Z_k^T$. Because Z_k can be postmultiplied by an orthogonal matrix if necessary, the near-singularity can cause one or more columns of Z_k to have very small norms. In this case equation (1.10) shows that, for most gradients $\nabla F(x_k)$, the search direction d_k contains only tiny multiples of such columns. Thus the near singularity is inherited by H_{k+1} if the BFGS formula (1.11) is applied. Further, the other updating formulae are no better because the relation (2.5) shows that the eigenvalues of H_{k+1} are bounded above by the corresponding eigenvalues of H_{k+1}^{BFGS}.

It has been found by numerical experiments that recovery from near-singularity usually occurs automatically, but many iterations are often required until $\nabla F(x_k)$ in equation (1.10) is nearly orthogonal to the large columns of Z_k. This observed property has some important implications. The main one concerns the remark in the previous paragraph that, in the nearly singular case, one may have to introduce multiplication by an orthogonal matrix in order that Z_k has a small column. In practice, however, it seems that small column vectors are generated automatically without the use of such multiplications, which must depend on the choice of Ω_k that is described in Section 3. An example of this phenomenon is given in [8], where the initial matrix Z_1 is singular but all its columns are large. When near-singularity is seen to occur, the technique of this section scales up the small columns of Z_{k+1} that are different from $\hat{z}_1$. It follows from the orthogonality of $\hat{z}_j$ to γ_k, given by formula (2.1), that these scalings preserve the quasi-Newton equation (1.13), but this remark alone does not justify the use of column scaling. Therefore we turn to some numerical tests.

Our implementation of column scaling is taken from a Fortran package for linearly constrained optimization calculations [10]. It depends on a parameter, ζ say, that is set to the value

$$\|\hat{z}_1\|_2 = \|\delta_k\|_2 / (\delta_k^T \gamma_k)^{1/2} \tag{4.2}$$

on the first iteration ($k=1$). Subsequent iterations update ζ, from ζ_{old} to ζ_{new} say, where ζ_{new} is the number in the interval

$$\sqrt{0.1\,\zeta_{\text{old}}\|\hat{z}_1\|_2} \leq \zeta_{\text{new}} \leq \sqrt{\zeta_{\text{old}}\|\hat{z}_1\|_2} \tag{4.3}$$

that is closest to ζ_{old}. For $j = 2, 3, \ldots, n$, the column $\hat{z}_j$ of Z_{k+1} is multiplied by the factor $\zeta_{\text{new}}/\|\hat{z}_j\|_2$ if and only if this factor is greater than one. We do not scale down large columns of Z_{k+1} because they do not seem to cause inefficiencies.

Our numerical results were obtained by running the Fortran software of Powell [10] with and without the scaling technique. We also compare the BFGS and DFP updating formulae, which was suggested by my colleague Arieh Iserles, and I am very grateful to him because the results are interesting. Only one subroutine of the package had to be adapted to provide these four variations. The calculations were run in double precision on a SUN 3/50 work-station, the stopping condition being a relative error of at most 10^{-10} in the Kuhn–Tucker residual vector.

A range of test problems was generated from the function of five variables

$$\Phi(x) = \sum_{j=0}^{100} [\,1 - e^{-jh}\,r(jh, x)\,]^2, \quad x \in \mathcal{R}^5, \tag{4.4}$$

where $h=0.05$ and $r(t, x)$ has the value

$$r(t, x) = \frac{p(t, x)}{q(t, x)} = \frac{x_1 + x_2 t + x_3 t^2}{1 + x_4(t-5) + x_5(t-5)^2}, \quad 0 \leq t \leq 5. \tag{4.5}$$

Thus Φ measures the relative error of a 2-2 rational approximation to the exponential function $\{e^t \mid 0 \leq t \leq 5\}$. Because the original purpose of the problems was to test the software for constrained optimization, we include the linear constraints

$$\left.\begin{array}{rcl} p(jh, x) & \geq & e^{jh}\,q(jh, x) \\ q(jh, x) & \geq & 10^{-5} \end{array}\right\}, \quad j = 0, 1, \ldots, 100, \tag{4.6}$$

Table 1

Diagonal scaling matrices

Elements of D		No scaling		With scaling	
D_{11}	D_{55}	BFGS	DFP	BFGS	DFP
10^{-6}	10^{-6}	133	$\star$	56	67
10^{-4}	10^{-4}	98	$\star$	52	62
10^{-2}	10^{-2}	53	198	45	60
1	1	38	69	38	50
10^{2}	10^{2}	39	42	39	44
10^{4}	10^{4}	39	40	39	47
10^{6}	10^{6}	39	40	39	47

on the components of x. The objective function itself is the expression

$$F(x) = \Phi(Dx), \quad x \in \mathcal{R}^5, \tag{4.7}$$

where D is a 5×5 positive diagonal matrix. We are going to make several choices of D, satisfying the condition that the diagonal elements form a geometric progression, in order that it is sufficient to tabulate only D_{11} and D_{55}. We start the iterations at the feasible point $x_1 = (D_{11}^{-1} \ D_{22}^{-1} \ 6D_{33}^{-1} \ 0 \ 0)^T$ with $Z_1 = I$.

Table 1 presents some numerical results when D is a multiple of the unit matrix, while Table 2 reports on some ill-conditioned choices of D. The main entries of the tables are the number of calls of a subroutine that provides $F(x)$ and $\nabla F(x)$ for any given value of x. A star, however, indicates that 500 calls of this subroutine were insufficient to achieve the required accuracy. We see that the BFGS formula is consistently superior to the DFP formula (excluding the last entry of Table 2), and that the scaling technique is highly successful at improving efficiency when some elements of D are much smaller than one.

Due to the relation

$$(\nabla^2 F(x))^{-1} = (D \, \nabla^2 \Phi(x) \, D)^{-1}, \tag{4.8}$$

small elements of D make it necessary for $\|Z\|_2$ to become large. Alternatively, because unadulterated variable metric algorithms are independent of linear transformations of the variables, provided that suitable changes are made to the starting conditions, we can take the view that $F(\cdot)$ is the well-scaled objective function $\{F(x) = \Phi(x) \mid x \in \mathcal{R}^5\}$ and that the initial approximation to the inverse of the second derivative matrix is $H_1 = Z_1 Z_1^T = D^2$. Thus small diagonal elements of D induce very small eigenvalues of H_1, so the middle two columns of the tables show the inefficiencies that are suggested by our discussion of near-singularities. We also see that the software can cope with condition numbers up to $\kappa(D) = 10^{20}$, which implies $\kappa(\nabla^2 F) \approx 10^{40}$. Of course they can be accommodated only because D is diagonal and because floating point arithmetic is employed. Nevertheless, it is encouraging that the methods with column scaling perform well in these cases, in view of the use of the Euclidean metric by the scaling technique.

Table 2

Ill-conditioned scaling matrices

Elements of D		No scaling		With scaling	
D_{11}	D_{55}	BFGS	DFP	BFGS	DFP
1	10^{-8}	71	⋆	52	93
1	10^{-4}	78	⋆	56	69
10^4	10^{-4}	49	⋆	44	51
10^{-8}	1	⋆	⋆	105	201
10^{-4}	1	112	⋆	80	98
10^{-4}	10^4	98	⋆	86	127
1	10^{-20}	⋆	⋆	102	135
10^{10}	10^{-10}	75	⋆	77	90
10^{20}	1	54	64	46	57
10^{-20}	1	⋆	⋆	120	193
10^{-10}	10^{10}	⋆	⋆	93	130
1	10^{20}	53	70	47	45

The success of the calculations when some elements of D are large initially also deserves comment. Now Z will tend to have some small columns, and we ask how this smallness is achieved. Typically $\|\gamma_k\|_2 \gg \|\delta_k\|_2$ occurs, and then, assuming that δ_k and γ_k are not pathologically close to orthogonality, equation (2.1) provides $\|\hat{z}_1\|_2 \ll 1$, but usually there is little change in the lengths of the other columns of the conjugate direction matrix. Thus the matrix can become small in a stable way, in contrast to several other implementations of updating formulae that achieve small matrix elements by cancellation. It may seem unsatisfactory that we expect only one column of Z to be reduced by each iteration, but, when the correction $H_{k+1} - H_k$ is of rank r, it is impossible for the updating formula to make more than r eigenvalues of the H matrix become less than any constant. Further, by considering the details of the BFGS formula (1.11), it can be proved that equation (2.5) shifts at most two of the eigenvalues of the H matrix through any constant threshold.

Some of the comparisons between the BFGS and DFP methods, shown in the "no scaling" columns of the tables, are typical of calculations that begin with a very poor second derivative approximation. It is particularly difficult for the DFP updating formula to recover from near-singularity of H because, if equation (4.1) is satisfied and if α_k yields the exact line search condition $\delta_k^T \nabla F(x_{k+1}) = 0$, then equation (1.15) implies the strict inequality

$$\nabla F(x_{k+1})^T H_{k+1} \nabla F(x_{k+1}) < \nabla F(x_k)^T H_k \nabla F(x_k). \tag{4.9}$$

Thus, if near-singularity of H_1 causes $\nabla F(x_1)^T H_1 \nabla F(x_1)$ to be small when $\nabla F(x_1)$ is large, and if every iteration makes an exact line search, then removal of the near-singularity is not possible until $\|\nabla F(x_k)\|_2$ becomes small. In this case, however, the theorem of Dixon [1] states that the BFGS and DFP methods are equivalent. It seems, nevertheless, that there is much scope for the scaling technique of this section to improve the performance of the DFP updating formula, perhaps because of the effects of linear constraints and practical line searches, and certainly because equation (2.5) shows that H_{k+1}^{DFP} has smaller elements than H_{k+1}^{BFGS}. This hope is realized very well in the last column of the tables.

A new family of updating methods has been presented that achieve the quasi-Newton equation (1.13). It has been shown that these methods are highly suitable in linearly constrained optimization calculations and that they allow a column scaling technique that overcomes some severe inefficiencies that can occur in variable metric algorithms. Similar remarks are made about the BFGS formula in [8] and [9], but then it was not known that the given advantages of working with conjugate direction matrices are enjoyed by a rank three family of updating methods.

References

[1] L.C.W. Dixon, "Quasi-Newton algorithms generate identical points", *Math. Programming*, Vol. 2 (1972), pp. 383–387.

[2] R. Fletcher, *Practical Methods of Optimization*, John Wiley & Sons (Chichester, 1987).

[3] P.E. Gill, W. Murray and M.H. Wright, *Practical Optimization*, Academic Press (New York, 1981).

[4] D. Goldfarb, "Factorized variable metric methods for unconstrained optimization", *Math. Comp.*, Vol. 30 (1976), pp. 796–811.

[5] D. Goldfarb, "Matrix factorizations in optimization of nonlinear functions subject to linear constraints", *Math. Programming*, Vol. 10 (1976), pp. 1–31.

[6] D. Goldfarb and A. Idnani, "A numerically stable dual method for solving strictly convex quadratic programs", *Math. Programming*, Vol. 27 (1983), pp. 1–33.

[7] S-P. Han, "Optimization by updated conjugate subspaces", in *Numerical Analysis: Pitman Research Notes in Mathematics Series 140*, eds. D.F. Griffiths and G.A. Watson, Longman Scientific & Technical (Burnt Mill, 1986), pp. 82–97.

[8] M.J.D. Powell, "Updating conjugate directions by the BFGS formula", *Math. Programming*, Vol. 38 (1987), pp. 29–46.

[9] M.J.D. Powell, "On a matrix factorization for linearly constrained optimization problems", Report DAMTP 1988/NA9, University of Cambridge (Cambridge, 1988) (to be published in *Proceedings of the IMA Conference on Applications of Matrix Theory*, Oxford University Press).

[10] M.J.D. Powell, "TOLMIN: a Fortran package for linearly constrained optimization calculations", Report DAMTP 1989/NA2, University of Cambridge (Cambridge, 1989).

T.F. RUSSELL
Eulerian-Langrangian Localized Adjoint Methods for Advection-Dominated Problems

ABSTRACT

Eulerian-Lagrangian localized adjoint methods (ELLAM), formulated in collaboration with M. A. Celia, R. E. Ewing, and I. Herrera, are a space-time finite-element methodology that builds on concepts of tracking advection-dominated flows (Eulerian-Lagrangian methods, or ELM) and of optimal test functions (adjoint methods). The ELLAM development yields a conservative scheme that treats boundary conditions systematically and generalizes ELM, thus overcoming the two principal shortcomings of ELM while maintaining the numerical advantages. A single formulation handles advection-diffusion and pure advection problems without upwinding or artificial boundary conditions. For the simplest case, this paper presents the formulation and some numerical results that confirm the potential of the approach.

1. INTRODUCTION

The importance of advection-diffusion problems in applications, and the difficulty of solving them numerically, have long been recognized. The vexing choice between artificially diffused results, non-physical oscillations, and impractically fine grids and time steps has plagued scientists and engineers for years. Eulerian-Lagrangian methods (ELM), developed independently in about the last decade by a variety of researchers and described in detail in Section 2, have offered considerable hope in resolving this dilemma. ELM have produced nonoscillatory answers without numerical diffusion, using long time steps on grids no finer than necessary to represent moving fronts. Unfortunately, ELM are generally not conservative, and it has not been clear how to deal accurately with boundary conditions in general situations. The former difficulty is of concern in essentially all applications, and the latter is of particular interest in such problems as semiconductor device modeling and contaminant transport in groundwater.

Recently, M. A. Celia, R. E. Ewing, I. Herrera, and the author have devised Eulerian-Lagrangian localized adjoint methods (ELLAM), a formulation that shows great promise in circumventing these drawbacks. (This group of four individuals will be referred to in this paper as the ELLAM group, or EG.) The purpose of this paper is to lay out the formulation from the author's perspective, including his view of its relationship to other work on ELM and other methods, and to present rudimentary numerical results. The other members of the EG made vital contributions to the ideas documented here, but they should not be held responsible for any opinions expressed, or for any errors, omissions, or misunderstandings. Attributions to the author below represent more an absolution of the other members than a claim by the author. Writings that more fairly reflect the consensus of the group will appear elsewhere [5,15].

Section 2 provides background on ELM from the author's point of view. Section 3 briefly discusses localized adjoint methods (LAM) as the author sees them, using some thoughts of the EG but not doing justice to the wide-ranging theory of Herrera or the numerical experience of Celia, Herrera, and others. In §4 we combine ELM and LAM into ELLAM in the interior of the spatial domain. The formulation is the work of the EG; the form of the presentation is the author's. One sees here how ELLAM can be reduced to ELM, so that it generalizes ELM. Section 5 develops ELLAM equations at an inlet boundary, establishing the techniques to handle Lagrangian characteristics that backtrack across the boundary. The basic development comes from the EG, while the backward Euler aspects and interpretive comments are the author's. Section 6 introduces the somewhat novel discretization of an outlet boundary and shows how this idea unites advection-diffusion and pure advection problems under one numerical roof; again, the basics are due to the EG, specific choices and comments to the author. Sections 7 and 8, dealing with the verification of mass conservation and the form of the equations to be solved, are due to the EG except for the case of piecewise-constant outlet fluxes. The numerical results in §9 are entirely the work of the author. Section 10 indicates future plans of the EG, §11 relates ELLAM to the work of others from the author's perspective, and §12 offers his brief conclusions.

To make the essential ideas clear, we shall present the formulation in the simplest possible context, which is the one-dimensional linear constant-coefficient advection-diffusion equation with Dirichlet or flux boundary conditions:

$$u_t + vu_x - Du_{xx} = f(x,t), \quad 0 < x < \ell, \quad 0 < t \le T, \qquad (1)$$
$$u(0,t) = g(t) \text{ or } vu(0,t) - Du_x(0,t) = g(t), \qquad 0 < t \le T, \qquad (2)$$
$$u(\ell,t) = h(t) \text{ or } vu(\ell,t) - Du_x(\ell,t) = h(t), \qquad 0 < t \le T, \qquad (3)$$
$$u(x,0) = u_0(x), \quad 0 \le x \le \ell. \qquad (4)$$

2. EULERIAN-LAGRANGIAN METHODS

These schemes have been developed over the last decade by many researchers with different viewpoints. In the mathematical literature, they have gone under the names of the modified method of characteristics (MMOC) [6,9,13,22], the transport-diffusion algorithm [21], and characteristic Galerkin methods [18,19,24]. The name of Eulerian-Lagrangian methods (ELM) has become popular in the engineering literature [1,20]. The methods have been highly successful in solving hyperbolic conservation laws [18,19], the Navier-Stokes equations [2,21,24], and problems arising from petroleum reservoir simulation [12,23], subsurface hydrology [20], and surface hydrology [1,16]. Despite the extensive literature, issues relating to treatment of boundary conditions and conservation of mass remain largely unresolved for ELM. Our next step is to derive a simple ELM for (1)–(4)

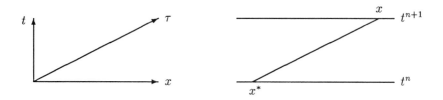

Figure 1: Characteristic direction and foot of characteristic

and indicate how these issues arise.

The principal idea of ELM is to view (1) in Lagrangian coordinates (for the pure advection equation $u_t + vu_x = 0$), in which it takes the form

$$\frac{Du}{Dt} - Du_{xx} = f, \tag{5}$$

where D/Dt denotes the total or substantial derivative. Alternatively, one can call this the space-time characteristic directional derivative

$$\frac{\partial}{\partial \tau} = v\frac{\partial}{\partial x} + \frac{\partial}{\partial t} \tag{6}$$

and write

$$u_\tau - Du_{xx} = f. \tag{7}$$

Note that $dx\, d\tau = dx\, dt$, i.e., integration in τ is measured by the amount covered in t. We shall use the latter notation; in either case, (1) has been reduced to a diffusion problem that can be solved accurately by standard procedures.

For this example we discretize (7) in "time" by a backward Euler approximation,

$$u_\tau^{n+1}(x) \approx \frac{u^{n+1}(x) - u^n(x - v\Delta t)}{\Delta t}, \tag{8}$$

where superscripts denote time steps and $\Delta t = t^{n+1} - t^n$. Let

$$x^* = x - v\Delta t, \tag{9}$$

i.e., let x^* be the point at the old time level n that advects to x at the new time level $n+1$ (see Figure 1). The points (x, t^{n+1}) and (x^*, t^n) are known respectively as the *head* and *foot* of the characteristic. We discretize in space by continuous piecewise-linear Galerkin finite elements as follows. Let $E > 0$ be an integer, $\Delta x = \ell/E$, and $x_i = i\Delta x$ for $i = 0, 1, \ldots, E$. The uniform grid is chosen only for simplicity of exposition, as the ideas of this paper are quite general. Let $\mathcal{M}$ be the finite-dimensional space of C^0 piecewise-linear functions with respect to this partition of $[0, \ell]$ (ignoring distinctions of essential and natural boundary conditions for the present), and let $w \in \mathcal{M}$ be a test function. Substitute (8) into (7), multiply by w, integrate over $[0, \ell]$, and (ignoring boundary terms) integrate by parts in the diffusion term to obtain

$$
\int_0^\ell U^{n+1}(x)w(x)\,dx + \Delta t \int_0^\ell DU_x^{n+1}(x)w_x(x)\,dx
$$
$$
= \int_0^\ell U^n(x^*)w(x)\,dx + \Delta t \int_0^\ell f^{n+1}(x)w(x)\,dx, \tag{10}
$$

where U is the numerical approximation of the true solution u. This is the C^0 piecewise-linear Galerkin MMOC for (1), which solves for $U^{n+1} \in \mathcal{M}$ such that (10) holds for all $w \in \mathcal{M}$. A finite-difference analogue can likewise be defined. For reference, we define the Courant number

$$
Cu = v\Delta t/\Delta x \tag{11}
$$

and the grid Peclet number

$$
Pe = v\Delta x/D. \tag{12}
$$

This approach offers many advantages. A solid theory exists [6,9,13,22], in which it is proved that for diffusion bounded below by a positive constant,

$$
\max_n \|u^n - U^n\|_{L^2} \le C(h^{r+1} + \Delta t); \tag{13}
$$

in (13), h is the maximum mesh diameter (in one or more than one space dimension) and r is the degree of the piecewise-polynomial trial and test functions. This result, by itself, is no different from what one obtains with standard Galerkin methods, but the constant C is greatly improved when MMOC is used for advection-dominated problems. In time, C depends on norms of u_{tt} with standard methods, but on the much smaller $u_{\tau\tau}$ with MMOC; thus, long, accurate time steps with large Courant numbers are possible. In space, standard Galerkin proofs are forced to inflate C in order to cover advection terms; MMOC proofs, in which advection is relegated to the right-hand side as in the x^* term of (10), need not do this. Hence there is a theoretical basis for expecting success with coarser grids, or large grid Peclet numbers, as well. If u is smooth, the theory extends to the zero-diffusion (hyperbolic) case, with h^{r+1} replaced by h^r [6].

Numerical work, cited above, bears out the theoretical predictions. Accurate, nonoscillatory results have been reported with $Cu \gg 1$ and $Pe \gg 2$. The success of $Cu \gg 1$ is readily understood. As for Pe, note that the width of a diffusing front in an exact solution of (1) is of order $\sqrt{Dt}$, so one would want $\Delta x = O(\sqrt{D})$ in the front. Standard methods demand $Pe \le 2$, or $\Delta x = O(D)$, to avoid oscillations, or else they increase D via some form of artificial or numerical diffusion. ELM have yielded oscillation-free answers, without numerical diffusion, for $\Delta x = O(\sqrt{D})$. Equivalently, a grid Peclet number $Pe = O(1/\sqrt{D})$ is feasible.

In addition to accuracy and efficiency in the discretization, note that the discrete equations arising from (10) are symmetric and positive definite. With advection on the right-hand side only, the system is better suited for iterative linear-solution techniques in multiple space dimensions. Note also that the left-hand side of (10) involves only standard terms, so that the problem at time level $n + 1$ can be solved on a static grid or on any adaptive grid chosen by the user. This comes about because the ELM track *backward* in time along characteristics; any irregularities due to advection are confined to the evaluation of the x^* integral in (10), which does not affect the grid. The ELM formulation is therefore straightforward to implement in more than one dimension, in contrast with forward-looking characteristic schemes that necessarily deform their grids.

Some questions raised by (10) and its analogues for more complex problems have been well discussed in the literature. Among these are the backtracking algorithm that determines x^* [1], and the numerical quadrature rule that approximates the associated integral [19] (or the finite-difference interpolation that defines $U^n(x^*)$ [1]). For the simple problem (1)–(4), these issues are trivial; one can easily carry out the calculations exactly, as we have done in the numerical results reported below. For complicated situations the state of the art is good, and so we do not address backtracking and quadrature (or interpolation) further in this paper.

Boundary conditions are another matter. If a backtracked characteristic crosses a boundary of the computational domain, as can be expected at an inlet boundary, it is not obvious what the meaning of x^* or $U^n(x^*)$ should be. If the boundary condition is Dirichlet, then a boundary value can be substituted for $U^n(x^*)$, but the time from the crossing to t^{n+1} is no longer Δt, and the implication of this has not been clear. Additionally, for a flux boundary condition, the value itself is not directly available. The tendency in the ELM literature has been to deal with these contingencies in an *ad hoc* manner, not always fully described. The ELLAM formulation yields a systematic approach in §§5–6 that leads to equations different from those published up to now. Clearly, if inlet boundary conditions are not treated as well as the rest of a problem, the advantages of ELM will be compromised. Resolution of this question is critical in many applications, e.g., semiconductor device modeling [8].

Perhaps the greatest drawback of ELM is their failure to conserve mass exactly. In (10), advection is tacitly in the x^* term, so x^* and the integral must be calculated exactly in order to conserve mass. For problems more difficult than (1)–(4), this is not practical. Conservation errors reported in the literature have generally been small but perceptible, and many potential users find this disconcerting. We shall indicate in §7 how the work of this paper can allow a conservative scheme to be devised.

3. LOCALIZED ADJOINT METHODS

The ELLAM formulation that treats boundary conditions and mass conservation for ELM uses ideas motivated by localized adjoint methods (LAM). We briefly describe such methods here, beginning with adjoint methods. Consider the partial differential equation in space or space-time,

$$Lu = f, \qquad x \in \Omega \text{ or } (x, t) \in \Omega. \tag{14}$$

Integrating against a test function w, we have the weak form

$$\int_\Omega Lu\, w\, d\Omega = \int_\Omega fw\, d\Omega. \tag{15}$$

If $L^*w = 0$ except at certain nodes or edges denoted by e_i, where L^* is the formal adjoint of L, and if $w = 0$ on the boundary $\partial\Omega$, then integration by parts leads to

$$\sum_i \int_{e_i} u\, L^*w\, d\Omega = \int_\Omega fw\, d\Omega. \tag{16}$$

The most familiar example would be $w(x) = G(x, x_k)$, the Green's function for node x_k, where $L^*w(x) = \delta(x - x_k)$, so that $u(x_k) = \int_\Omega fw\, d\Omega$. Thus the Green's function, as test function, extracts the point value of the solution; other test functions can be chosen to extract other information. Herrera has built an extensive theory around this concept; see [14] for a list of references. The theory can deal with situations where distributions do not apply, such as u and w both being discontinuous. Related ideas known as the H^{-1}-Galerkin method appear in [7].

Unfortunately, the Green's function is global, as are other test functions of the type just described that can immediately extract information. Hence these are not suitable for practical computations. To maintain the formalism to the extent possible, we force w to be local but allow $L^*w = 0$ to be violated at more points. For example, in one space dimension a chapeau function satisfies $L^*w = 0$ except at 3 points, where its derivative has jumps; instead of extracting the point value of the solution, one extracts the familiar relation between the values of the solution at the 3 nodes. What interests us here is the application of this concept in space-time, e.g., for (1), where we want a test function w that satisfies

$$L^*w = -w_t - vw_x - Dw_{xx} = 0. \tag{17}$$

It is clear that (17) can be achieved in a multitude of ways. By taking $w_t = 0$ and $vw_x + Dw_{xx} = 0$, Celia $et\ al.$ [4] obtain so-called optimal test functions in space that are static in time. These are upwinded by an amount that varies with the Peclet number; see [4] for details and a guide to other work in this area. The resulting methods are Eulerian, though nonstandard, so they are subject to the Peclet-number (in the sense that they are numerically diffusive) and Courant-number limitations that ELM are intended to circumvent. In §4, a different way of satisfying (17) leads to the formulation of this paper.

4. INTERIOR ELLAM FORMULATION

In accordance with the idea of the characteristic directional derivative, we satisfy (17) by considering

$$w_\tau = w_t + vw_x = 0, \qquad Dw_{xx} = 0. \tag{18}$$

In space, for fixed time, we obtain the chapeau function (which violates $Dw_{xx} = 0$ at 3 points); in time, the test function is constant along characteristics. Being constant in τ, to localize the test function we must make it discontinuous at t^n and t^{n+1}, the beginning

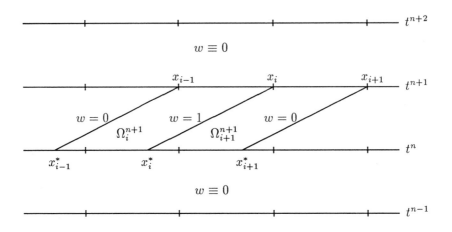

Figure 2: Interior space-time test function w_i^{n+1}

and end of the time step under consideration. We define the test function w_i^{n+1} as in Figure 2. Let Ω_i^{n+1} and Ω_{i+1}^{n+1} denote the two parallelograms on which w_i^{n+1} is nonzero. Since we will be working only with the time step from t^n to t^{n+1} in what follows, we drop the superscript $n+1$ on w and Ω.

For the time being, we assume that $\Omega_i \cup \Omega_{i+1}$ does not meet the boundary of Ω; boundary cases are treated in §§5–6. A weak form of (1) is

$$\int \int_{\Omega_i \cup \Omega_{i+1}} (u_t + vu_x - Du_{xx} - f)w_i \, dx \, d\tau = 0. \tag{19}$$

Noting that $u_t + vu_x = u_\tau$ and that w_i is independent of τ,

$$\int \int_{\Omega_i \cup \Omega_{i+1}} (u_t + vu_x)w_i \, dx \, d\tau = \int \int u_\tau \, d\tau \, w_i \, dx$$

$$= \int_{x_{i-1}}^{x_{i+1}} (u^{n+1}(x) - u^n(x^*))w_i(x, t^{n+1}) \, dx. \tag{20}$$

Another way of viewing (20) is as an integration by parts in τ, where $(w_i)_\tau = 0$ and the last member of (20) constitutes the resulting boundary terms at t^n and t^{n+1}. For the diffusion term, integration by parts in x yields

$$\int \int_{\Omega_i \cup \Omega_{i+1}} -Du_{xx}w_i \, dx \, d\tau = \int \int_{\Omega_i \cup \Omega_{i+1}} Du_x(w_i)_x \, dx \, d\tau, \tag{21}$$

where the terms at the left edge of Ω_i and the right edge of Ω_{i+1} vanish because w_i vanishes. Putting (19) through (21) together, we have the interior ELLAM equation

$$\int_{x_{i-1}}^{x_{i+1}} u^{n+1}(x)w_i(x, t^{n+1}) \, dx + \int \int_{\Omega_i \cup \Omega_{i+1}} Du_x(w_i)_x \, dx \, d\tau$$

$$= \int_{x_{i-1}}^{x_{i+1}} u^n(x^*)w_i(x, t^{n+1}) \, dx + \int \int_{\Omega_i \cup \Omega_{i+1}} fw_i \, dx \, d\tau. \tag{22}$$

212

If we now use backward Euler time integration *along characteristics* (backward Euler in τ), replacing the τ integrals in the D and f terms by Δt times the value at the head of the characteristic, and if we seek the numerical solution $U^{n+1} \in \mathcal{M}$ (note that backward Euler τ integration implies that U needs only to be known at the discrete time levels) using (22), we obtain precisely (10). That is, *in the interior, ELLAM with backward Euler τ integration is MMOC with chapeau functions.*

In deriving (22), we used smoothness of u, constancy of w_i along characteristics, and continuity of w_i. Thus, MMOC with any continuous element, such as higher-order piecewise polynomials, can be represented via ELLAM. Higher-order test functions do not satisfy $Dw_{xx} = 0$, but this is no longer critical. The LAM framework motivated the formulation presented here and will continue to be useful in constructing generalizations, but there appears to be no reason to adhere to it slavishly. Indeed, Varoğlu and Finn [25] came to a related formulation independently; the relationship to this work will be elaborated in §11.

So far we have not developed anything new, but the existing theory for MMOC tells us that ELLAM is likely to be on sound mathematical footing. New procedures arise when we move on in §§5–6 to space-time elements that meet the spatial boundary. To match the form of equations to come, we confine the backward Euler version of (22) to the single element Ω_i and write

$$\int_{x_{i-1}}^{x_i} u^{n+1}(x)w(x,t^{n+1})\,dx + \Delta t \int_{x_{i-1}}^{x_i} Du_x^{n+1}(x)w_x(x,t^{n+1})\,dx + \text{IET}$$
$$= \int_{x_{i-1}}^{x_i} u^n(x^*)w(x,t^{n+1})\,dx + \Delta t \int_{x_{i-1}}^{x_i} f^{n+1}(x)w(x,t^{n+1})\,dx, \tag{23}$$

where IET denotes an interior edge term that is canceled by an adjacent element (Ω_{i-1} if $w = w_{i-1}$, Ω_{i+1} if $w = w_i$); this term need not be computed or stored. Note that $w_{i-1} + w_i \equiv 1$ on Ω_i, assuring mass conservation on Ω_i when the w_{i-1} and w_i equations are summed. Note also that w_i, for example, is nonzero on Ω_i and Ω_{i+1}, so the actual ELLAM equation for w_i is (23) with $w = w_i$, plus (23) on Ω_{i+1} with $w = w_{(i+1)-1}$.

5. INLET BOUNDARY

At the inlet boundary, we encounter four different types of elements, as shown in Figure 3. We consider only the first in detail here, since the second has the form of the first with $\tilde{x}_0 = x_i$ (where $\tilde{x}_0^* = x_0$), and the third and fourth have the form of the first and second, respectively, with $i = 1$. Referring to Fig. 3, the test function w will be 1 on the edge from t_{i-1}^* to x_{i-1} and 0 on the edge from x_i^* to x_i (w_{i-1}), or the reverse (w_i). Note that, in either case, w is nonzero on the edge from t^n to t_{i-1}^*. When $i = 1$ (the third and fourth types in Fig. 3), w will be identically 1 if the inlet boundary condition is Dirichlet; we discuss this further below.

Proceeding as before, the analogue of (20) is ˙

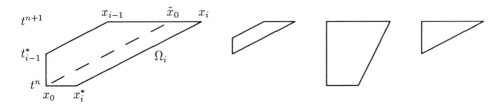

Figure 3: Four types of elements meeting an inlet boundary

$$\int \int_{\Omega_i} (u_t + vu_x) w \, dx \, d\tau$$
$$= \int_{x_{i-1}}^{x_i} u^{n+1}(x) w(x, t^{n+1}) \, dx - \int_{x_{i-1}}^{\tilde{x}_0} u(x_0, t^*(x)) w(x, t^{n+1}) \, dx$$
$$- \int_{\tilde{x}_0}^{x_i} u^n(x^*) w(x, t^{n+1}) \, dx, \tag{24}$$

where $t^*(x)$ refers to the time at which the characteristic traced backward from x meets the boundary. The analogue of (21) is

$$\int \int_{\Omega_i} -D u_{xx} w \, dx \, d\tau = \int \int_{\Omega_i} D u_x w_x \, dx \, d\tau + \int_{t^n}^{t^*_{i-1}} D u_x(x_0, t) w(x_0, t) \, dt + \text{IET}. \tag{25}$$

If the inlet boundary has a flux condition, note that

$$\int_{x_{i-1}}^{\tilde{x}_0} u(x_0, t^*(x)) w(x, t^{n+1}) \, dx = \int_{t^n}^{t^*_{i-1}} vu(x_0, t) w(x_0, t) \, dt \tag{26}$$

and combine (19), (24), (25), and (26) to see that

$$\int_{x_{i-1}}^{x_i} u^{n+1}(x) w(x, t^{n+1}) \, dx + \int \int_{\Omega_i} D u_x w_x \, dx \, d\tau + \text{IET}$$
$$= \int_{t^n}^{t^*_{i-1}} (vu - D u_x)(x_0, t) w(x_0, t) \, dt + \int_{\tilde{x}_0}^{x_i} u^n(x^*) w(x, t^{n+1}) \, dx$$
$$+ \int \int_{\Omega_i} fw \, dx \, d\tau. \tag{27}$$

Substituting (2) into (27) and applying backward Euler integration along characteristics in the D and f terms, we have

$$\int_{x_{i-1}}^{x_i} u^{n+1}(x) w(x, t^{n+1}) \, dx + \int_{x_{i-1}}^{x_i} D u_x^{n+1}(x) w_x(x, t^{n+1}) \Delta t(x) \, dx + \text{IET}$$
$$= \int_{t^n}^{t^*_{i-1}} g(t) w(x_0, t) \, dt + \int_{\tilde{x}_0}^{x_i} u^n(x^*) w(x, t^{n+1}) \, dx$$
$$+ \int_{x_{i-1}}^{x_i} f^{n+1}(x) w(x, t^{n+1}) \Delta t(x) \, dx, \tag{28}$$

where $\Delta t(x) = t^{n+1} - t^*(x)$ if $x < \tilde{x}_0$, taking account of the reduced elapsed time along a characteristic that meets the boundary. This is the inlet-boundary ELLAM equation

214

for a flux boundary condition. As usual for a natural boundary condition, $U^{n+1}(x_0)$ will be solved for as an unknown; corresponding to this degree of freedom is the test function w_0 that is 1 at (x_0, t^{n+1}) and 0 on the characteristic back from x_1 (see third and fourth types in Fig. 3). For a Neumann condition, $-Du_x(x_0, t) = g(t)$, (28) holds with an extra left-hand-side term consisting of the negative of the left-hand side of (26); $u(x_0, t^*(x))$ in this term can be expressed in terms of the known $U^n(x_0)$ and the unknown $U^{n+1}(x_0)$. This would appear inaccurate in a highly advective problem with large Courant number (many parallel characteristics meeting the boundary), but it is unlikely that a Neumann condition would be physically appropriate in such a case.

We treat the case of a Dirichlet condition slightly differently. To obtain (28) we first integrated by parts in x and then used backward Euler integration in τ. If we take these steps in the reverse order, only the diffusion term (the only term to which both steps apply) is altered. Instead of (25), we have

$$
\int_{x_{i-1}}^{x_i} -Du_{xx}^{n+1}(x)w(x, t^{n+1})\Delta t(x)\, dx
$$
$$
= \int_{x_{i-1}}^{x_i} Du_x^{n+1}(x)[w_x(x, t^{n+1})\Delta t(x) + w(x, t^{n+1})(\Delta t)_x(x)]\, dx + \text{IET}, \tag{29}
$$

since the edge term at x_0 has $\Delta t(x_0) = 0$. Observe that $(\Delta t)_x = 1/v$ for $x < \tilde{x}_0$ and $(\Delta t)_x = 0$ for $x > \tilde{x}_0$, substitute $g(t)$ for $u(x_0, t)$, and move the g integration from $[t^n, t_{i-1}^*]$ to $[x_{i-1}, \tilde{x}_0]$, yielding

$$
\int_{x_{i-1}}^{x_i} u^{n+1}(x)w(x, t^{n+1})\, dx + \int_{x_{i-1}}^{x_i} Du_x^{n+1}(x)w_x(x, t^{n+1})\Delta t(x)\, dx
$$
$$
+ \int_{x_{i-1}}^{\tilde{x}_0} Du_x^{n+1}(x)w(x, t^{n+1})\frac{1}{v}\, dx + \text{IET}
$$
$$
= \int_{x_{i-1}}^{\tilde{x}_0} g(t^*(x))w(x, t^{n+1})\, dx + \int_{\tilde{x}_0}^{x_i} u^n(x^*)w(x, t^{n+1})\, dx
$$
$$
+ \int_{x_{i-1}}^{x_i} f^{n+1}(x)w(x, t^{n+1})\Delta t(x)\, dx \tag{30}
$$

instead of (28). The factor of $1/v$ causes no trouble, since the integration is over an interval of length at most $v\Delta t$. As in standard finite-element methods, the Dirichlet condition is essential, so $U^{n+1}(x_0)$ is assigned from boundary data and is not solved for. With no degree of freedom at x_0, and asking that test functions sum to 1 for mass conservation, the only test function on Ω_1 must be $w_1 \equiv 1$ as noted above.

It is of interest to compare (23), (28), and (30). The x^* term in (23) has been replaced by the t^* and x^* terms in (30). Thus, in the Dirichlet case, the integral at the feet of the characteristics "goes around the corner" (see Fig. 3). A similar situation holds in the flux case of (28), except that the boundary segment integrates the known flux instead of the dependent variable itself. In either case, from the hyperbolic point of view, both types of "initial data" are integrated by going around the corner. The other major change from (23) to (28) and (30) is the x-dependent Δt in the principal diffusion term; this seems quite appropriate, as the diffusion at each point is weighted by the length of time over which it acts. The effect of applying backward Euler first and then integrating by parts in the Dirichlet case is to put the $Du_x w$ term at level $n+1$, where it conveniently fits as part of the implicit solution, instead of at the inlet boundary, where it is not convenient to evaluate it. In the flux case, $Du_x w$ is left at the boundary, where it fits as part of the known flux.

We compare this boundary formulation to previous approaches in the literature, insofar as we can interpret them. It appears to us that the usual approach to the Dirichlet problem, say in the context of finite differences, has been to backtrack from grid points at t^{n+1}, obtaining values at the feet of characteristics by interpolation of the t^n solution if the boundary is not crossed, or by the boundary datum if it is. These values are assigned to the grid points at t^n, and then a diffusion problem is solved over the step Δt. Thus, diffusion operates over the full Δt everywhere, in contrast with our x-dependent Δt. In the more careful MMOC formulation of Douglas et al. [8], the x-dependent Δt is put in the denominator in (8); this leads to terms like those in (30), except that each integrand is divided by $\Delta t(x)$ and the $1/v$ term is absent. Our formulation, which is integrated in space and time, is closest to the physical origins of the differential equations; individual terms can be interpreted as masses (weighted by test functions) associated with storage, interior flux, boundary flux, or sources over a time step. Boundary conditions, and the flux condition in particular, fit in naturally in accordance with their physical meaning. It is not clear how, or if, the literature has dealt with flux conditions. We view the development above as the manner in which MMOC can accommodate characteristics that cross the boundary.

If the τ integration is not backward Euler, (28) and (30) will have additional $Du_x w_x \Delta t$ terms on the inlet boundary. For a flux condition, Du_x can be expressed in terms of the unknown U and the known flux. If the boundary condition is Dirichlet, Du_x could be a discrete derivative of U, or an additional degree of freedom for an unknown $U_x^{n+1}(x_0)$ could be introduced. In the latter case, Ω_1 would have the two test functions w_0 and w_1 as for a flux boundary, instead of the single $w_1 \equiv 1$. These ideas will be elaborated in [5] and are not discussed further here.

Figure 4 shows the test functions, through w_2, for the different boundary conditions, where $1 < Cu < 2$ is chosen for illustration.

6. DISCRETIZED OUTLET BOUNDARY AND HYPERBOLIC CASE

If $Cu \gg 1$, we see that many spatial degrees of freedom cross the outlet boundary in a time step. To preserve the information in these degrees of freedom, it seems appropriate to discretize the outlet boundary in time. Let $c = [Cu]$, the integer part of the Courant number, and for $j = E, E+1, \ldots, E+c-1$ set $t_j = t^{n+1} - (j-E)\Delta x/v$, with $t_{E+c} = t^n$. Thus $[t^n, t^{n+1}]$ is partitioned into c intervals, backward in time, with the first $c-1$ of length $\Delta t/Cu$. The last, up to twice the size of the others, is of length $((Cu)+1)\Delta t/Cu$, where $(Cu) = Cu - c$ is the fractional part of Cu. Alternatively, we could put in the node

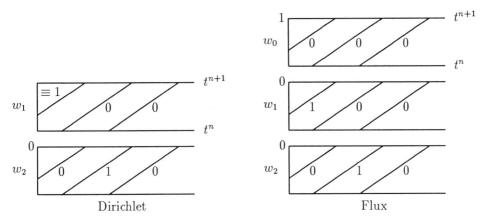

Dirichlet   Flux

Figure 4: Test functions near an inlet boundary

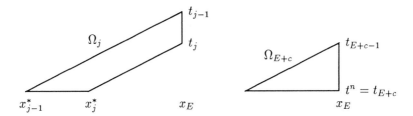

Figure 5: Two types of elements meeting an outlet boundary

$t_{E+c} = t^{n+1} - c\Delta x/v$ and, if $(Cu) > 0$, take $t_{E+c+1} = t^n$. We refer to these as the c and $c + 1$ cases, respectively. From here on, we use the c case in the exposition. The types of elements that arise are shown in Figure 5. Since the second has the form of the first with $t_j = t^n$, we discuss only the first in detail. The test function w will be 1 on the edge from x_{j-1}^* to t_{j-1} and 0 on the edge from x_j^* to t_j (w_{j-1}), or the reverse (w_j). Note that w_E is nonzero on Ω_E, the last parallelogram of the type in Fig. 2, and on Ω_{E+1}, the first trapezoid of the type in Fig. 5. Because the solution at $(x_E, t_{E+c}) = (x_E, t^n)$ is known from the previous time level, we do not solve for an unknown corresponding to t_{E+c}; accordingly, the second element Ω_{E+c} in Fig. 5 has the single test function $w_{E+c-1} \equiv 1$ instead of two.

The analogue of (20) is now, since $v\, dt = dx$,

$$\int\int_{\Omega_i} (u_t + vu_x)w\, dx\, d\tau = \int_{t_j}^{t_{j-1}} (u(x_E, t) - u(x^*(t), t^n))w(x_E, t)v\, dt, \qquad (31)$$

where $(x^*(t), t^n)$ is on the characteristic backtracked from (x_E, t). The analogue of (21) is

$$\int\int_{\Omega_i} -Du_{xx}w\, dx\, d\tau = \int\int_{\Omega_i} Du_x w_x\, dx\, d\tau - \int_{t_j}^{t_{j-1}} Du_x(x_E, t)w(x_E, t)\, dt + \text{IET}. \qquad (32)$$

217

Combining (19), (31), and (32), and using backward Euler in τ, we have

$$\int_{t_j}^{t_{j-1}} (vu - Du_x)(x_E, t)w(x_E, t)\, dt$$

$$+ \int_{t_j}^{t_{j-1}} Du_x(x_E, t)w_x(x_E, t)(t - t^n)v\, dt + \text{IET}$$

$$= \int_{t_j}^{t_{j-1}} u^n(x^*(t))w(x_E, t)v\, dt + \int_{t_j}^{t_{j-1}} f(x_E, t)w(x_E, t)(t - t^n)v\, dt. \tag{33}$$

For a flux outlet condition, we add a $(vu - Du_x)w_x(t - t^n)v$ term to both sides of (33), substitute from (3), and see that

$$\int_{t_j}^{t_{j-1}} u(x_E, t)w_x(x_E, t)(t - t^n)v^2\, dt + \text{IET}$$

$$= \int_{t_j}^{t_{j-1}} u^n(x^*(t))w(x_E, t)v\, dt - \int_{t_j}^{t_{j-1}} h(t)w(x_E, t)\, dt$$

$$+ \int_{t_j}^{t_{j-1}} h(t)w_x(x_E, t)(t - t^n)v\, dt + \int_{t_j}^{t_{j-1}} f(x_E, t)w(x_E, t)(t - t^n)v\, dt. \tag{34}$$

With a Neumann condition, h in (34) is replaced by $vu + h$, leading to (34) with the integral on the left-hand side replaced by $\int_{t_j}^{t_{j-1}} u(x_E, t)w(x_E, t)v\, dt$. In either case, the unknowns on Ω_j are $U(x_E, t_{j-1})$ and $U(x_E, t_j)$. For a Dirichlet boundary, subtract vuw from both sides of (33) and obtain

$$\int_{t_j}^{t_{j-1}} Du_x(x_E, t)w_x(x_E, t)(t - t^n)v\, dt - \int_{t_j}^{t_{j-1}} Du_x(x_E, t)w(x_E, t)\, dt + \text{IET}$$

$$= \int_{t_j}^{t_{j-1}} u^n(x^*(t))w(x_E, t)v\, dt - \int_{t_j}^{t_{j-1}} h(t)w(x_E, t)v\, dt$$

$$+ \int_{t_j}^{t_{j-1}} f(x_E, t)w(x_E, t)(t - t^n)v\, dt. \tag{35}$$

Here the appropriate unknowns will be fluxes, with u itself being known. If U is represented by piecewise-linear trial functions, we could consider linear ones for U_x at the outlet, but recalling the expected loss of one order of accuracy in passing from U to U_x, piecewise constants seem more suitable. This has been borne out by the numerical results reported in §9. With this choice, Ω_j has the single unknown $U_x(x_E, t)$, $t_j < t < t_{j-1}$. Finally, consider the purely hyperbolic case, $D = 0$. Then (33) becomes

$$\int_{t_j}^{t_{j-1}} u(x_E, t)w(x_E, t)v\, dt$$

$$= \int_{t_j}^{t_{j-1}} u^n(x^*(t))w(x_E, t)v\, dt + \int_{t_j}^{t_{j-1}} f(x_E, t)w(x_E, t)(t - t^n)v\, dt, \tag{36}$$

with unknowns $U(x_E, t_{j-1})$ and $U(x_E, t_j)$ on Ω_j. The same equation, (33), handles all of the cases represented by (34) through (36); in particular, no artificial boundary condition

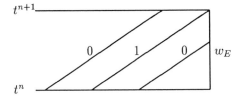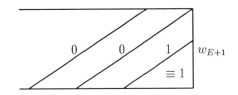

Figure 6: Test functions near an outlet boundary

is needed in the hyperbolic case, even though upwinding is not used in the treatment of advection.

Comparing (23) and (33), we see that the $[x_{i-1}, x_i]$ terms of (23) have $[t_j, t_{j-1}]$ analogues in (33), with an additional diffusion term on the left-hand side of (33) arising from integration by parts. Dual to integrating old information around the corner at the inlet boundary, as in (30), here we are solving for new information around the corner (x_E, t^{n+1}). The ELLAM formulation treats the inlet and the previous time level, which together constitute hyperbolic data, as data, and the new time level and outlet, which comprise hyperbolic unknowns, as unknowns. This helps to explain why the development is well-suited to advection-dominated problems and passes readily to the purely hyperbolic case. The formulation also handles diffusion-dominated problems; if $v = 0$, then $x^* = x$, the test functions are independent of t, the elements of Fig. 5 (and hence (33)) are dropped, and ELLAM reduces to a standard finite-element method for $u_t - D u_{xx} = f$. The only step that is not completely straightforward in this involves the $1/v$ term in (30), which collapses to $D u_x^{n+1}(x_0) w(x_0, t^{n+1}) \Delta t$, the same boundary term resulting from integration by parts in a standard method.

For the c case with $2 < Cu < 3$, $c = 2$, Figure 6 shows the last two test functions w_E and w_{E+1}.

7. CONSERVATION OF MASS

Unlike the ELM developed previously, ELLAM conserves mass. To see this, note that the inlet-boundary, interior, and outlet-boundary test functions have been set up so that their sum is identically 1 on $[x_0, x_E] \times [t^n, t^{n+1}]$. It follows that, selecting appropriate equations from (23), (28), (30), (34), (35), and (36) according to boundary conditions, the sum will lead to (following (19))

$$
\begin{aligned}
0 &= \int_{t^n}^{t^{n+1}} \int_{x_0}^{x_E} (u_t + v u_x - D u_{xx} - f) \, dx \, dt \\
&= \int_{x_0}^{x_E} u^{n+1}(x) \, dx - \int_{x_0}^{x_E} u^n(x) \, dx - \int_{t^n}^{t^{n+1}} \int_{x_0}^{x_E} f \, dx \, dt \\
&\quad + \int_{t^n}^{t^{n+1}} (vu - D u_x)(x_E, t) \, dt - \int_{t^n}^{t^{n+1}} (vu - D u_x)(x_0, t) \, dt, \qquad (37)
\end{aligned}
$$

which is precisely the statement of global mass conservation. To obtain (37) exactly in an

$$\begin{pmatrix}
\text{x} & \text{x} & & & & & & \\
\text{x} & \text{x} & \text{x} & & & & & \\
& \text{x} & \text{x} & \text{x} & & & & \\
& & \text{x} & \text{x} & 0 & & & \\
\hline
& & & \text{x} & \text{x} & 0 & & \\
& & & & \text{x} & \text{x} & 0 & \\
& & & & & \text{x} & \text{x} &
\end{pmatrix}$$

Figure 7: Form of discrete matrix for Dirichlet boundaries

implementation, some care is needed in the consistent evaluation of integrals; for example, the x^* integrals in the ELLAM equations must be computed in such a way that they sum to the u^n integral in (37).

8. TRIAL FUNCTIONS AND MATRIX STRUCTURE

To this point we have not specified the space-time trial functions in the ELLAM formulation, except for occasional allusions to solving for certain unknowns U. With backward Euler integration in τ, it is sufficient to define the trial functions at the discrete time levels t^n only. The most natural choice, in view of the test functions, is continuous piecewise-linear elements, which we have used in the numerical results of §9. As noted in §4, nothing prevents consideration of higher-order test and trial functions. If the τ integration were other than backward Euler, trial functions would need to be defined in space-time. Here the natural idea is to have the same elements at t^{n+1} as in the backward Euler case, and then demand that a trial function be linear in τ along a characteristic backtracking in time; the unknown at (x, t^{n+1}) and the known value at (x^*, t^n) determine the trial function on the characteristic.

The matrix structure resulting from the ELLAM equations and the above trial functions depends on the boundary conditions. For flux inlet and outlet boundaries, the situation is straightforward: rows correspond to the test functions $w_0, w_1, \ldots, w_{E+c-1}$, columns to the unknowns $U^{n+1}(x_0), U^{n+1}(x_1), \ldots, U^{n+1}(x_E), U(x_E, t_{E+1}), \ldots, U(x_E, t_{E+c-1})$, and the matrix is tridiagonal. A Dirichlet inlet boundary removes w_0 and $U^{n+1}(x_0)$, with the matrix still tridiagonal. In either case, $U(x_E, t_{E+c}) = U(x_E, t^n)$ is known and closes the linear system. A Dirichlet outlet boundary has a more noticeable effect, since the unknowns become $U^{n+1}(x_1), \ldots, U^{n+1}(x_{E-1}), (U_x)_E, \ldots, (U_x)_{E+c-1}$, where $(U_x)_j$ is the piecewise-constant value of $U_x(x_E, t)$ for $t_{j+1} < t < t_j$. The w_{E-1} equation involves $U^{n+1}(x_{E-2})$, $U^{n+1}(x_{E-1})$, and $U^{n+1}(x_E)$, the last of which is known and moves to the right-hand side. The w_E equation is the sum of two parts: the first, from (23), has $U^{n+1}(x_{E-1})$ and $U^{n+1}(x_E)$, while the second, from (30), has $(U_x)_E$. For $E+1 \leq j \leq E+c-1$, the w_j equation, with both parts from (30) on Ω_j and Ω_{j+1}, has $(U_x)_{j-1}$ and $(U_x)_j$. Thus, the matrix is as depicted in Figure 7, which shows the case $E = 5$, $c = 3$, Dirichlet boundaries.

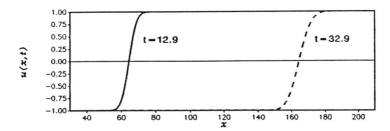

Figure 8: Exact initial and final solutions of test problem

Because of the zero in the upper-right block in Fig. 7, the Dirichlet equations can be solved for $U^{n+1}(x_1)$, ..., $U^{n+1}(x_{E-1})$ independently of any boundary results, just as in standard finite elements. The lower-bidiagonal structure of the lower-right corner shows that each $(U_x)_j$ can then be obtained in turn, independent of all succeeding ones. If $U_x(x_E, t)$ is represented by piecewise-linear instead of piecewise-constant elements, with unknowns $U_x(x_E, t_j)$ for $E \leq j \leq E + c - 1$, the w_j equation couples $U_x(x_E, t_{j-1})$, $U_x(x_E, t_j)$, and $U_x(x_E, t_{j+1})$. This fills in the 0's in the lower-right corner of Fig. 7, and the known $U_x(x_E, t_{E+c}) = U_x(x_E, t^n)$ closes the system. It is still the case that $U^{n+1}(x_1)$, ..., $U^{n+1}(x_{E-1})$ are independent of the outlet fluxes, but each flux is no longer independent of succeeding ones.

9. NUMERICAL RESULTS

Since the MMOC has been extensively tested in earlier interior work [11,12,23], our principal purpose here is to demonstrate the ability of ELLAM to handle boundary conditions. To this end, consider (1) with $f = 0$, and (4) with $u_0(x) = 1$ for $x > 0$, $u_0(x) = -1$ for $x < 0$. The exact solution of this problem is

$$u(x, t) = \text{erf}\left(\frac{x - vt}{2\sqrt{Dt}}\right). \tag{38}$$

We obtained a variety of boundary-value problems by cutting off the spatial and temporal domains, selecting different types of boundary conditions, and assigning $g(t)$ and $h(t)$ in (2) and (3) such that (38) is the exact solution. To avoid quadrature errors, ELLAM integrals involving g or h were evaluated by high-order Gauss rules, and x^* integrals at the feet of characteristics were calculated by high-order rules in the first time step, when $u^n(x^*)$ is known analytically, and by exact integration thereafter, when $U^n(x^*)$ is a piecewise polynomial. As a result, all computations conserved mass to essentially within rounding error. To avoid initial singularities, which would require grid refinement and are not our main interest, we chose initial time $t^0 > 0$. The results reported below used $v = 5$, $D = 0.5$, $t^0 = 12.9$, and final time $t^N = 32.9$. The initial and final exact solutions are shown in Figure 8. Three spatial domains were considered: $[30, 210]$, denoted by Int, for which the front remains in the interior; $[90, 210]$, called In, where the front must cross

Run	Domain	Inlet	Outlet	$\Delta x(Pe)$	$\Delta t(Cu)$	$\|u - U\|_{L^2}(t^N)$	L^2 projection
1	Int	Dir	Flux	3(30)	$10(16\frac{2}{3})$	5.16E−2	1.93E−2
2	Int	Dir	Flux	3(30)	$2(3\frac{1}{3})$	2.33E−2	1.93E−2
3	Int	Dir	Flux	0.6(6)	$2(16\frac{2}{3})$	1.46E−2	7.35E−4
4	Int	Dir	Flux	0.6(6)	$0.08(\frac{2}{3})$	9.10E−4	7.35E−4
5	In	Dir	Flux	3(30)	$10(16\frac{2}{3})$	3.32E−2	1.93E−2
6	In	Dir	Flux	3(30)	$2(3\frac{1}{3})$	2.28E−2	1.93E−2
7	In	Dir	Flux	0.6(6)	$2(16\frac{2}{3})$	1.06E−2	7.35E−4
8	In	Dir	Flux	0.6(6)	$0.08(\frac{2}{3})$	7.89E−4	7.35E−4
9	In	Flux	Flux	3(30)	$10(16\frac{2}{3})$	3.31E−2	1.93E−2
10	In	Flux	Flux	3(30)	$2(3\frac{1}{3})$	2.19E−2	1.93E−2
11	In	Flux	Flux	0.6(6)	$2(16\frac{2}{3})$	1.05E−2	7.35E−4
12	In	Flux	Flux	0.6(6)	$0.08(\frac{2}{3})$	8.36E−4	7.35E−4
13	Out	Dir	Dir	3(30)	$10(16\frac{2}{3})$	4.17E−2**	2.29E−2**
14	Out	Dir	Dir	3(30)	$2(3\frac{1}{3})$	2.02E−2*	1.33E−2*
15	Out	Dir	Dir	0.6(6)	$2(16\frac{2}{3})$	3.82E−3*	5.17E−4*
16	Out	Dir	Dir	0.6(6)	$0.08(\frac{2}{3})$	6.24E−4*	5.17E−4*
17	Out	Dir	Flux	3(30)	$10(16\frac{2}{3})$	4.47E−2**	2.29E−2**
18	Out	Dir	Flux	3(30)	$2(3\frac{1}{3})$	1.42E−2*	1.33E−2*
19	Out	Dir	Flux	0.6(6)	$2(16\frac{2}{3})$	4.95E−3*	5.17E−4*
20	Out	Dir	Flux	0.6(6)	$0.08(\frac{2}{3})$	5.65E−4*	5.17E−4*
21	Out	Dir	Dir	3(30)	$0.4(\frac{2}{3})$	2.68E−2*	1.33E−2*
22	Out	Dir	Dir	0.6(6)	$0.4(3\frac{1}{3})$	9.05E−4*	5.17E−4*

*At $t = 24.9$ **At $t = 22.9$

Table 1: Numerical results on the domain $[x_0, x_E]$

the inlet boundary; and $[30, 120]$, or Out, with the front crossing the outlet. The initial front width, expressed as $2/q$, where q is the maximum slope (or almost equivalently as the distance between the points with values -0.8 and $+0.8$), is about 9.0 units; the final width is about 14.4.

Previous work has shown that MMOC with linear elements is oscillation-free if at least 3 intervals discretize a front, i.e., $\Delta x \leq 2/3q$ for this problem. Accordingly, our base case here was $\Delta x = 3$, with grid Peclet number $Pe = v\Delta x/D = 30$. To check convergence rates we also used $\Delta x = 0.6$ $(Pe = 6)$. The base time step was $\Delta t = 2$, for a base-case Courant number of $Cu = v\Delta t/\Delta x = 10/3$. For $\Delta x = 3$, we ran $\Delta t = 0.4, 2, 10$ $(Cu = 2/3, 10/3, 50/3)$; for $\Delta x = 0.6$, $\Delta t = 0.08, 0.4, 2$ $(Cu = 2/3, 10/3, 50/3)$ were used. Base-case boundary conditions were Dirichlet at the inlet and flux at the outlet; flux at the inlet was also run for the domain In, and Dirichlet at the outlet was run for Out.

Table 1 displays some of the results for the usual domain (not the discretized outlet boundary). The next-to-last column shows the L^2 error of the computed solution at the final time for Int and In, or at a time when the front is crossing the boundary for Out.

The last column contains the L^2 error of the L^2 projection of the exact solution into the grid, i.e., the best value that could appear in the next-to-last column. In the Int cases, Runs 1 through 4, ELLAM is effectively MMOC, and familiar behavior [11] is observed. Comparison of Runs 2 and 4 reveals the expected $O(\Delta x^2 + \Delta t)$ convergence, Run 2 shows that most of the error can be spatial even with $Cu > 1$, and Run 1 yields reasonable results even with a very large time step. Run 3, on the other hand, is dominated by truncation in the characteristic τ direction.

The In cases, Runs 5 through 12, show clearly that the formulation can advect the front across the inlet with no loss of accuracy. Indeed, the approximations are actually a bit better than their Int counterparts. Presumably, this occurs because the analytical boundary data of In have a shorter time to advect and incur errors than the corresponding initial data of Int. The improvement is greatest when Δt is large, hinting that the favorable influence of boundary data is of longer duration in this instance. These results strongly suggest that ELLAM provides the "correct" way of dealing with inflow boundaries in MMOC, and in ELM in general. The Out cases, Runs 13 through 20, also show no significant deterioration due to the outlet boundary approximation. Each error relates to its L^2 projection error about as well as the Int errors do, with the possible exception of Run 14. But even the Run 14 error, which covers part of the front, is less than the Int error of Run 2, covering the whole front, so this does not appear to be of concern. Note that the results in the domain are affected by the discretized outlet boundary when the outlet has a flux condition (recall §8).

We turn next to the discretized outlet boundary, which we discuss only briefly here. It is significant only for the Out cases. For a flux condition, where we solve for $U^{n+1}(x_E)$, $U(x_E, t_{E+1})$, ..., $U(x_E, t_{E+c-1})$, the results were good, matching well with interior U values in corresponding Int runs. Some small oscillations, which do not appear in corresponding Int runs, were observed at the head of the front. The magnitude of these oscillations was of the order of 0.01, 0.001, and 0.0001 for $\Delta t = 2, 0.4$, and 0.08, respectively; it seemed insensitive to Δx. These results are useful and encouraging, though further analysis of the oscillations is warranted.

With a Dirichlet outlet, where fluxes are solved for, the situation is more mixed. Runs 13 through 16, 21, and 22 in Table 1 were made with both piecewise-linear and piecewise-constant U_x, and both the c and $c + 1$ cases. For piecewise-linear U_x, among the cases with $Cu > 1$, Runs 13, 14, and 15 showed sizable oscillations of U_x in time within a time step, while Run 22 was free of this. Run 21, with $Cu < 1$, solved for just one U_x per time step, but this showed an extra peak and valley before and after the passage of the front; Run 16 did not oscillate in this way. It turns out that the homogeneous equations for the piecewise-linear Dirichlet outlet boundary are nearly solved by an oscillatory vector of $+1$'s and -1's; it fails only in the first and last equations. Note that this vector represents no net flux through each outlet boundary element, since the integral over the outlet cancels the $+1$ and -1. This may be responsible for the oscillatory tendency when $Cu > 1$ and bears further investigation, perhaps with an analogue of upwinding along the outlet. In any event, the piecewise-constant equations do not suffer from this, and Runs 13, 14, and 15 (as well as 16 and 22) produced smooth results with them. However, Run 21 still had extra peaks and valleys; only the finer grid and time step of Run 16 eliminated this. At this preliminary stage, piecewise constants perform well within a step, but the long-time picture needs future clarification.

Given the use of piecewise constants, the c case appears to be distinctly preferable to the $c + 1$ case. If (Cu) is small, i.e., Cu is slightly greater than an integer, the $c + 1$ case gives rise to a small element Ω_{E+c+1}. It then turns out that the w_{E+c} equation has matrix entries of order (Cu), but a right-hand side potentially of order 1, and $(U_x)_{E+c}$ can be anomalous. The right-hand side, as in (35), involves the difference between the approaching front u^* and the prescribed Dirichlet data h. Even in the absence of a genuine boundary layer, errors in $(U_x)_{E+c}$ can be magnified. It can be shown that use of the c case instead of the $c + 1$ case replaces $(U_x)_{E+c-1}$ and $(U_x)_{E+c}$ by their weighted average and removes this difficulty. All of these points regarding the discretized outlet boundary will be pursued in more depth and detail in subsequent research.

10. EXTENSIONS

The simple problem (1)–(4) is, of course, not what we actually had in mind in developing ELLAM. The formulation should perhaps be viewed more as an approach or a methodology than as a method, since the ideas are very general and choices must be made to apply them to a particular problem. In future work, the EG expects to consider a variety of extensions of the framework presented here. Multiple space dimensions, variable coefficients, nonlinear equations with reaction terms, coupled systems in which advective velocities are solutions of continuity equations (as is typical in petroleum recovery, groundwater flow and transport, and semiconductor device modeling), and incorporation of dynamic local grid refinement are some of these.

In principle, we do not believe that any of these extensions present fundamental obstacles. For example, when the velocity is variable or unknown, we cannot expect to backtrack characteristics exactly. Hence, the space-time elements will only be approximations of those with characteristic edges. In MMOC, analogous approximations caused mass-conservation errors. In ELLAM, by contrast, this will manifest itself in a residual advective term that remains on the left-hand side of the weak form of the problem; i.e., advection will not fall entirely into x^* as it did in (20). This is foreshadowed, in a different context, in the research of Espedal and Ewing [10] that formulates MMOC with a nonlinear flux function. By keeping this new term, we expect to maintain the conservative property of the scheme. Since the size of the term depends on the error of backtracking, it should usually be small, and the numerical advantages of removing advection in the simpler case should persist in this case.

The other avenue of further investigation is a theoretical convergence analysis. The known error estimates for MMOC should serve as a basis for this.

11. RELATIONSHIP TO OTHER WORK

We are aware of three other uses of space-time elements for advection-dominated problems, the finite element method incorporating characteristics (FEMIC) of Varoḡlu and Finn [25], the streamline diffusion (SD) method of Brooks and Hughes [3], and work of several French researchers to be discussed below. FEMIC is in many ways similar to ELLAM, in that it uses space-time elements with edges oriented along characteristics, and its test functions are like those presented here. However, FEMIC is a forward-tracking, not a backtracking, scheme. This gives rise to the usual difficulties of distorted Lagrangian grids in more than one dimension, which backtracking methods avoid. In one dimension, at each time step FEMIC allows a triangular space-time element at an inlet boundary that essentially introduces one new grid point; to avoid loss of resolution, this effectively limits FEMIC to Courant numbers of the order of 1, where an important feature of ELM is the ability to use $Cu \gg 1$. Finally, in the step analogous to (20), FEMIC integrates by parts in t instead of τ; this appears to make no difference in the interior, but does matter at the boundary. For instance, for (1)–(4) with $Cu = 1$, the u_τ terms in the FEMIC analogue of the w_1 equation from (30) are influenced by the Dirichlet boundary value $u^n(x_0) = g(t^n)$, but not by $u^{n+1}(x_0) = g(t^{n+1})$. In (30), ELLAM includes information for all times on this boundary. This is another major difference, since the boundary treatment is one of the principal purposes of ELLAM. Thus, FEMIC and ELLAM are superficially similar, but in substance very different.

The SD scheme, analyzed over the last several years by Johnson and coworkers ([17] and references therein), is a very effective procedure for hyperbolic conservation laws that apparently adds the "right" artificial diffusion: enough to assure convergence to the correct physical solution while minimizing loss of accuracy due to smearing. In an analogue of (13), SD leads to the spatial term $h^{r+1/2}$ when the solution is smooth, compared to the MMOC results of h^{r+1} when D is bounded below and h^r when it is not. ELLAM, as formulated here, has no artificial diffusion added; the theory suggests that incorporating the SD term would be helpful when the Peclet number is very large or infinite. On the other hand, if D is significant, the artificial term appears to be a disadvantage. Hence, there should be a threshold at which the SD term could be switched on or off. Johnson, using space-time elements not aligned with characteristics, has recommended and theoretically substantiated $D = O(h)$ as a threshold. However, (38) shows a front width of $O(\sqrt{D})$, and ELM such as MMOC have converged optimally with this as noted in §2. If a front reaches a boundary layer, it sharpens to $O(D)$. This hints that, if space-time elements are aligned with characteristics, a sharp threshold would be $D = O(h)$ in a boundary layer, and $D = O(h^2)$ elsewhere. Theoretical verification of this awaits further research.

After a first draft of this paper was written, we became aware of some related investigations in France [2]. In this work, space-time test functions satisfying the adjoint equation $w_t + v w_x = 0$, as in (18), were used. This led to a development similar to ours; indeed, it appears that this is how the French came to study ELM [21], and they applied their concepts in more complicated settings, such as Navier-Stokes flows, than we have done here. Our main objective here is generalization of ELM to treat boundary conditions and conserve mass, so we comment briefly on the French contributions in those directions. Boundary conditions were treated by assuming that diffusive influx was negligible, extending the space-time domain outside the boundary, and replacing boundary integrals with integrals outside the domain. Diffusion operated over the full Δt everywhere, as in the *ad hoc* approaches mentioned in §5. The reported numerical results did not involve significant boundary behavior. For nonzero diffusion, this boundary treatment causes mass-conservation errors; it was noted in [2] that mass is exactly conserved in the case of pure advection. Thus, the combination of characteristics with finite elements via the local adjoint equation is not new to the ELLAM work, and the notion that this helps to conserve mass is likewise not new. The systematic boundary formulation, and resulting general conservative property, do appear to be new.

12. SUMMARY AND CONCLUSIONS

The general ideas of Eulerian-Lagrangian localized adjoint methods (ELLAM) have been illustrated here for the simplest possible advection-diffusion problem. Using space-time finite elements, the formulation extends Eulerian-Lagrangian methods (ELM) such as the modified method of characteristics (MMOC) in a manner that conserves mass and treats boundary conditions systematically and accurately, without *ad hoc* assumptions. Discretization of the outlet boundary allows a single formulation to handle the advection-diffusion and purely hyperbolic problems, with no need for upwinding or artificial boundary conditions. Numerical results confirm that ELLAM preserves the numerical advantages of ELM conservatively in the context of boundary-value problems. Results on the discretized outlet boundary are accurate for non-Dirichlet conditions, but computed fluxes for Dirichlet conditions show some extra peaks and valleys in time as a front passes the outlet with coarse time steps.

ACKNOWLEDGMENTS

This research was supported in part by the National Science Foundation under grant DMS-8821330. It is hoped that the invaluable contributions of the other members of the ELLAM group (M. A. Celia, R. E. Ewing, I. Herrera) have been adequately cited in the text. We are grateful to R. Glowinski for informing us of [2].

References

[1] A. M. Baptista, "Solution of advection-dominated transport by Eulerian-Lagrangian methods using the backwards method of characteristics," Ph. D. Thesis, Massachusetts Institute of Technology, 1987.

[2] J. P. Benqué and J. Ronat, "Quelques difficultés des modèles numériques en hydraulique," *Computing Methods in Applied Sciences and Engineering, V*, R. Glowinski and J. L. Lions, eds., North-Holland, Amsterdam, 1982, pp. 471–494.

[3] A. Brooks and T. J. R. Hughes, "Streamline upwind Petrov-Galerkin formulations for convection dominated flows with particular emphasis on the incompressible Navier-Stokes equations," *Comp. Meth. Appl. Mech. Engng., 32* (1982), 199–259.

[4] M. A. Celia, I. Herrera, and E. T. Bouloutas, "Adjoint Petrov-Galerkin methods for multi-dimensional flow problems," *Proceedings of the Seventh International Conference on Finite Element Methods in Flow Problems*, Huntsville, AL, April 3–7, 1989, pp. 953–958.

[5] M. A. Celia, T. F. Russell, I. Herrera, and R. E. Ewing, "An Eulerian-Lagrangian localized adjoint method for the advection-diffusion equation," in preparation.

[6] C. N. Dawson, T. F. Russell, and M. F. Wheeler, "Some improved error estimates for the modified method of characteristics," *SIAM J. Numer. Anal.*, to appear.

[7] J. Douglas, Jr., T. Dupont, H. H. Rachford, Jr., and M. F. Wheeler, "Local H^{-1} Galerkin and adjoint local H^{-1} Galerkin procedures for elliptic equations," *RAIRO Anal. Numér., 11* (1977), 3–12.

[8] J. Douglas, Jr., I. Martínez-Gamba, and M. C. J. Squeff, "Simulation of the transient behavior of a one-dimensional semiconductor device," *Mat. Aplic. Comp., 5* (1986), 103–122.

[9] J. Douglas, Jr., and T. F. Russell, "Numerical methods for convection-dominated diffusion problems based on combining the method of characteristics with finite element or finite difference procedures," *SIAM J. Numer. Anal., 19* (1982), 871–885.

[10] M. S. Espedal and R. E. Ewing, "Characteristic Petrov-Galerkin subdomain methods for two-phase immiscible flow," *Comp. Meth. Appl. Mech. Engng., 64* (1987), 113–135.

[11] R. E. Ewing and T. F. Russell, "Multistep Galerkin methods along characteristics for convection-diffusion problems," *Advances in Computer Methods for Partial Differential Equations – IV*, R. Vichnevetsky and R. S. Stepleman, eds., IMACS, Rutgers Univ., 1981, pp. 28–36.

[12] R. E. Ewing, T. F. Russell, and M. F. Wheeler, "Simulation of miscible displacement using mixed methods and a modified method of characteristics," SPE 12241, *Proceedings of the Seventh SPE Symposium on Reservoir Simulation*, Society of Petroleum Engineers, Dallas, 1983, pp. 71–81.

[13] R. E. Ewing, T. F. Russell, and M. F. Wheeler, "Convergence analysis of an approximation of miscible displacement in porous media by mixed finite elements and a modified method of characteristics," *Comp. Meth. Appl. Mech. Engng., 47* (1984), 73–92.

[14] I. Herrera, "Unified formulation of numerical methods. I. Green's formulas for operators in discontinuous fields," *Numer. Meth. PDE, 1* (1985), 25–44.

[15] I. Herrera, R. E. Ewing, M. A. Celia, and T. F. Russell, in preparation.

[16] F. M. Holly, Jr., and A. Preissmann, "Accurate calculation of transport in two dimensions," *J. Hydraulics Div., Proc. Amer. Soc. Civil Engrs., 103* (1977), No. HY11, 1259–1277.

[17] C. Johnson and A. Szepessy, "On the convergence of a finite element method for a nonlinear hyperbolic conservation law," *Math. Comp., 49* (1987), 427–444.

[18] K. W. Morton, "Generalized Galerkin methods for hyperbolic problems," *Comp. Meth. Appl. Mech. Engng., 52* (1985), 847–871.

[19] K. W. Morton, A. Priestley, and E. Süli, "Stability of the Lagrange-Galerkin method with non-exact integration," M^2AN, *22* (1988), 625–653.

[20] S. P. Neuman, "An Eulerian-Lagrangian numerical scheme for the dispersion-convection equation using conjugate space-time grids," *J. Comp. Phys., 41* (1981), 270–294.

[21] O. Pironneau, "On the transport-diffusion algorithm and its application to the Navier-Stokes equations," *Numer. Math., 38* (1982), 309–332.

[22] T. F. Russell, "Time stepping along characteristics with incomplete iteration for a Galerkin approximation of miscible displacement in porous media," *SIAM J. Numer. Anal., 22* (1985), 970–1013.

[23] T. F. Russell, M. F. Wheeler, and C. Y. Chiang, "Large-scale simulation of miscible displacement by mixed and characteristic finite element methods," *Mathematical and Computational Methods in Seismic Exploration and Reservoir Modeling*, W. E. Fitzgibbon, ed., SIAM, Philadelphia, 1986, pp. 85–107.

[24] E. Süli, "Convergence and nonlinear stability of the Lagrange-Galerkin method for the Navier-Stokes equations," *Numer. Math., 53* (1988), 459–483.

[25] E. Varoḡlu and W. D. L. Finn, "Finite elements incorporating characteristics for one-dimensional diffusion-convection equation," *J. Comp. Phys., 34* (1980), 371–389.

Thomas F. Russell
Computational Mathematics Group
Department of Mathematics
University of Colorado at Denver
1200 Larimer Street, Campus Box 170
Denver, Colorado 80204-5300
U. S. A.

228

D.C. SORENSON

Numerical Aspects of Divide and Conquer Methods for Eigenvalue Problems

Abstract

This paper discusses some numerical aspects of the divide and conquer technique for the computation of the eigensystem of a symmetric matrix. A brief analysis of the numerical properties and sensitivity to round off error is presented and a remedy is proposed for avoiding numerical difficulties associated with computing orthogonal eigenvectors when eigenvalues are clustered. We show how to explicitly overlap the initial reduction to tridiagonal form with the parallel computation of the eigensystem of the tridiagonal matrix. The algorithm is therefore able to exploit parallelism at all levels of the computation and is well suited to a variety of architectures.

1. Introduction

The symmetric eigenvalue problem is one of the most fundamental problems of computational mathematics. In [3] a new parallel algorithm was presented for the symmetric tridiagonal eigenvalue problem. A surprising result of the work in [3] was that the parallel algorithm developed there, even when run in serial mode, is significantly faster than the previously best sequential algorithm on large problems, and is effective on moderate size (order ≥ 30) problems when run in serial mode.

In [3] the difficulty of computing numerically orthogonal eigenvectors was discussed but not fully resolved. Recent progress has been made to overcome this numerical problem. This involves two approaches. One is to use extended precision arithmetic. The other is to recast the evaluation of the secular function to avoid round off difficulties. These results will be presented in greater detail in a joint paper with P. Tang [9] that is in preparation. In this paper the second approach is briefly discussed.

The problem we consider is the following: Given a real $n \times n$ symmetric matrix A, find all of the eigenvalues and corresponding eigenvectors of A. It is well known [8] that under these assumptions

$$A = QDQ^T , \quad \text{with } Q^TQ = I , \tag{1.1}$$

so that the columns of the matrix Q are the orthonormal eigenvectors of A and

Work supported in part by the Applied Mathematical Sciences subprogram of the Office of Energy Research, U.S. Department of Energy under Contracts W-31-109-Eng-38, DE-AC05-840R21400 and DE-FG02-85ER25001.

229

$D = diag\,(\delta_1, \delta_2, ..., \delta_n)$ is the diagonal matrix of eigenvalues. The standard algorithm for computing this decomposition is first to use a finite algorithm to reduce A to tridiagonal form using a sequence of Householder transformations, and then to apply a version of the QR-algorithm to obtain all the eigenvalues and eigenvectors of the tridiagonal matrix[8].

Once this tridiagonal form has been obtained the algorithm described in[3] may be used to find the eigenvalues and eigenvectors in parallel. The method given in [3] is based upon a divide and conquer algorithm suggested by Cuppen[2]. A fundamental tool used to implement this algorithm is a method that was developed by Bunch, Nielsen, and Sorensen[1] for updating the eigensystem of a symmetric matrix after modification by a rank one change. This rank-one updating method was inspired by some earlier work of Golub[5] on modified eigenvalue problems. The basic idea of the new method is to use rank-one modifications to tear out selected off-diagonal elements of the tridiagonal problem in order to introduce a number of independent subproblems of smaller size. The subproblems are solved at the lowest level using the subroutine TQL2 from EISPACK [7] and then results of these problems are successively glued together using the rank-one modification routine SESUPD that we have developed based upon the ideas presented in [1].

For the sake of completeness we briefly describe the tridiagonal algorithm but refer the reader to [3] for details. First we review the partitioning of the tridiagonal problem into smaller problems by rank-one tearing. Then we describe the numerical algorithm for gluing the results back together. The new method for calculating solutions to the secular equation is introduced. We present numerical results for this method and then briefly discuss overlapping the tridiagonal computations with the initial reduction to tridiagonal form. Throughout this paper we adhere to the convention that capital Roman letters represent matrices, lower case Roman letters represent column vectors, and lower case Greek letters represent scalars. A superscript T denotes transpose. All matrices and vectors are real, but the results are easily extended to matrices over the complex field.

2. Partitioning by Rank-One Tearing

The crux of the algorithm is to divide a given problem into two smaller subproblems. To do this, we consider the symmetric tridiagonal matrix

$$T = \begin{bmatrix} T_1 & \beta e_k e_1^T \\ \beta e_1 e_k^T & T_2 \end{bmatrix} = \begin{bmatrix} \hat{T}_1 & 0 \\ 0 & \hat{T}_2 \end{bmatrix} + \theta\beta \begin{bmatrix} e_k \\ \theta^{-1}e_1 \end{bmatrix} (e_k^T, \theta^{-1}e_1^T) \qquad (2.1)$$

where $1 \le k \le n$ and e_j represents the $j-th$ unit vector of appropriate dimension. The $k-th$ diagonal element of T_1 has been modified to give $\hat{T}_1$ and the first diagonal element of T_2 has been modified to give $\hat{T}_2$. Potential numerical difficulties associated with cancellation may be avoided through the appropriate choice of θ. If the diagonal entries to be modified are of the same sign then $\theta = \pm 1$ is chosen so that $-\theta\beta$ has this sign and cancellation is avoided. If the two diagonal entries are of opposite sign, then the sign of θ is chosen so that $-\theta\beta$ has the same sign as one of the elements and the magnitude of θ is chosen to avoid severe loss of significant digits when $\theta^{-1}\beta$ is subtracted from the other. This is perhaps a minor detail, but it does allow the partitioning to be selected solely on the basis of position and without regard to numerical considerations.

Now we have two smaller tridiagonal eigenvalue problems to solve. According to equation (1.1) we compute the two eigensystems

$$\hat{T}_1 = Q_1 D_1 Q_1^T , \quad \hat{T}_2 = Q_2 D_2 Q_2^T .$$

This gives

$$T = \begin{bmatrix} Q_1 D_1 Q_1^T & 0 \\ 0 & Q_2 D_2 Q_2^T \end{bmatrix} + \theta\beta \begin{bmatrix} e_k \\ \theta^{-1} e_1 \end{bmatrix} (e_k^T , \theta^{-1} e_1^T) \tag{2.2}$$

$$= \begin{bmatrix} Q_1 & 0 \\ 0 & Q_2 \end{bmatrix} \left(\begin{bmatrix} D_1 & 0 \\ 0 & D_2 \end{bmatrix} + \theta\beta \begin{bmatrix} q_1 \\ \theta^{-1} q_2 \end{bmatrix} (q_1^T , \theta^{-1} q_2^T) \right) \begin{bmatrix} Q_1^T & 0 \\ 0 & Q_2^T \end{bmatrix}$$

where $q_1 = Q_1^T e_k$ and $q_2 = Q_2^T e_1$. The problem at hand now is to compute the eigensystem of the interior matrix in equation (2.2). A numerical method for solving this problem has been provided in [1] and we shall discuss this method in the next section.

It should be fairly obvious how to proceed from here to exploit parallelism. One simply repeats the tearing on each of the two halves recursively until the original problem has been divided into the desired number of subproblems and then the rank one modification routine may be applied from bottom up to glue the results together again.

3. The Updating Problem

The general problem we are required to solve is that of computing the eigensystem of a matrix of the form

$$\hat{Q}\hat{D}\hat{Q}^T = D + \rho z z^T \tag{3.1}$$

where D is a real $n \times n$ diagonal matrix, ρ is a nonzero scalar, and z is a real vector of order n. It is assumed without loss of generality that z has Euclidean norm 1.

We seek a formula for an eigen-pair for the matrix on the right hand side of (3.1). In [3] such a formula is derived under the assumptions that $D = diag(\delta_1, \delta_2, \cdots, \delta_n)$ with $\delta_1 < \delta_2 < \cdots < \delta_n$, and that no component ζ_i of the vector z is zero. In Section 4 we discuss how this may always be arranged. We just note here that if λ is a root of the equation

$$1 + \rho z^T (D - \lambda I)^{-1} z = 0 \tag{3.2}$$

and

$$q = \theta(D - \lambda I)^{-1} z \tag{3.3}$$

231

then q ,λ is such an eigen-pair, i.e they satisfy the relation

$$(D + \rho z z^T)q = \lambda q.$$

In (3.3) the scalar θ may be chosen so that $\|q\| = 1$ to obtain an orthonormal eigensystem.

If we write equation (3.2) in terms of the components ζ_i of z then λ must be a root of the equation

$$f(\lambda) \equiv 1 + \rho \sum_{j=1}^{n} \frac{\zeta_j^2}{\delta_j - \lambda} = 0. \tag{3.4}$$

Under our assumptions this equation has precisely n roots, one in each of the open intervals (δ_j, δ_{j+1}) $j = 1,2,...n-1$ and one to the right of δ_n if $\rho > 0$ or one to the left of δ_1 if $\rho < 0$. We construct the eigenvectors corresponding to each of these roots using formula (3.3).

An excellent numerical method was developed in [1] which took advantage of this structure, to find the roots of the secular equation and as a by-product to compute the eigenvectors to full accuracy. However, even though this method is satisfactory for most problems, it still may fail when there are extremely difficult situations in the secular equation. Difficult problems arize when two adjacent roots δ_j and δ_{j+1} are close and the corresponding weights ζ_j^2 and ζ_{j+1}^2 are small and of the the same order of magnitude. This situation defeats the deflation process described in the following section.

In an iterative method for computing a root λ, we require rapid convergence as well as reliable calculation of f in order to satisfy the stringent stopping criterion

(i) $|f(\lambda)| \le \eta \max(|\delta_1|, |\delta_2|)$ and (ii) $|\tau| \le \eta \min(|\delta_i - \lambda|, |\delta_i - \lambda|)$,

where λ is the current iterate and τ is the last iterative correction that was computed. The condition on τ is very stringent. The purpose of such a stringent stopping criteria will be clarified in the next section when we discuss orthogonality of computed eigenvectors. Let it suffice at this point to say that condition (i) assures a small residual and that (ii) assures orthogonal eigenvectors.

This has been a brief description of the rank-one updating scheme. Full theoretical details are available in [1,3].

4. Deflation and Orthogonality of Eigenvectors

At the outset of this discussion we made the assumption that the diagonal elements of D were distinct and that no component of the vector z was zero. These conditions are not satisfied in general, so deflation techniques must be employed to ensure their satisfaction. This deflation can occur in two ways in exact arithmetic, either through zero components of z or through multiple eigenvalues. In order to obtain an algorithm suitable for finite precision arithmetic, we must refine these notions to include "nearly zero" components of z and "nearly equal" eigenvalues. Analysis of the first situation is straightforward. The second situation can be quite delicate in certain pathological cases, however, so we shall discuss it here in some detail. Deflation can have a very dramatic effect upon the amount of work required in our parallel method. As first observed by Cuppen[2], there can be significant deflation in the updating process as the original matrix is rebuilt from the sub-problems.

It is straightforward to see that when $z^T e_i = 0$ the i-th eigenvalue and corresponding eigenvector of D will be an eigen-pair for $D + \rho z z^T$. Let us ask the question "when is an eigen-pair of D a good approximation to an eigen-pair for the modified matrix?". This question is easily answered. Recall that $\|z\| = 1$, so

$$\|(D + \rho z z^T)e_i - \delta_i e_i\| = |\rho \zeta_i| \|z\| = |\rho \zeta_i|.$$

Thus we may accept this eigen-pair as an eigen-pair for the modified matrix whenever

$$|\rho \zeta_i| \le tol$$

where $tol > 0$ is our error tolerance. Typically $tol = macheps \, (\eta \|A\|)$, where $macheps$ is machine precision and η is a constant of order unity. However, at each stage of the updating process we may use

$$tol = macheps \, \eta \, (\max(|\delta_1|, |\delta_n|) + |\rho|)$$

since at every stage this will represent a bound on the spectral radius of a principal subma-trix of A. Let us now suppose there are two eigenvalues of D separated by ε so that $\varepsilon = \delta_{i+1} - \delta_i$. Consider the 2×2 submatrix

$$\begin{bmatrix} \delta_i & \\ & \delta_{i+1} \end{bmatrix} + \rho \begin{bmatrix} \zeta_i \\ \zeta_{i+1} \end{bmatrix} (\zeta_i \, , \, \zeta_{i+1})$$

of $D + \rho z z^T$ and let us construct a Givens transformation to introduce a zero in one of the two corresponding components in z. Then

$$\begin{bmatrix} \gamma & \sigma \\ -\sigma & \gamma \end{bmatrix} \left(\begin{bmatrix} \delta_i & \\ & \delta_{i+1} \end{bmatrix} + \rho \begin{bmatrix} \zeta_i \\ \zeta_{i+1} \end{bmatrix} (\zeta_i \, , \, \zeta_{i+1}) \right) \begin{bmatrix} \gamma & -\sigma \\ \sigma & \gamma \end{bmatrix}$$

$$= \begin{bmatrix} \hat{\delta}_i & \\ & \hat{\delta}_{i+1} \end{bmatrix} + \rho \begin{bmatrix} \tau \\ 0 \end{bmatrix} (\tau \, , \, 0) + \varepsilon \sigma \gamma \begin{bmatrix} 0 & 1 \\ 1 & 0 \end{bmatrix}$$

where $\gamma^2 + \sigma^2 = 1$, $\hat{\delta}_i = \delta_i \gamma^2 + \delta_{i+1} \sigma^2$, $\hat{\delta}_{i+1} = \delta_i \sigma^2 + \delta_{i+1} \gamma^2$ and $\tau^2 = \zeta_i^2 + \zeta_{i+1}^2$. Now, if we put G_i equal to the $n \times n$ Givens transformation constructed from the identity by replac-ing the appropriate diagonal block with the 2×2 rotation just constructed, we have

$$G_i (D + \rho z z^T) G_i^T = \hat{D} + \rho \hat{z} \hat{z}^T + E_i \qquad (4.1)$$

with $e_i^T \hat{z} = \tau$ and $e_{i+1}^T \hat{z} = 0$, $e_i^T \hat{D} e_i = \hat{\delta}$, $e_{i+1}^T \hat{D} e_{i+1} = \hat{\delta}$, and

$$\| E_i \| = |\varepsilon \sigma \gamma|.$$

If we choose, instead, to zero out the i-th component of z then we simply apply G_i on the right and its transpose on the left in equation(4.1) to obtain the desired similar result. The only exception to the previous result is that the sign of the matrix E_i is reversed. Of course this deflation is only done when

$$|\epsilon\gamma\sigma| \leq tol$$

is satisfied.

The result of applying all of the deflations is to replace the updating problem (3.1) with one of smaller size. When appropriate, this is accomplished by applying similarity transformations consisting of several Givens transformations. If G represents the product of these transformations the result is

$$G(D + \rho zz^T)G^T = \begin{bmatrix} D_1 - \rho z_1 z_1^T & 0 \\ 0 & D_2 \end{bmatrix} + E \, ,$$

where

$$\| E \| \leq tol \, \eta_2$$

with η_2 of order unity. The cumulative effect of such errors is additive and thus the final computed eigensystem $\hat{Q}\hat{D}\hat{Q}^T$ satisfies

$$\| A - \hat{Q}\hat{D}\hat{Q}^T \| \leq \eta_3 tol$$

where η_3 is again of order 1 in magnitude. The reduction in size of $D_1 - \rho z_1 z_1^T$ over the original rank 1 modification can be spectacular in certain cases. The effects of such deflation can be dramatic, for the amount of computation required to perform the updating is greatly reduced.

Let us now consider the possible limitations on orthogonality of eigenvectors due to nearly equal roots. Our first result (proved in [3]) is a perturbation lemma that will indicate the inherent difficulty associated with nearly equal roots.

LEMMA (4.2) Let

$$q_\lambda^T \equiv (\frac{\zeta_1}{\delta_1 - \lambda}, \frac{\zeta_2}{\delta_2 - \lambda}, \cdots \frac{\zeta_n}{\delta_n - \lambda}) \left[\frac{\rho}{f'(\lambda)} \right]^{1\!/\!2}, \tag{4.3}$$

where f is defined by formula (3.4). Then for any $\lambda, \mu \notin \{\delta_i : i = 1,...,n\}$

$$|q_\lambda^T q_\mu| = \frac{1}{|\lambda - \mu|} \frac{|f(\lambda) - f(\mu)|}{\left[f'(\lambda)f'(\mu) \right]^{1\!/\!2}}. \tag{4.4}$$

Note that in (4.3) q_λ is always a vector of unit length, and the set of n vectors selected by setting λ equal to the roots of the secular equation is the set of eigenvectors for $D + \rho zz^T$. Moreover, equation (4.4) shows that those eigenvectors are mutually

234

orthogonal whenever λ and μ are set to distinct roots of f. Finally, the term $|\lambda - \mu|$ appearing in the denominator of (4.4) sends up a warning that it may be difficult to attain orthogonal eigenvectors when the roots λ and μ are close. We wish to examine this situation now. In [3] it was shown that, as a result of the deflation process, the $\{\delta_i\}$ are sufficiently separated, and the weights ζ_i are uniformly large enough that the roots of f are bounded away from each other. Because of this, we can expect to be able to compute the differences $\delta_j - \lambda$ to high relative accuracy. This is quite important with regards to orthogonality of the computed eigenvectors as the following lemma shows.

LEMMA (4.5) *Suppose that* $\hat{\lambda}$ *and* $\hat{\mu}$ *are numerical approximations to exact roots* λ *and* μ *of* f. *Assume that these roots are distinct and let the relative errors for the quantities* $\delta_i - \lambda$ *and* $\delta_i - \mu$ *be denoted by* θ_i *and* η_i *respectively. That is, the computed quantities*

$$\delta_i - \hat{\lambda} = (\delta_i - \lambda)(1 + \theta_i) \quad \text{and} \quad \delta_i - \hat{\mu} = (\delta_i - \mu)(1 + \eta_i), \tag{4.6}$$

for $i = 1,2,...,n$. *Let* $q_{\hat{\lambda}}$ *and* $q_{\hat{\mu}}$ *be defined according to formula (4.3) using the computed quantities given in (4.6). If* $|\theta_i|, |\eta_i| \le \varepsilon \ll 1$, *then*

$$|q_{\hat{\lambda}}^T q_{\hat{\mu}}| = |q_{\lambda}^T E q_{\mu}| \le \varepsilon(2+\varepsilon) \left[\frac{1+\varepsilon}{1-\varepsilon} \right]^2,$$

with E *a diagonal matrix whose* i-*th diagonal element is*

$$E_{ii} = \frac{\theta_i + \eta_i + \theta_i \eta_i}{(1 + \theta_i)(1 + \eta_i)} \left[\frac{f'(\lambda)f'(\mu)}{f'(\hat{\lambda})f'(\hat{\mu})} \right]^{1/2}.$$

A proof of this result is given in [3]. It shows that orthogonality can be assured whenever it is possible to provide small relative errors when computing the differences $\delta_j - \lambda$. Since, as we mentioned in Section 3, these quantities are updated with the iterative corrections to λ and since deflation has guaranteed that the differences $\delta_j - \lambda$ cannot be arbitrarily small it will be possible in theory to provide small relative errors.

5. Reformulating the Evaluation of f

In this section we reformulate the evaluation of the function f using the perturbation result of Lemma (4.2) above. The purpose of this reformulation is to avoid the problems of trying to compute an exceptionally accurate f. We make use of the fact derived in the proof of Lemma (4.2) that for any λ, $\mu \notin \{\delta_j : 1 \le j \le n\}$

$$f(\lambda) = f(\mu) + (\lambda - \mu)(f'(\lambda)f'(\mu))^{1/2} q_{\lambda}^T q_{\mu} \tag{5.1}$$

where q_{λ} is given by (4.3). Note that if the eigenvalue $\lambda_{i-1} \in (\delta_{i-1}, \delta_i)$ is known, then we can find the root λ_i in the interval (δ_i, δ_{i+1}) by solving $h(\lambda) = 0$ where

$$h(\lambda) \equiv q_{\lambda}^T q_{\mu}, \quad \text{with } \mu = \lambda_{i-1}. \tag{5.2}$$

The main point here is that we do not need to calculate the difference $\lambda - \mu$ that appears in (5.1). It is available implicitly. This means that the function f may be evaluated accurately to a resolution of

$$f(\lambda) = (\lambda - \mu)(f'(\lambda)f'(\mu))^{1/2}O\,(macheps)$$

This is quite important with respect to obtaining numerical orthogonality of the computed eigenvectors. In [1,3] this is accomplished by adding the iterative corrections τ directly to the differences $\Delta_j \equiv \delta_j - \lambda$ and then the components of the corresponding eigenvector are ζ_j / Δ_j. Since the iterative corrections are added directly to these differences, the stringent requirement of Lemma (4.5) can be met because as convergence takes place the iterative corrections τ are able to iteratively correct the inaccurate trailing digits in the differences Δ_j. However, this is only possible if f can be evaluated accurately enough to assure correct leading digits in the iterative correction τ.

6. Iterations Based on Rational Functions

In this sections we discuss some special properties of rational functions that are naturally related to the reformulation of the evaluation of f developed in Section 5. The function h in (5.2) lends itself to some interesting numerical schemes due to its special structure. The following lemma applies when λ and μ are in adjacent intervals as specified in (5.2).

LEMMA (6.1) Suppose

$$\Delta_1 < \Delta_2 < \cdots < \Delta_i \leq 0 < \Delta_{i+1} < \cdots < \Delta_n$$

and let

$$\phi(\tau) \equiv \sum_{j \neq i} \frac{\eta_j}{\Delta_j - \tau}$$

where $\eta_j \Delta_j > 0$ *for* $j \neq i$, $\eta_i > 0$. *Then*

 i) $\phi(\tau)$ *is convex on* (Δ_i, Δ_{i+1}) ,

 ii) $\dfrac{1}{\phi(\tau)}$ *is concave on* (Δ_i, Δ_{i+1})

 iii) Newton's method applied to finding a root of

$$\frac{\eta_i}{\phi(\tau)} + (\Delta_i - \tau) = 0$$

 converges monotonically and quadratically from any initial $\tau^o \in (\tau^*, \Delta_{i+1})$
 where τ^* *satisfies*

$$\frac{\eta_i}{\phi(\tau^*)} + (\Delta_i - \tau^*) = 0.$$

This Lemma is directly applicable to h as specified in (5.2) because the hypothesis are satisfied. However, even though Lemma (6.1) is naturally of interest, our numerical experience would indicate that the monotonicity is not obtained in the face of round of due to finite precision arithmetic and this can lead to difficulty in finding roots in the proper intervals when there are clusters. Newton's method will ultimately require safeguarding and this will prevent the rapid convergence that will eventually be required. There is a method that will serve our needs however. Instead of dealing with (5.2) directly, we consider the equivalent problem of solving

$$-\frac{\eta_i}{(\Delta_i - \tau)} = \phi(\tau).$$

The method is based upon approximating $\phi(\tau)$ near $\tau = 0$ by

$$\hat{\phi}(\tau) \equiv \frac{\Delta^2 \phi'(0)}{\Delta - \tau} + \phi(0) - \Delta \phi'(0) \approx \phi(\tau)$$

Note that the interpolation conditions

$$\phi(0) = \hat{\phi}(0) \ , \ \phi'(0) = \hat{\phi}'(0)$$

hold and that $\hat{\phi}(\tau)$ has a pole at Δ . We take

$$\Delta = \begin{cases} \Delta_{i+1} \text{ if } \phi'(0) \le 0 \\ \Delta_{i-1} \text{ if } \phi'(0) > 0 \end{cases}$$

Then the approximation to the solution τ^* is computed by solving

$$\frac{\eta_i}{(\Delta_i - \tau)} + \frac{\Delta^2 \phi'(0)}{\Delta - \tau} + \phi(0) - \Delta \phi'(0) = 0. \tag{6.2}$$

This equation may of course be solved via computing the solution of a quadratic equation. It can be shown that the resulting iteration converges globally and quadratically.

Providing initial approximations which provide upper and lower bounds on the root $\tau^* \in (\Delta_i, \Delta_{i+1})$ is a simple but important matter. Note that

$$f(\lambda) = \frac{\zeta_i^2}{\delta_i - \lambda} + \frac{\zeta_{i+1}^2}{\delta_{i+1} - \lambda} + \psi(\lambda)$$

where

$$\psi(\lambda) \equiv 1 + \sum_{j \ne i, i+1} \frac{\zeta_j^2}{\delta_j - \lambda}.$$

Now, $\psi'(\lambda) > 0$ for λ in (δ_i, δ_{i+1}) and thus

$$\psi(\delta_i) < \psi(\lambda) < \psi(\delta_{i+1}).$$

If $f(\lambda^*) = 0$ then

$$-\psi(\delta_{i+1}) < \frac{\zeta_i^2}{\delta_i} - \lambda^* + \frac{\zeta_{i+1}^2}{\delta_{i+1}} - \lambda^* < -\psi(\delta_i).$$

Solving

$$\frac{\zeta_i^2}{(\delta_i - \lambda)} + \frac{\zeta_{i+1}^2}{(\delta_{i+1} - \lambda)} = -\psi(\delta_j)$$

for the root in (δ_i, δ_{i+1}) gives a lower bound λ^l when $j = i + 1$ and an upper bound λ^u when $j = i$ so that

$$\delta_i < \lambda^l < \lambda^* < \lambda^u < \delta_{i+1}.$$

These bounds have been used in the original code developed by Sorensen to implement the algorithm developed in [1].

These approximations can be used to construct an iterative algorithm to find zeros of f.

Algorithm (6.3)

1) Given the root μ in (δ_{i-1}, δ_i) construct an initial guess λ to the root in $(\delta_i, \delta_i + 1)$;

2) Put $\Delta_j = \delta_j - \lambda$ and put $\eta_j = \dfrac{\zeta_j^2}{(\delta_j - \mu)}$ for $j = 1, n$

3) for $k = 1, 2, \ldots$ until "convergence"

    3.1) Solve (6.2) for τ;

    3.2) $\Delta_j \leftarrow \Delta_j - \tau$ for $j = 1, n$;

    3.3) $\lambda \leq \lambda + \tau$.

This iteration is quite successful as will be shown in the following section. However, it is quite serial in nature due to the requirement of knowing the eigenvalue in the previous interval in order to compute one in the current interval. However, it is straightforward to obtain simple a-priori bounds on the distance between adjacent λ_j during the deflation stage which may be used to partition the root finding problems. Unfortunately, such a partitioning may not lead to a suitable load balance for parallel processing and this is a subject for further research.

7. Some Numerical Results

In this section the results of a numerical experiment will be discussed which supports the adoption of the iteration outlined in the previous section. Performance of the algorithm has been reported elsewhere [3]. Therefore, attention will be restricted to a single

238

pathological example that has caused the technique developed in [3] to fail. The example is

$$A = \begin{bmatrix} W_{21} & E^T & . & & 0 \\ E & W_{21} & . & & . \\ 0 & E & . & & . \\ . & & . & . & E^T \\ 0 & 0 & & E & W_{21} \end{bmatrix}$$

where W_{21} is the symmetric tridiagonal matrix of order 21 with diagonal elements $(10, 9, \cdots, 1, 0, 1, 2, \cdots, 10)$ and sub-diagonal (super-diagonal) elements equal to 1. The matrix E is of the form

$$E = \begin{bmatrix} 0 & \varepsilon \\ 0 & 0 \end{bmatrix}$$

where epsilon is usually chosen to be a small number. The matrix W_{21} is an example devised by Wilkinson to illustrate difficulties algorithms might have with nearly equal eigenvalues. The matrix has pairs of extremely close eigenvalues. The matrix A can be made arbitrarily large of order $k*21$ by adjoining k copies of W_{21}. The resulting matrix will have eigenvalues in k clusters. Eigenvalues within a cluster will be separated by no more than $O(\varepsilon)$.

A sample sub-problem of order 48 (after deflation) that arose in the divide and conquer step had the following weights and poles, near the value 9.201 :

poles	$\mid \zeta_i \mid$
9.210678644276706	$0.72(10^{-5})$
9.210678644392496	$0.57(10^{-6})$
9.210678644408787	$0.77(10^{-6})$
9.210678644420730	$0.54(10^{-6})$
9.210678647333125	$0.93(10^{-9})$
9.210678650133307	$0.69(10^{-5})$
9.210678650243938	$0.58(10^{-6})$
9.210678650260574	$0.77(10^{-6})$
9.210678650272200	$0.54(10^{-6})$

The new method (called TREEQL) was tested on the above example for $k = 2, 3, \cdots, 7$. with two values of $\varepsilon = 10^{-8}$ and $\varepsilon = 10^{-14}$. Previously, the method developed in [1,3] would fail to give full orthogonality of the eigenvectors on some of these examples. However, the new method was able to resolve all of these extremely difficult problems to full accuracy providing a matrix of eigenvectors Q and diagonal matrix of eigenvalues Λ such that

$$\|AQ - Q\Lambda\|_F < \eta\|A\|_F \; , \quad \|Q^T Q - I\|_F < \eta$$

where η is of the order of *macheps* (i.e η ranged between 10^{-14} and 10^{-15} in 64 bit arithmetic). These results were compatible with the accuracy of TQL2 from EISPACK. Interestingly enough, both algorithms were run in serial mode with TREEQL executing 3 - 13 times faster than TQL2. The computations were done on a SUN4 workstation in Fortran using double precision.

8. Reduction to Tridiagonal Form and Related Issues

This algorithm is designed to work in conjunction with Householder's reduction of a symmetric matrix to tridiagonal form. In this standard technique a sequence of $n-1$ Householder transformations is applied to form

$$A_{k+1} = (I - \alpha_k w_k w_k^T)A_k(I - \alpha_k w_k w_k^T) \; ,$$

with

$$A_k = \begin{bmatrix} T_k & 0 \\ 0 & \hat{A}_k \end{bmatrix},$$

where T_k is a tridiagonal matrix of order $k-1$. See [7,8,10] for details. We note here that we can form $(I - \alpha ww^T)A(I - \alpha ww^T)$ using the following algorithm.

Algorithm 8.1
1. $v = Aw$
2. $y^T = v^T - \alpha(w^T v)w^T$
3. Replace A by $A - \alpha vw^T - \alpha wy^T$.

Steps 1 and 3 may all be parallelized because A may be partitioned into blocks of columns $A = (A_1, A_2, \cdots, A_j)$ and each of the contributions corresponding to these columns may be carried out independently in steps 1 and 3. For example in step 3,

$$A_i \leftarrow A_i - \alpha v w_i^T - \alpha w y_i^T$$

where the vectors w and y have been partitioned in a corresponding manner. In step 2 the partial results will have to be stored in temporary locations until all are completed and then they may be added together to obtain the final result. Note also that when the calculations are arranged this way advantage may be taken of vector operations when they are available.

Algorithm 8.1 has some disadvantages when incorporated into the reduction of a symmetric matrix to tridiagonal form. At the k-th stage of the reduction we have

$$\begin{bmatrix} T^{(k)} & \beta e_k e_1^T \\ \beta e_1 e_k^T & A^{(k)} \end{bmatrix}$$

The reduction is advanced one step through the application of Algorithm 5.1 to the submatrix $A^{(k)}$. First, a fork-join synchronization construct is imposed since the matrix-vector product requires the entire matrix $A^{(k)}$ to be in place before this product can be completed, so that no portion of Step 2 may begin until Step 1 is finished. Second, due to symmetry, only the lower triangle of $A^{(k)}$ need be computed and this implies that vector lengths shorten during the computation.

When used in conjunction with the rank one tearing scheme, these drawbacks may be overcome. Suppose the final result of this decomposition is T and that this matrix would be partitioned into ($T_1, T_2, \ldots, T_m$) by the rank one tearing if it were known. It is not necessary to wait until the entire reduction is completed, for $T^{(k)}$ represents a leading principal submatrix of T. Thus, as soon as the first sub-tridiagonal matrix T_1 is exposed the process of computing its eigensystem may be initiated. Similarly, as soon as T_2 is exposed its eigensystem may be computed. Then a rank one update may occur, and so on. In this way a number of independent processes may be spawned early on and the number of such processes ready to execute will remain above a reasonable height throughout the course of the computation. This technique has been discussed in detail in [4] and will not be repeated here.

9. Conclusions and Remarks

We have presented a new method for solving the secular equation in the divide and conquer update step. This new technique is able to resolve pathological cases that were not possible before. Another possibility is to use simulated extended precision arithmetic. This is being developed along with the technique presented here and will be presented in detail in [9]. The author wishes to acknowledge that most of this work was done while he was a member of the Mathematics and Computer Science Division of Argonne National Labororatory. Also, the author wishes to acknowledge the support of the IBM Scientific Center in Bergen, Norway where these ideas were developed during a delightful summer visit.

10. References

[1] J.R. Bunch, C.P. Nielsen, and D.C. Sorensen, *Rank-One Modification of the Symmetric Eigenproblem,* Numerische Mathematik 31, pp. 31-48, 1978.

[2] J.J.M. Cuppen *A Divide and Conquer Method for the Symmetric Tridiagonal Eigenproblem,* Numerische Mathematik 36, pp. 177-195, 1981.

[3] J.J. Dongarra and D.C. Sorensen, *A Fully Parallel Algorithm for the Symmetric Eigenproblem,* SIAM J. Sci. Stat. Comput. 8, pp.139-154, 1987.

[4] J.J. Dongarra and D.C. Sorensen, *On the Implementation of A Fully Parallel Algorithm for the Symmetric Eigenprobl em,* SPIE Vol 696 , Advanced Algorithms and Architectures for Signal Processing, pp. 45-53 (1986).

[5] G.H. Golub, *Some Modified Matrix Eigenvalue Problems,* SIAM Review, 15, pp. 318-334, 1973.

[6] I.C.F. Ipsen and E.R. Jessup, *Solving the Symmetric Tridiagonal Eigenvalue Problem on the Hypercube,* Research Report 548, Department of Computer Science, Yale University, 1987, (To Appear SIAM J. Sci. Stat. Comput.).

[7] B.T. Smith, J.M. Boyle, J.J. Dongarra, B.S. Garbow, Y. Ikebe, V.C. Klema, and C.B. Moler, *Matrix Eigensystem Routines - EISPACK Guide,* Lecture Notes in Computer Science, Vol. 6, 2nd edition, Springer-Verlag, Berlin, 1976.

[8] G.W. Stewart, *Introduction to Matrix Computations,* Academic Press, New York, 1973.

[9] D.C. Sorensen and P. Tang, *On the Orthogonality of Eigenvectors Computed by a Divide and Conquer Method,* (In Preparation).

[10] J.H. Wilkinson, *The Algebraic Eigenvalue Problem,* Clarendon Press, Oxford, 1965.

D. C. Sorensen
Department of Mathematical Sciences
Rice University
P.O. Box 1829
Houston, Texas 77251-1829
USA

A. SPENCE, P.J. ASTON and W. WU

Bifurcation and Stability Analysis in Nonlinear Equations Using Symmetry Breaking in Extended Systems

1. INTRODUCTION

Consider the parameter dependent nonlinear system

$$\dot{x} + f(x,\underline{\lambda}) = 0 \qquad x \in H, \ \underline{\lambda} \in \mathbf{R}^p \tag{1.1}$$

where $f \in C^3(H \times \mathbf{R}^p, H)$ with H an n-dimensional real Hilbert space and $p=1$ or 2. Usually the first step in the analysis is to find out some information about the steady state solutions of (1.1), that is, the solutions of

$$f(x,\underline{\lambda}) = 0, \qquad x \in H, \quad \underline{\lambda} \in \mathbf{R}^p. \tag{1.2}$$

When $p=1$ then solution "paths" in $H \times \mathbf{R}$ can often be approximated using numerical continuation codes of which there are now several, for example, ALCON [4], AUTO [5] and PITCON [16]. When $p=2$ solution "surfaces" in $H \times \mathbf{R}^2$ can be approximated by repeated computation of one dimensional "slices", though this might prove expensive. Much depends on the actual requirements of the application and in many cases alternative strategies exist; for example, one might compute paths of singular points (see [10], [11] and [16]).

Often the next step in the analysis of (1.1) is to determine some stability information about the steady states and, in particular, to discover when the system loses stability as one of the parameters varies. Commonly this will be decided from information about the spectrum of the $n \times n$ Jacobian matrix of $f(x,\underline{\lambda})$, given by

$$f_x(x,\underline{\lambda}) := \left[\frac{\partial f_i}{\partial x_j}(x,\underline{\lambda}) \right], \qquad f_x^0 := f_x(x_0,\underline{\lambda}_0),$$

and with eigenvalues denoted by σ_k, $k=1,\ldots,n$. It is well known (see for example, [6]) that if $\mathrm{Re}(\sigma_k) > 0$, $\forall k$, then $(x,\underline{\lambda})$ is a linearly stable steady state solution, but if at least one eigenvalue satisfies $\mathrm{Re}(\sigma_k) < 0$ then $(x,\underline{\lambda})$ is unstable. The two typical situations where a system can lose stability are

a) when a simple real eigenvalue of f_x crosses into the unstable half plane ($z \in \mathbf{C}$, $\mathrm{Re}(z) < 0$), and

b) when a complex conjugate pair of eigenvalues of f_x cross the imaginary axis into the unstable half plane.

We refer to case a) as *steady state bifurcation* and to case b) as *Hopf bifurcation*, and both are well understood, at least under the simplest nondegeneracy assumptions.

In this paper we look at two more complicated situations. The first is discussed in Section 3 and involves the interaction of Hopf and steady state bifurcation at a so called *Takens–Bogdanov point*, which provides an illustration of *mode interaction* ([7], Chapter XIX). The second is discussed in Section 4 and involves bifurcation to rotating waves when f satisfies an $O(2)$ equivariance condition. These two apparently different phenomena are unified by the fact that the important bifurcation results are obtained using steady state bifurcation theory in the presence of a reflectional symmetry (see Section 2), and that both involve a zero eigenvalue of $f_x(x,\lambda)$ which has algebraic multiplicity two but geometric multiplicity one.

This paper summarises results initially presented in [2] and [18] though we do include two extensions to the theory in [18]. The main aim however is to illustrate the unified treatment of the two different cases and the usefulness and power of symmetry-breaking bifurcation theory. This work is motivated by the desire to obtain reliable numerical algorithms and so the emphasis is on a theoretical analysis which produces readily computable quantities to provide dependable tests for the detection of various types of bifurcation phenomena.

Finally we note that though the theory is given for a finite dimensional H, much extends to the infinite dimensional case and these points are discussed in [2] and [18].

2. BIFURCATION IN THE PRESENCE OF A REFLECTIONAL SYMMETRY

In this section we present some results for steady state bifurcation in the presence of reflectional Z_2 symmetry which may be found in [3], [19] and [20] and which will be used in Sections 3 and 4. We present the background theory as a series of theorems, lemmas etc. without proofs, which may be found in the quoted literature. With H an n-dimensional Hilbert space we consider the nonlinear problem

$$f(x,\lambda)=0, \qquad f:=H\times\mathbf{R}\to H, \tag{2.1}$$

where $f\in C^3(H\times\mathbf{R},H)$ satisfies the following reflectional equivariance condition:

There exists an orthogonal $S\in L(H)$ with $S\neq I$, $S^2=I$, and

$$f(Sx,\lambda)=Sf(x,\lambda). \tag{2.2}$$

This is the simplest example of an equivariance condition (see Section 4 for another example and [7] for the general theory) and since the 2 element group $\{I,S\}$ is isomorphic to $Z_2=\{1,-1\}$, we often call (2.2) a Z_2-equivariance condition.

First we note that there is a natural decomposition of H into symmetric and antisymmetric components, $H=H_s\oplus H_a$, where $x\in H_s$ implies that $Sx=x$ and $x\in H_a$ implies that $Sx=-x$. It follows from (2.2) that for $\lambda\in\mathbf{R}$ and $x\in H_s$,

a) $f(x,\lambda), f_\lambda(x,\lambda)\in X_s$,

b) $f_x(x,\lambda)Sv=Sf_x(x,\lambda)v$, which leads to $f_x(x,\lambda):H_s\to H_s$ and $f_x(x,\lambda):H_a\to H_a$

and so we may introduce the restrictions $f_x(x,\lambda)|_{H_s}$ and $f_x(x,\lambda)|_{H_a}$, $\qquad(2.3)$

c) $\forall v,w\in H_a$ or $\forall v,w\in H_s$, $\quad f_{xx}(x,\lambda)vw\in H_s$.

The main assumption in the bifurcation theory is that $(x_0,\lambda_0)\in H\times\mathbf{R}$ is a *simple singular point* of (2.1) with $x_0\in H_s$, that is

$\qquad a)\quad f(x_0,\lambda_0)=0,$

$\qquad b)\quad \mathrm{Null}\,(f_x^0)=\mathrm{span}\,\{\varphi_1\},\qquad\qquad\qquad\qquad\qquad(2.4)$

$\qquad\qquad \mathrm{Null}\,((f_x^0)^*)=\mathrm{span}\,\{\psi_1\},$

where * denotes the adjoint operator.

The first result, which follows from (2.3b) is that either $S\varphi_1=\varphi_1$ or $S\varphi_1=-\varphi_1$. In this paper we shall consider only the second case, that is, the *symmetry–breaking* case with

$$x_0\in H_s,\quad \varphi_1\in H_a.\qquad\qquad(2.5)$$

Since $\varphi_1\notin H_s$, $f_x(x_0,\lambda_0)|_{H_s}$ is nonsingular and the Implicit Function Theorem shows that there exists a smooth path of solutions of (2.1), $(x(\lambda),\lambda)\in H_s\times\mathbf{R}$, with $x(\lambda_0)=x_0$. Other results are that $\psi_1\in H_a$ and

$$\psi_1^T x=0,\qquad \forall x\in H_s,\qquad\qquad(2.6)$$

and so from (2.3a,c)

$$\psi_1^T f_\lambda^0=0,\qquad \psi_1^T f_{xx}^0\varphi_1\varphi_1=0.\qquad\qquad(2.7)$$

The first important result in the bifurcation analysis of (2.1) is the following theorem (see [20]) on the existence of symmetry-breaking bifurcating solutions near (x_0,λ_0).

THEOREM 1. Assume (2.2), (2.4), (2.5) and that

$$b_f:=\psi_1^T(f_{x\lambda}^0\varphi_1+f_{xx}^0\varphi_1 v_0)\neq 0\qquad\qquad(2.8)$$

where

$$f_x^0 v_0+f_\lambda^0=0,\qquad v_0\in X_s.\qquad\qquad(2.9)$$

Then (x_0,λ_0) is a simple Z_2-symmetry-breaking pitchfork bifurcation point of (2.1). ∎

The precise form of the solution paths of (2.1) near (x_0,λ_0) is given by the following Lemma (see [3]):

LEMMA 2. a) The path of symmetric solutions near (x_0,λ_0) has the following local parameterisation:

$$\lambda=\lambda_0+t, \quad x=x_0+tv_0+o(t).\tag{2.10}$$

b) The path which breaks the symmetry, the asymmetric path, has the following local parameterisation:

$$\lambda=\lambda_0+\delta t^2+o(t^2), \quad x=x_0+t\varphi_1+t^2(\delta v_0+\tfrac{1}{2}w_0)+o(t^2)\tag{2.11}$$

where

$$f_x^0 w_0+f_{xx}^0\varphi_1\varphi_1=0, \quad w_0\in H_s,\tag{2.12}$$

$$d_f:=\psi_1^T(f_{xx}^0\varphi_1\varphi_1\varphi_1+3f_{xx}^0\varphi_1 w_0)\tag{2.13}$$

and

$$\delta:=-\frac{d_f}{6b_f}. \quad\blacksquare\tag{2.14}$$

These parameterisations give rise to the familiar "pitchfork" bifurcation diagram (see [3], Figure 1). Further analysis (see remark 3 in Section 3 of [19]) provides

LEMMA 3. Assume the zero eigenvalue of f_x^0 is algebraically simple, that is,

$$\psi_1^T\varphi_1\neq0.\tag{2.15}$$

Then near (x_0,λ_0) we may write the eigenvalue problem for $f_x(x,\lambda)$ corresponding to the smallest eigenvalue as

$$f_x(x(\lambda),\lambda)\varphi(\lambda)=\sigma(\lambda)\varphi(\lambda), \quad \varphi^T(\lambda)\varphi(\lambda)=1\tag{2.16}$$

for $x(\lambda)\in H_s$, $\varphi(\lambda)\in H_a$, with $\sigma(\lambda_0)=0$, $\varphi(\lambda_0)=\varphi_1$, and $\varphi(\lambda)$, $\sigma(\lambda)$ smooth functions of λ. $\blacksquare$

Note that in (2.16) we could replace $f_x(x(\lambda),\lambda)$ by $f_x(x(\lambda),\lambda)|_{H_a}$. Also a similar result holds for the left eigenvector ψ, that is, $\psi=\psi(\lambda)$ with $\psi(\lambda_0)=\psi_1$.

A useful corollary, obtained by differentiating (2.16) and then evaluating at $\lambda=\lambda_0$ is:

COROLLARY 4. $\sigma'(\lambda_0)=b_f/\psi_1^T\varphi_1$

and hence the simple eigenvalue $\sigma(\lambda)$ of $f_x(x(\lambda),\lambda)$, $x(\lambda)\in H_s$, crosses zero at $\lambda=\lambda_0$ with nonzero velocity provided (2.8) holds. $\blacksquare$

This result has some important implications. First we may prove local stability results such as: If points on the symmetric path $(x(\lambda),\lambda)$ are linearly stable for $\lambda<\lambda_0$, then the points on the symmetric path for $\lambda>\lambda_0$ are linearly unstable. Second it follows that the determinant of $f_x(x(\lambda),\lambda)$ (and of $f_x(x(\lambda),\lambda)|_{H_a}$) changes sign at $\lambda=\lambda_0$ and this fact can be used in algorithms for the detection of a bifurcation point.

Finally in this section we remark that once a symmetry-breaking bifurcation point has been detected and computed then we may use the information from (2.11) to "jump" onto the asymmetric path. A simple approach is to use the fact that the tangent to the asymmetric path at (x_0,λ_0) is $(\varphi_1,0)$ and to use this information in the pseudo-arc length algorithm [12].

3. ANALYSIS OF A TAKENS-BOGDANOV POINT

One of the simplest cases of mode interaction arises in the two parameter problem

$$\frac{dx}{dt}+f(x,\lambda,\alpha)=0, \quad \lambda,\alpha\in\mathbf{R}, \tag{3.1}$$

where $f:H\times\mathbf{R}^2\to H$ has no symmetry or, at least, no symmetry which is broken. The steady state equation is

$$f(x,\lambda,\alpha)=0, \quad x\in H, \quad \lambda,\alpha\in\mathbf{R}, \tag{3.2}$$

and, for fixed α, the simplest steady state bifurcation found in (3.2) is a fold (turning or limit) point as indicated by A in Figure 1 (see [12], [14] and [16]). Hopf bifurcation [8], [10] is also possible for fixed α and might occur anywhere on the path of solutions of $f(x,\lambda,\alpha_0)=0$, though for illustrative purposes we assume it occurs at H in Figure 1. (Note that, unlike the fold point situation, the geometric form of the path of steady states provides no information about the presence of a Hopf bifurcation, and that is one reason why Hopf bifurcations are so difficult to detect in large systems when direct eigenvalue calculation is impractical.) If now α is allowed to vary the solution surface of (3.2) contains a path of fold points (the fold curve) and a path of Hopf bifurcation points (see Figure 1). The point at which these two paths meet is called a *Takens–Bogdanov* point (point O in Figure 1), and this is an example of a steady state/Hopf mode interaction which has been well analysed in the dynamical systems literature ([9], p364). We shall present an analysis of the solutions of (3.2) near this point using the symmetry-breaking bifurcation theory of Section 2. Some stability results are obtained very simply but we note that the more complicated dynamical phenomena of homoclinic orbits are not found by our approach.

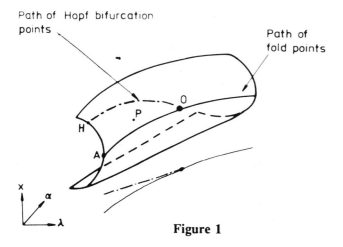

Path of Hopf bifurcation points

Path of fold points

Figure 1

Schematic diagram of a fold curve with a Takens-Bogdanov point and path of Hopf bifurcation points. If P is a stable solution of (1.2) then only points in the region of the solution surface containing P are stable. In parameter space the paths meet tangentially.

First let us state the assumptions on f. Assume that

    a) $f(x_0,\lambda_0,\alpha_0)=0$,

    b) Null $(f_x^0)=$ span $\{\varphi_1\}$,   Null $((f_x^0)^*)=$ span $\{\psi_1\}$,

    c) $\psi_1^T\varphi_1=0$,   (3.3)

    d) $l^T\varphi_1\neq0$, $l\in H$,

    e) $f_x^0\zeta_1+\varphi_1=0$, $l^T\zeta_1=0$,

    f) $\psi_1^T\zeta_1\neq0$,

that is, f_x^0 has an *algebraically double* zero eigenvalue which is *geometrically simple*. Note that in (3.3d) l is a normalising element (see [14]). In addition we assume

$$\psi_1^T f_\lambda^0\neq0, \qquad \psi_1^T f_{xx}^0\varphi_1\varphi_1\neq0. \qquad (3.4)$$

which are the usual quadratic fold point nondegeneracy conditions (see [14]).

We shall analyse the solutions of $f(x,\lambda,\alpha)=0$ near (x_0,λ_0,α_0) using the extended system for a Hopf bifurcation point [8], [10] which can be written as

$$F(z,\alpha)=0, \quad F:Z\times\mathbf{R}\to Z, \quad Z:=H\times H\times\mathbf{R}\times H\times\mathbf{R},$$

$$F(z,\alpha):=\begin{bmatrix} f(x,\lambda,\alpha) \\ f_x(x,\lambda,\alpha)a-\omega b \\ l^T a-1 \\ f_x(x,\lambda,\alpha)b+\omega a \\ l^T b \end{bmatrix}, \quad z=(x,a,\lambda,b,\omega). \tag{3.5}$$

Note that, for α fixed, the 2nd and 4th equations tell us that f_x has a pure imaginary eigenvalue $i\omega$ with eigenvector $a+ib$ and the 3rd and 5th equations normalise a and b for an appropriately chosen l. As α varies then (z,α), the solution of (3.5), can represent a path of Hopf bifurcation points of (3.1). What is not quite so obvious is that, for a fixed α, fold points can also satisfy (3.5). To see this put $b=0$ and $\omega=0$ so that (3.5) reduces to the requirement that f_x has a zero eigenvalue with eigenvector a. Therefore $(z(\alpha),\alpha)$ can also represent a path of fold points. Thus in one sense fold points are degenerate Hopf bifurcation points and it is precisely this feature which enables us to use the symmetry-breaking bifurcation analysis. The main elements in the analysis are to show that F satisfies a reflectional equivariance condition of the form (2.2) and that the Takens-Bogdanov point defined by (3.3) corresponds to a symmetry-breaking pitchfork bifurcation of (3.5) with the path of fold points being the symmetric path.

The first lemma provides the reflectional symmetry result.

LEMMA 5. If S is defined on Z by

$$Sz:= (x,a,\lambda,-b,-\omega), \quad z=(x,a,\lambda,b,\omega)$$

then $F(z,\alpha)$ defined by (3.5) satisfies the reflectional equivariance condition (2.2). The symmetric elements have the form $(x,a,\lambda,0,0)\in Z_s:=H\times H\times\mathbf{R}\times\{0\}\times\{0\}$ and the antisymmetric elements have the form $(0,0,0,b,\omega)\in Z_a:=\{0\}\times\{0\}\times\{0\}\times H\times\mathbf{R}$. ∎

The next Lemma verifies that (2.4), (2.5) and (2.15) hold for $F(z_0,\alpha_0)=0$.

LEMMA 6. With F given by (3.5), and $z_0=(x_0,\varphi_1,\lambda_0,0,0)$ satisfying (3.3) we have

a) $F(z_0,\alpha_0)=0,$

b) $\text{Null}(F_z(z_0,\alpha_0)) = \text{span}\{\Phi_1\}, \Phi_1=(0,0,0,\zeta_1,1),$ $$\tag{3.6}$$
$\text{Null}((F_z(z_0,\alpha_0))^*) = \text{span}\{\Psi_1\}, \Psi_1=(0,0,0,\psi_1,0),$

c) $z_0\in Z_s, \quad \Phi_1\in Z_a,$

d) $\Psi_1^T\Phi_1\neq 0.$ ∎

The only condition left to investigate is (2.8). Direct calculation (see [18], Lemma 2.2) shows that

$$\Psi_1^T[F_{z\alpha}^0\Phi_1+F_{zz}^0\Phi_1 V_0]=\frac{d}{d\alpha}(\psi_1^T\varphi_1)$$

where $F_z^0V_0+F_\alpha^0=0$, and so condition (2.8) reduces to a condition on the nondegeneracy of (3.3c), in the sense that it requires that $\psi_1^T\varphi_1$ cross though zero at $\alpha=\alpha_0$ with nonzero velocity. This fact can be used in the numerical detection of the Takens-Bogdanov point.

The results so far are summarised in:

THEOREM 7. Assume (3.3) and (3.4) and that $F(z,\alpha)$ is given by (3.5). If

$$\frac{d}{d\alpha}(\psi_1^T\varphi_1)\neq0 \tag{3.7}$$

then $F(z,\alpha)=0$ has a simple symmetry-breaking bifurcation point at $\alpha=\alpha_0$. The symmetric solutions represent a path of fold points and at (x_0,λ_0,α_0) a path of Hopf bifurcation points bifurcates from the path of fold points. ∎

Note that in the schematic Figure 1 no antisymmetric component is plotted and so the Hopf path does not exhibit the familiar pitchfork shape, however the asymmetric branch is double in the sense that for each solution (x,a,λ,b,ω) there is a conjugate solution $(x,a,\lambda,-b,-\omega)$ but the conjugate solution represents the same Hopf bifurcation point of (3.1).

In a stability analysis of steady state solutions of (3.1) near a Takens-Bogdanov point one can show the following: If in Figure 1 P represents a stable steady-state solution then all points in the same region as P repsent stable steady state solutions. However points on the lower path of the fold curve, and on the other side of the Hopf path represent unstable steady state solutions.

The reader is referred to [1] for applications of the above theory in the explanation of a complicated stability interchange in 5-cell flows in the finite Tayor problem. Numerical and experimental results are also presented there. A different approach to the analysis of a Takens-Bogdanov point is also given in [17] with supporting numerical results.

We conclude this section with two extensions of results in [18]. The first is given by

COROLLARY 8. In (λ,α)-space the paths of fold points and Hopf points are tangential at their point of contact (see Figure 1).

Proof. Direct calculation shows that V_0 and W_0 given by $F_z^0V_0+F_\alpha^0=0$, $F_z^0W_0+F_{zz}^0\Phi_1\Phi_1=0$ have the form $V_0=(*,*,v_3,0,0)$, $W_0=(*,*,0,0,0)$. Thus (2.10) gives $\alpha=\alpha_0+t$, $\lambda=\lambda_0+tv_3+o(t)$ as the parameterisation of the path of fold points in (λ,α)-space. Clearly at (λ_0,α_0), $d\lambda/d\alpha=v_3$. Similarly (2.11) gives, using the form of Φ_1 given by (3.6b),

$\alpha=\alpha_0+\delta t^2+o(t^2)$, $\lambda=\lambda_0+\delta v_3 t^2+o(t^2)$, as the parameterisation of the path of Hopf bifurcation points in (λ,α)-space, which gives a one-sided derivative at (λ_0,α_0) with the same value as above. ∎

The second result not in [18] relates the eigenpairs of $F_z(z(\alpha),\alpha)$, $z(\alpha)\in Z_s$ with the eigenpairs of $f_x(x,\lambda,\alpha)$ along the path of fold points. Clearly along $z(\alpha)\in Z_s$ we have $f_x(x,\lambda,\alpha)\varphi_1(\alpha)=0$. The next Lemma provides information about the way a second eigenvalue of $f_x(x,\lambda,\alpha)$ crosses through zero at $\alpha=\alpha_0$, which may be useful in a stability analysis.

LEMMA 9. Along the path of fold points there exist φ_2 and σ_2 which are smooth functions of α and which satisfy

$$f_x(x,\lambda,\alpha)\varphi_2(\alpha)=\sigma_2(\alpha)\varphi_2(\alpha)$$

with $\sigma_2(\lambda_0)=0$ and $\varphi_2(\lambda_0)=\varphi_1$. Provided (3.7) holds then $\sigma_2{}'(\alpha_0)\neq 0$.

Proof. From Lemma 3 we know that for $z(\alpha)\in Z_s$, there exists $\Phi(\alpha)$ and $\sigma_2(\alpha)$, say, which are smooth functions of α such that

$$F_z(z(\alpha),\alpha)|_{z_a}\Phi(\alpha)=\sigma_2(\alpha)\Phi(\alpha) \tag{3.8}$$

with $\sigma_2(\alpha_0)=0$. Also from Corollary 4 and Theorem 7 we have that $\sigma_2{}'(\alpha_0)\neq 0$ provided (3.7) holds.

In "component wise" notation (3.8) is equivalent to

$$\begin{bmatrix} f_x(x,\lambda,\alpha) & \varphi_1(\alpha) \\ l^T & 0 \end{bmatrix}\begin{bmatrix} \zeta(\alpha) \\ 1 \end{bmatrix}=\sigma_2(\alpha)\begin{bmatrix} \zeta(\alpha) \\ 1 \end{bmatrix}$$

with $\Phi(\alpha)=(0,0,0,\zeta(\alpha),1)$. Thus

$$(f_x(x,\lambda,\alpha)-\sigma_2(\alpha)I)\zeta(\alpha)+\varphi_1(\alpha)=0,$$

and if we define $\varphi_2(\alpha):=-f_x(x,\lambda,\alpha)\zeta(\alpha)$ then

$$f_x(x,\lambda,\alpha)\varphi_2(\alpha)=f_x(x,\lambda,\alpha)[-\sigma_2(\alpha)\zeta(\alpha)+\varphi_1(\alpha)]=\sigma_2(\alpha)\varphi_2(\alpha)$$

where we see that $(\varphi_2(\alpha),\sigma_2(\alpha))$ is an eigenpair for $f_x(x,\lambda,\alpha)$ with $(\varphi_2(\alpha_0),\sigma_2(\alpha_0))=(\varphi_1,0)$. The result follows. ∎

4. ROTATING WAVES IN EQUATIONS WITH O(2)-SYMMETRY

In this section we consider the one parameter problem

$$\frac{dX}{dt}+f(X,\lambda)=0, \qquad \lambda\in\mathbf{R}, \quad X\in H \tag{4.1}$$

where H is an n-dimensional real Hilbert space, with the corresponding steady state equation

$$f(X,\lambda)=0, \tag{4.2}$$

and where f is equivariant with respect to an action of the group $O(2)$, that is,

$$\gamma f(x,\lambda)=f(\gamma x,\lambda), \quad \forall\gamma\in O(2), \quad \forall x\in H. \tag{4.3}$$

As usual $O(2)$ denotes the group generated by the rotations r_α, $\alpha\in[0,2\pi)$ and a reflection s (see Chapter 2, [21]). A rotating wave solution of (4.1) is given by

$$X(t):=r_{ct}x, \tag{4.4}$$

for $c\in\mathbf{R}$ and x independent of time. This section describes the main idea in [2] though that paper provides a deeper account and also contains a numerical example. The aim of this section is to show how the problem of the analysis of a rotating wave bifurcation point can be put into the reflectional symmetry-breaking setting of Section 2.

First we need to provide some preliminary information about the consequences of the action of the group $O(2)$ in H. As in Section 2 we make use of the subspaces

$$H_s:=\{x\in H:sx=x\}, \quad H_a:=\{x\in H:sx=-x\},$$

where s is the reflection in $O(2)$. Without loss of generality we may assume that the inner product on H is $O(2)$ invariant, that is,

$$(\gamma x)^T(\gamma y)=x^Ty \quad \forall\gamma\in O(2), \quad \forall x,y\in H,$$

which induces

$$x^Ty=0 \quad \forall x\in H_s, \quad y\in H_a.$$

Finally we introduce the linear operator A defined by

$$A:=r_0', \quad \text{where} \quad r_\alpha'x:=\frac{d}{d\alpha}(r_\alpha x), \tag{4.5}$$

(here A is bounded on H since the group action is smooth on finite dimensional H [13]) and we can show that

$$a) \quad r_\alpha'=r_\alpha A=Ar_\alpha, \quad b) \quad sA=-As. \tag{4.6}$$

The first important results about rotating wave solutions are given in the next Lemma:

LEMMA 10. a) A rotating wave solution $(r_{ct}x,\lambda)$ satisfies (4.1) if and only if (x,c,λ) satisfies

$$f(x,\lambda)+cAx=0. \tag{4.7}$$

b) If (x,c,λ) satisfies (4.7) then

$$(f_x(x,\lambda)+cA)Ax=0.$$

Proof. Part a) is proved by substitution and use of (4.6a) (cf. Lemma 1.1, [15]); part b) follows by differentiating the equivariance condition (4.3) with respect to α and setting $\alpha=0$. ∎

This lemma shows that the rotating wave solutions satisfy the *steady state* equation (4.7), which enables the analysis and computation of rotating waves to be performed using standard steady state theory. However result b) shows that, provided $Ax \neq 0$, the linearisation of (4.7) is always singular at a solution point. This is not surprising since one consequence of (4.3) and (4.6a) is that if (x,c,λ) solves (4.7) then so does $(r_\alpha x, c, \lambda)$, providing a continuum of *conjugate* solutions, called the *orbit*, which has tangent Ax at (x,c,λ).

In order to eliminate this nonuniqueness we use the common approach of adding a phase condition (or normalising equation) to (4.7). In this paper we use the condition $l^T x = 0$, where l is a fixed element in H, though in [2] we discuss a more general phase condition (see also [10], where the numerical performance of various phase conditions is discussed). So we obtain the equation on which we shall base our analysis:

$$F(y,\lambda)=0, \quad F:Y\times R \to Y, \quad Y:=H\times R,$$

$$F(y,\lambda):= \begin{bmatrix} f(x,\lambda)+cAx \\ l^T x \end{bmatrix} \quad y=(x,c). \tag{4.8}$$

Note that

$$F_y(y,\lambda)= \begin{bmatrix} f_x(x,\lambda)+cA & Ax \\ l^T & 0 \end{bmatrix},$$

and though the upper left hand block of F_y is singular it is possible that the whole matrix will be nonsingular. In particular, if $l^T Ax \neq 0$ then $(Ax,0)$ will not be a null vector of F_y.

Let us assume that $x=0$ is a trivial $O(2)$-symmetric steady state solution of (4.2) for all λ. It is well known [21] that steady state symmetry-breaking pitchfork bifurcation occurs from the trivial solution resulting in nontrivial solution branches lying in H_s. We shall be concerned with bifurcations to rotating wave solutions occurring on these nontrivial branches. To be precise we assume that f and l satisfy (cf. (3.3))

a) $f(x_0,\lambda_0)=0, \quad x_0 \in H_s, \quad x_0 \neq 0,$

b) $\text{Null}(f_x^0)=\text{span}\{Ax_0\}, \quad \text{Null}((f_x^0)^*)=\text{span}\{\psi_1\},$

c) $\psi_1^T Ax_0=0,$ (4.9)

d) $l \in H_a, \quad l^T Ax_0 \neq 0,$

e) $f_x^0 \zeta_1 + Ax_0=0, \quad l^T \zeta_1=0,$

f) $\psi_1^T \zeta_1 \neq 0.$

Before starting the bifurcation analysis we note that if $x_0 \in H_s, \, x_0 \neq 0$, then from (4.6b),

$$Ax_0 \in H_a \tag{4.10}$$

and (see [2])

$$Ax_0 \neq 0. \tag{4.11}$$

Now we can proceed with the symmetry-breaking bifurcation theory for (4.8) and our first step is obviously to set up the symmetry operator on Y. This is done in the following Lemma.

LEMMA 11. If S is defined on $Y = H \times \mathbf{R}$ by

$$Sy := (sx, -c), \qquad y = (x, c),$$

where s is the reflection in $O(2)$, then $F(y, \lambda)$ defined by (4.8) satisfies the reflectional equivariance condition (2.2). The symmetric elements have the form $y_s := (x_s, 0) \in Y_s := H_s \times \{0\}$, and antisymmetric elements have the form $(x_a, c) \in Y_a := H_a \times \{\mathbf{R}\}$. ∎

Clearly for λ fixed a symmetric solution, x_s say, is a steady state solution of (4.2) lying in H_s. As λ varies then $x_s(\lambda)$ represents a smooth path of steady state solutions in $H_s \times \{\mathbf{R}\}$ since $f_x(x, \lambda)|_{H_s}$ is nonsingular using (4.10), but this result immediately translates to: $y_s(\lambda) = (x_s(\lambda), 0) \in Y_s$ represents a path of symmetric solutions of $F(y, \lambda) = 0$ since $F_y(y, \lambda)|_{Y_s}$ is nonsingular.

Next we need to show that under (4.9), $F(y, \lambda) = 0$ satisfies the conditions (2.4), (2,5) and (2.15). This is given in the next Lemma.

LEMMA 12. Let F be given by (4.8), and $y_0 = (x_0, 0) \in Y_s$ satisfy $F(y_0, \lambda_0) = 0$. Assume (4.9) holds. Then

    a)   Null$(F_y(y_0, \lambda_0)) = \text{span}\{\Phi_1\}$, $\Phi_1 = (\zeta_1, 1) \in Y_a$,

         Null$(F_y(y_0, \lambda_0)^*) = \text{span}\{\Psi_1\}$, $\Psi_1 = (\psi_1, 0)$,

    b) $\Psi_1^T \Phi_1 \neq 0$. ∎

To find out whether or not bifurcation occurs we need to check (2.8). After some manipulation we find that

$$\Psi_1^T [F_{y\lambda}^0 \Phi_1 + F_{yy}^0 \Phi_1 V_0] = \frac{d}{d\lambda}(\psi_1^T A x_0) \tag{4.12}$$

and so condition (2.8) reduces to a condition on the nondegeneracy of (4.9c) (cf. 3.7). Thus the following theorem holds:

THEOREM 13. Assume (4.9) and that $F(y,\lambda)$ is given by (4.8). If

$$\frac{d}{d\lambda}(\psi_1^T A x_0) \neq 0 \qquad\qquad (4.13)$$

then $F(y,\lambda)=0$ has a simple symmetry-breaking bifurcation point at $\lambda=\lambda_0$. The symmetric solutions represent a path of steady state solutions of $f(x,\lambda)=0$, $x\in H_s$, and at (x_0,λ_0) a path of rotating wave solutions bifurcates from the steady state branch. ■

It is clear from the above description that the analysis for the bifurcating rotating waves has close similarities with that for a Takens-Bogdanov point. We shall not explore this relationship further but end with some remarks about stability.

Lemma 10 shows that if (x,λ) is a steady state solution of (4.2) then $f_x(x,\lambda)Ax=0$ and so the Jacobian matrix has an eigenvalue, $\sigma_1(\lambda)$ say, fixed at zero. This requires the introduction of a different concept of stability, called *orbital stability*, though in the discussion of gain or loss of stability the essential point is that we ignore $\sigma_1(\lambda)$ (p87, [7]). To produce orbital stability results we need to find out when a second eigenvalue, σ_2 say, crosses the imaginary axis. As one might expect we can prove a result similar to that given by Lemma 9:

LEMMA 14. Along the path of steady states in H_s there exist smooth functions of λ, $\varphi_2(\lambda)$ and $\sigma_2(\lambda)$, such that

$$f_x(x,\lambda)\varphi_2(\lambda)=\sigma_2(\lambda)\varphi_2(\lambda)$$

with $\sigma_2(\lambda_0)=0$, $\varphi_2(\lambda_0)=Ax_0$. Provided (4.13) holds then $\sigma_2{}'(\lambda_0)\neq 0$. ■

Thus the type of stability result we can prove is the following:

LEMMA 15. An orbitally stable path of steady state solutions loses stability at a bifurcation to rotating waves provided (4.13) holds. ■

We end by noting that the form of the solutions projected onto parameter space can be obtained using an approach similar to that in Corollary 8. We refer the reader to [2] for numerical results illustrating this and other bifurcation phenomena for a spectral Galerkin discretization of the Kuramoto-Sivashinski equation.

ACKNOWLEDGEMENT

The authors acknowledge the support of the SERC in the pursuance of this research.

REFERENCES

1. Anson, D.K., Mullin, T. and Cliffe, K.A. A numerical investigation of a new solution in the Tayor vortex problem (to appear in *J. Fluid Meth.*), 1989.

2. Aston, P.J., Spence, A. and Wu, W. Bifurcation to rotating waves in equations with O(2) symmetry (submitted to *SIAM J. Appl. Maths*), 1989.

3. Brezzi, F., Rappaz, J. and Raviart, P.A. Finite dimensional approximations of nonlinear problems. Part III. Simple bifurcation points, *Numer. Math.* 38, (1981) 1-30.

4. Deuflehard, P., Fiedler, B. and Kunkel, P. Efficient numerical path following beyond critical points. *SIAM J. Numer. Anal.* 24, (1987) 912-927.

5. Doedel, E.J. AUTO: a program for the automatic bifurcation analysis of autonomous systems, *Congressus Numerantium,* 30, (1981) 265-284.

6. Golubitsky, M. and Schaeffer, D. *Singularities and Groups in Bifurcation Theory, Appl. Math. Sci.* 51, *Springer, New York,* 1986.

7. Golubitsky, M., Stewart, I.N. and Schaeffer, D.G. *Singularities and Groups in Bifurcation Theory, Vol. 2. Appl. Math. Sci.* 69, *Springer, New York,* 1988.

8. Griewank, A. and Reddien, G. The calculation of Hopf points by a direct method. *IMA J. Numer. Anal.* 3, (1983) 295-304.

9. Guckenheimer, J. and Holmes, P. *Nonlinear Oscillations, Dynamical Systems, and Bifurcations of Vector Fields. Springer, New York,* 1983.

10. Jepson, A.D. and Keller, H.B. Steady state and periodic solution paths: their bifurcations and computations, in Küpper, T. Mittelmann, H.D. and Weber, H. eds., *Numerical Methods for Bifurcation Problems. ISNM 70 Birkhäuser, Basel,* (1984) 219-246.

11. Jepson, A.D. and Spence, A. Folds in solutions of two parameter systems and their calculation. Part I. *SIAM J. Numer. Anal.* 22, (1985) 347-368.

12. Keller, H.B. Numerical solution of bifurcation and nonlinear eigenvalue problems, in *Rabinowitz, P.H. ed. Applications of Bifurcation Theory, Academic Press, New York,* (1977) 359-384.

13. Knapp, A.W. *Representation Theory of Semisimple Groups: An Overview Based on Examples. Princeton University Press, Princeton, N.J.* 1986.

14. Moore, G. and Spence, A. The calculation of turning points of nonlinear equations. *SIAM J. Numer. Analysis,* 17, (1980) 567-576.

15. Renardy, M. Bifurcation from rotating waves. *Arch. Rat. Mech. Anal.,* 79, (1982) 49-84.

16. Rheinboldt, W.C. *Numerical Analysis of Parameterized Nonlinear Equations. Wiley-Interscience, New York,* 1986.

17. Roose, D. Numerical computation of origins for Hopf bifurcation in a two parameter problem, in Küpper, T., Seydel, R. and Troger, H. eds., *Bifurcation: Analysis, Algorithms, Applications, ISNM 79, Birkhäuser, Basel,* 1987.

18. Spence A., Cliffe, K.A. and Jepson, A.D. A note on the calculation of paths of Hopf bifurcations. *JCAM* 26, (1989) 125-131.

19. Werner, B. Regular systems for bifurcation points with underlying symmetries, in Küpper, T., Mittelmann, H.D. and Weber, H. eds., *Numerical methods in Bifurcation Problems, ISNM 70, Birkhäuser, Basel,* 1984.

20. Werner, B. and Spence, A. The computation of symmetry-breaking bifurcation points. *SIAM J. Numer. Anal.,* 21, (1984) 388-399.

21. Vanderbauwhede, A. *Local Bifurcation and Symmetry. Research Notes in Mathematics* 75, *Pitman, London,* 1982.

A. Spence and P.J. Aston
School of Mathematical Sciences
University of Bath
Bath BA2 7AY
United Kingdom

W. Wu
Department of Mathematics
University of Jilin
Changchun, China

G.A. WATSON
Computing the Structured Singular Value, and Some Related Problems

1. Introduction

Let M be a real $n \times n$ matrix, and let $\mathbf{k} \in Z_+^m$ be a vector with positive integer components summing to n. Define

$$S = \{\Delta \in \mathbb{R}^{n \times n} : \Delta = \text{block diag } (\Delta_1, \ldots, \Delta_m), \ \Delta_i \in \mathbb{R}^{k_i \times k_i}, \ i = 1,2,..,m \},$$

and consider the problem

$$\text{find } \Delta \in S \text{ to minimize } \|\Delta\| \text{ subject to } \det(I + M\Delta) = 0, \tag{1.1}$$

where the norm is the l_2 norm or spectral norm (the largest singular value). Throughout this paper, unadorned norms on matrices will refer to this norm, and those on vectors will refer to the l_2 vector norm. Now define $\mu(M)$ by

$$\mu(M) = \|\Delta^*\|^{-1}, \tag{1.2}$$

where Δ^* solves (1.1), with $\mu(M) = 0$ if there is no feasible solution (a possibility which we will exclude). For the special case when $m = 1$, it can be shown that

$$\mu(M) = \|M\|,$$

in other words the largest singular value of M, and in the more general situation described above, $\mu(M)$ is referred to as the *structured singular value*. This concept was introduced by Doyle [4] as a tool for the analysis and synthesis of feedback systems with structured uncertainties. In this context M is in fact a complex matrix, but for convenience we will restrict attention here to the real case: the appropriate extensions to permit the treatment of the corresponding complex problems are readily available.

In his paper, Doyle gives an algorithm for computing the structured singular value for the special case when $m \leq 3$. The algorithm can also be applied in more general cases, but there is then no guarantee of success. The approach requires the solution of the problem:

$$\underset{D \in T}{\text{minimize}} \|DMD^{-1}\| \tag{1.3}$$

where

$$T = \{D \in \mathbb{R}^{n \times n} : D = \text{block diag } \{d_1 I_{k_1}, ..., d_m I_{k_m}\}, \ d_i \in \mathbb{R}^+, \ i = 1,..,m \}.$$

258

Now for any $D \in T$ and any $\Delta \in S$ which is feasible in (1.1), there exists $\mathbf{x}$, $\mathbf{x}^T \mathbf{x} = 1$ such that

$$(I + DMD^{-1}\Delta)\mathbf{x} = 0,$$

so that

$$\|\Delta\|^{-1} \leq \|DMD^{-1}\|,$$

and

$$\mu(M) \leq \inf_{D \in T} \|DMD^{-1}\|. \tag{1.4}$$

Doyle [4] establishes the result that equality holds in (1.4) if $m \leq 3$ or if the minimum in (1.3) corresponds to the largest singular value of DMD^{-1} being simple. This shows that the structured singular value can often be computed by solving (1.3), which may be interpreted as an unconstrained optimization problem (if D is replaced by e^D in the objective function) with effectively $m - 1$ unknowns. It is interesting also that, although (1.1) is not a convex problem, $\|e^D M e^{-D}\|^2$ is a convex function of D ([16]), so that any local solution is also global. In the next section we consider how (1.3) may be solved; in fact we consider a more general problem without the special structure assumed here.

2. Minimizing the spectral norm of a matrix

Let $A: \mathbb{R}^s \to \mathbb{R}^{p \times q}$ where $p \geq q$, and the components of A are continuously differentiable functions of $\mathbf{a} \in \mathbb{R}^s$ in some region of interest. Consider the problem

$$\text{find } \mathbf{a} \in \mathbb{R}^s \text{ to minimize } \|A(\mathbf{a})\|. \tag{2.1}$$

Problems of this type arise in control engineering (see, for example, Polak and Wardi[14]). Let $A_j(\mathbf{a}) = \dfrac{\partial A(\mathbf{a})}{\partial a_j}$, $j = 1, .., s$, and for given $\mathbf{a} \in \mathbb{R}^s$, let the singular value decomposition of A be

$$A = U \Sigma V^T,$$

where $\Sigma = \text{diag} \{\sigma_1 = ... = \sigma_k > \sigma_{k+1} \geq \cdots \geq \sigma_q\}$ and U and V are orthogonal matrices. Let $\mathbf{u}_i$ ($\mathbf{v}_i$) denote the *ith* column of U (V), let U_1 (V_1) denote the first k columns of U (V) (these of course are not uniquely defined), and let $\partial\|.\|$ denote the subdifferential of $\|.\|$, so that $G \in \partial\|A\|$ if and only if

(i) $\|G\|^* \leq 1$, where $\|.\|^*$ denotes the dual norm,

(ii) trace $(G^T A) = \|A\|$.

By standard analysis carried over from the vector case (see, for example Rockafellar [15]) the following results are readily established.

Lemma 1 (Directional derivative)

$$\lim_{\gamma \to 0+} \frac{\|A(\mathbf{a} + \gamma \mathbf{d})\| - \|A(\mathbf{a})\|}{\gamma} = \max_{G \in \partial\|A\|} \sum_{j=1}^{s} d_j \, \text{trace}(G^T A_j).$$

Lemma 2 (Necessary conditions for a solution) Let $\mathbf{a} \in I\!R^s$ solve (2.1). Then there exists $G \in \partial\|A\|$ such that

$$\text{trace}(G^T A_j) = 0, \ j=1,..,s. \tag{2.2}$$

If A is a convex function, then this condition is sufficient as well as necessary.

Lemma 3 (Ziętak [20], Berens and Finzel [1]) $G \in \partial\|A\|$ if and only if $G = U_1 \Lambda V_1^T$, where Λ is a $k \times k$ symmetric positive semi-definite matrix with trace $(\Lambda) = 1$.

Define

$$H_j = \frac{1}{2}(U_1^T A_j V_1 + V_1^T A_j^T U_1), \ j=1,..,s. \tag{2.3}$$

Then if $k = 1$, G is unique and corresponds to the matrix of partial derivatives of $\|A\|$ with respect to the components of A, and further

$$H_j = \mathbf{u}_1^T A_j \mathbf{v}_1 = \frac{\partial}{\partial a_j}\|A\|, \ j=1,..,s.$$

Theorem 1 The following conditions are all equivalent to (2.2).

(i) There exists a $k \times k$ symmetric, positive semi-definite matrix Λ, with trace$(\Lambda)=1$, such that

$$\text{trace}(\Lambda H_j) = 0, \ j=1,2,..,s.$$

(ii)

$$0 \in \text{conv}(\nabla), \text{where}$$

$$\nabla = \{\mathbf{d} \in I\!R^s : d_j = \mathbf{c}^T H_j \mathbf{c}, \ j=1,..,s, \ \mathbf{c}^T \mathbf{c} = 1\}, \tag{2.4}$$

and conv denotes the convex hull.

(iii)

$$\lambda_{\text{max}}(\sum_{j=1}^{s} d_j H_j) \geq 0 \ \text{ for all } \mathbf{d} \in I\!R^s,$$

where $\lambda_{\text{max}}(.)$ denotes the largest eigenvalue.

Proof That (2.2) and (i) are equivalent follows using Lemma 3.

(i) => (ii)

Using a singular value decomposition of Λ, it follows from (i) that

$$\text{trace}(\sum_{i=1}^{k} \gamma_i \mathbf{c}_i \mathbf{c}_i^T H_j) = 0, \ j=1,..,s,$$

where $c_i^T c_i = 1$, $\gamma_i \geq 0$, $i = 1,..,k$ and $\sum_{i=1}^{k} \gamma_i = 1$. Thus

$$\sum_{i=1}^{k} \gamma_i c_i^T H_j c_i = 0, \quad j=1,..,s,$$

and (ii) follows.

(ii) => (iii)

Assume that (iii) does not hold. Then

$$\lambda_{\max}(\sum_{j=1}^{s} d_j H_j) < 0 \quad \text{for some } d \in I\!R^s.$$

This contradicts (ii).

(iii) => (i)

Assume that (i) does not hold. Then there exists a descent direction at a. Consider the directional derivative at a. Using Theorem 1 this is

$$\max_{\substack{\Lambda \text{ pos semi-def} \\ \text{trace}(\Lambda)=1}} \sum_{j=1}^{s} d_j \text{trace}(\Lambda H_j)$$

$$= \max_{\substack{c_i^T c_i = 1, \gamma_i \geq 0 \\ \sum \gamma_i = 1}} \sum_{j=1}^{s} d_j \text{trace}(\sum_{i=1}^{k} \gamma_i c_i c_i^T H_j)$$

$$= \max_{c^T c = 1} c^T (\sum_{j=1}^{s} d_j H_j) c$$

$$= \lambda_{\max}(\sum_{j=1}^{s} d_j H_j).$$

Thus (iii) is contradicted and the result is proved. □

It is always possible to compute descent directions away from a minimum. Let d solve:

$$\text{minimize } \|d\|. \qquad (2.5)$$
$$d \in \text{conv}(\nabla)$$

Then either $d = 0$, in which case (2.2) is satisfied, or

$$\lambda_{\min}(\sum_{j=1}^{s} d_j H_j) > 0,$$

so that $-d$ is a descent direction. Thus the analogue of steepest descent may be employed (although the solution of (2.5) is obviously not a finite problem when $k > 1$).

At a point a where the largest singular value σ_1 is simple, it is not difficult to calculate first and second derivatives of σ_1 with respect to the components of a. Specifically, for all $i,j = 1,..,s$,

$$\frac{\partial \sigma_1}{\partial a_i} = \mathbf{u}_1^T A_i \mathbf{v}_1,$$

and, if A_{ij} denotes the matrix of second derivatives of A with respect to $a_i a_j$, then

$$\frac{\partial^2 \sigma_1}{\partial a_i \partial a_j} = \sum_{l=2}^{q} \frac{\sigma_1(\alpha_{1il}\alpha_{1jl} + \alpha_{li1}\alpha_{lj1}) + \sigma_l(\alpha_{1il}\alpha_{lj1} + \alpha_{li1}\alpha_{1jl})}{\sigma_1^2 - \sigma_l^2} + \mathbf{u}_1^T A_{ij} \mathbf{v}_1,$$

where

$$\alpha_{ijk} = \mathbf{u}_i^T A_j \mathbf{v}_k,$$

for all relevant i,j,k. It follows that these derivatives are defined only if σ_1 is simple. If this is so at a solution to (2.1), then Newton's method is available, and thus a second order convergence rate is possible.

Now let k^* be the multiplicity of the largest singular value of A at the solution (assumed unique). At an approximation to the solution, for each l, $l=1,..,s$ define the $k^* \times k^*$ symmetric matrices G_l by

$$(\sigma_i + \sigma_j)(G_l)_{ij} = \sigma_i \mathbf{u}_i^T A_l \mathbf{v}_j + \sigma_j \mathbf{u}_j^T A_l \mathbf{v}_i, \quad i,j=1,..,k^*.$$

Clearly at a point where $\sigma_1 = .. = \sigma_{k^*}$,

$$G_l = H_l, \quad l=1,..,s.$$

Then the following system of equations must be satisfied at a solution of (2.1):

$$\text{trace}(\Lambda G_l) = 0, \quad l=1,..,s, \tag{2.6}$$

$$\sigma_i(A) - h = 0, \quad i=1,..,k^*, \tag{2.7}$$

$$1 - \text{trace}(\Lambda) = 0, \tag{2.8}$$

where, additionally, Λ is positive semi-definite. This is a total of (k^*+s+1) equations for $\tfrac{1}{2}k^*(k^*+1)+s+1$ unknowns. Even if derivatives are assumed to exist, there are (if $k^* > 1$) insufficient equations to define the Newton step. In the context of inverse eigenvalue problems, Friedland, Nocedal and Overton [7], [8] point out that the difficulty in defining the Newton step when there is mutiplicity in the target eigen-values arises from the fact that the eigenvectors corresponding to the multiple eigen-values are not uniquely defined, and they show that the correct way to implement the Newton iteration is to impose suitable normalization conditions on the relevant eigen-vectors. Using the relationship between the singular vectors of A and the eigenvectors of $A^T A$, the analogous modification needed here is to provide an additional $\tfrac{1}{2}k^*(k^*-1)$ equations by imposing the requirement that the Newton step $\mathbf{d}$ in the $\mathbf{a}$ variables should satisfy the normalization conditions on the singular vectors

$$\sum_{l=1}^{s} d_l(\sigma_i \mathbf{u}_i^T A_l \mathbf{v}_j + \sigma_j \mathbf{u}_j^T A_l \mathbf{v}_i) = 0, \quad i,j,=1,..,k^*; \; i<j. \tag{2.9}$$

Define $\mathbf{l}_1 = (\Lambda_{11}, \ldots , \Lambda_{k^*k^*})^T$, $\mathbf{l}_2 = (\Lambda_{12},.....)^T \in I\!R^{\frac{1}{2}k^*(k^*-1)}$. Then the system of equations which gives the full Newton step is

$$
\begin{bmatrix}
J & X & 2C & 0 \\
X^T & 0 & 0 & -e \\
C^T & 0 & 0 & 0 \\
0 & -e^T & 0 & 0
\end{bmatrix}
\begin{bmatrix}
\mathbf{d} \\
\delta l_1 \\
\delta l_2 \\
\delta h
\end{bmatrix}
=
\begin{bmatrix}
-X l_1 - 2C l_2 \\
h e - \sigma \\
0 \\
e^T l_1 - 1
\end{bmatrix},
\tag{2.10}
$$

where $\mathbf{e} = (1,..,1)^T$, $\sigma = (\sigma_1, \ldots, \sigma_{k^*})^T$, J is the matrix of partial derivatives of (2.6) with respect to the components of $\mathbf{a}$, $X \in \mathbb{R}^{s \times k^*}$ has elements $(G_l)_{ii}$, $l=1,..,s$, $i=1,..,k^*$ and $C \in \mathbb{R}^{s \times (\frac{1}{2} k^* (k^* - 1))}$ has elements $(G_l)_{ij}$, $l=1,..,s$, $i,j=1,..,k^*$, $i < j$. The system includes (2.9), modified so that the left hand side of each equation is divided by $\sigma_i + \sigma_j$.

Now let W be an $s \times s$ symmetric matrix, let $h = \sigma_1(A)$ and consider the quadratic programming problem:

$$\text{minimize } p + \tfrac{1}{2} \mathbf{d}^T W \mathbf{d}$$

$$
\text{subject to } \sum_{l=1}^{s} d_l \mathbf{u}_i^T A_l \mathbf{v}_i - p \le h - \sigma_i(A), \quad i=1,...,k^*,
\tag{2.11}
$$

$$
\sum_{l=1}^{s} d_l (\sigma_i \mathbf{u}_i^T A_l \mathbf{v}_j + \sigma_j \mathbf{u}_j^T A_l \mathbf{v}_i) = 0, \quad i,j=1,..,k^*; \ i<j.
$$

At a solution to this problem, there exist Lagrange multipliers $\lambda_1 \ge 0$ and λ_2 such that

$$W \mathbf{d} + X \lambda_1 + C \lambda_2 = 0, \tag{2.12}$$

$$e^T \lambda_1 = 1. \tag{2.13}$$

Thus if $W = J$ and equality holds in all the constraints of (2.11), then $\mathbf{d}$ is the Newton step in $\mathbf{a}$ for solving (2.6), (2.7) and (2.8), which also satisfies (2.9). The Lagrange multipliers are such that

$$\lambda_1 = l_1 + \delta l_1,$$

$$\lambda_2 = 2(l_2 + \delta l_2).$$

Motivation for (2.11) can also be obtained by considering the problem (2.1) posed in the form of the minimax problem:

$$\text{minimize } h \text{ subject to } \sigma_i(A) \le h, \quad i=1,2,..,q.$$

Locally, it is sufficient to solve:

$$\text{minimize } h \text{ subject to } \sigma_i(A) \le h, \quad i=1,2,..,k^*, \tag{2.14}$$

where the constraints may be considered to hold as equalities at the solution. The quadratic programming problem (2.11) may then be interpreted as a modification of the subproblem arising when a sequential quadratic programming method is applied to (2.14): the matrix W is related to the Hessian matrix of the Lagrangian function for the problem. The use of (2.11) also provides a means of finitely generating descent directions at **a**, as is shown by the following theorem.

Theorem 2 Let W be positive definite and let p, **d** solve the quadratic programming problem (2.11) at **a**, where k^* is assumed to be a guess at the correct final value, and $\sigma_i(A) = h$, $i=1,...,k \le k^*$. Then $p \le 0$, and if $p < 0$, **d** is a descent direction for $\|A\|$ at **a**.

Proof The objective function is clearly non-positive at a solution, so that $p \le 0$ if W is positive definite. Assume that $p < 0$. Then **d**$\ne 0$, and using the definition of H_j, it is a consequence of the constraints that $\sum_{j=1}^{s} d_j H_j$ is a diagonal matrix with negative diagonal elements. It follows from the alternative expression for the directional derivative derived in the proof of Theorem 1 that **d** is a descent direction as required. $\square$

Based on this result, it is possible to devise a descent algorithm for (2.1). If a sequence of descent steps is taken along directions **d** solving (2.11), then assuming that the step-lengths are bounded away from zero, the coresponding sequence of values of p may be driven to zero. An alternative approach is to impose bounds on the size of the solution **d** at each iteration, in a conventional trust region strategy. Now assume that W is positive definite, **a** is such that $\sigma_i(A) = h$, $i=1,...,k^*$, and (2.11) is solved by $p = 0$, so that **d** $= 0$. Then from (2.12), (2.13) it follows that there is a symmetric $s \times s$ matrix Λ, with non-negative diagonal elements summing to 1, such that

$$\text{trace}(\Lambda H_j) = 0, \quad j=1,...,s.$$

Specifically, the diagonal elements of Λ are formed by the components of λ_1, and the off-digonal elements by 0.5 times the appropriate components of λ_2. If Λ is in addition positive semi-definite, then it follows that **a** satisfies necessary conditions for a solution. Otherwise, let ω be a negative eigenvalue of Λ with corresponding eigenvector **x**, $\mathbf{x}^T \mathbf{x} = 1$. Then if **d**,$p$ satisfy the constraints,

$$\sum_{l=1}^{s} d_l \mathbf{u}_i^T A_l \mathbf{v}_i - p = \omega x_i^2, \quad i=1,...,k^*,$$

$$\sum_{l=1}^{s} d_l (\mathbf{u}_i^T A_l \mathbf{v}_j + \mathbf{u}_j^T A_l \mathbf{v}_i) = 2\omega x_i x_j, \quad i,j=1,...,k^*; \ i<j,$$

it is readily established that $p = -\omega^2$. In addition,

$$\sum_{j=1}^{s} d_j H_j = \omega \mathbf{x}\mathbf{x}^T - \omega^2 I,$$

so that **d** is a descent direction.

Some preliminary numerical experiments have been carried out with an algorithm based on the use of (2.11) for solving (2.1). The idea has been to generate constraints based on the largest k' singular values of the current matrix A (initially there is normally just one constraint based on σ_1). On evidence of singular values coalescing to σ_1

as the computation proceeds, k' is increased, and of course the additional equality constraints introduced (except when $s = 1$, as this would immediately terminate the algorithm). The matrix W was taken as the exact Hessian matrix when $k' = 1$, otherwise a multiple of the unit matrix. This simple strategy may be quite effective, but of course will also just give a first order convergent algorithm in many cases. Notice that if

$$\tfrac{1}{2}k^*(k^*+1) = s+1, \qquad (2.15)$$

then locally a second order rate of convergence is to be expected *independently* of the choice of W, for we are simply applying the correct form of Newton's method to (2.7) (see Friedland, Nocedal and Overton [8]). If these conditions are not satisfied, however, the choice of W will crucially affect the performance of any algorithm. Progress of the algorithm can be measured by the size of p solving (2.11), and termination required this value to be sufficiently close to zero. Some numerical experiments were carried out with a trust region algorithm. Bounds

$$-\tau \le d_j \le \tau, \ j=1,...,s,$$

were introduced into (2.11), the value of τ being changed in a systematic way in a manner commonly used in conventional optimization methods: descent can then be achieved even if W is not positive definite. The following simple example illustrates the progress of the method in a situation where a second order convergence rate can be obtained in the presence of multiple singular values at a solution to (2.1).

Example 1 Let

$$A = \begin{bmatrix} 2+a_2 & a_1-3a_2 \\ a_2-a_1 & 1+a_1 \end{bmatrix}.$$

The progress of the method is shown in Table 1, where k' refers to the number of constraints in the guessed active set. The number of inequality constraints (guessed active set) is increased to 2, and the extra equality constraint introduced, when $\sigma_1 - \sigma_2 < 0.1$. The matrix Λ at the final iterate is

$$\Lambda = \begin{bmatrix} 0.75 & 0.25 \\ 0.25 & 0.25 \end{bmatrix},$$

showing that the conditions of Theorem 2 are satisfied.

Other problems have been tackled when the relation (2.15) does not hold. Limited numerical experience shows that the algorithm can successfully drive the value of the largest singular value to a minimum, although without the correct choice of W the rate of convergence can be slow.

Example 2 Let

$$A = \begin{bmatrix} 2+c_3 & c_1-3c_2 \\ c_3-c_1 & 1+c_1+c_2 \end{bmatrix}.$$

The progress of the method is shown in Table 2. Three more iterations were required to give 5 decimal place accuracy in the coefficients, and the matrix Λ is then

$$\Lambda = \begin{bmatrix} 0.664398 & -0.141437 \\ -0.141437 & 0.335602 \end{bmatrix},$$

showing that the conditions of Theorem 2 are satisfied.

i	k'	τ	a_1	a_2	σ_1	σ_2
0			0.5	0.5	2.768660	1.354446
1	1	1.0	-0.5	-0.5	1.825141	0.410927
2	1	1.0	-1.5	-0.622381	1.634017	0.618733
3	1	4.0	-1.82516	-0.991416	1.532799	1.167984
4	1	4.0	-1.87736	-0.907375	1.463747	1.214712
5	1	4.0	-1.95742	-0.997909	1.442903	1.354055
6	2	4.0	-2.00040	-1.00030	1.414869	1.414055
7	2	4.0	-2.00000	-1.00000	1.414214	1.414214

Table 1

i	k'	τ	a_1	a_2	a_3	σ_1	σ_2
0			-2.0	0.0	-2.0	2.236068	0.0
1	1	0.25	-1.75	-0.25	-1.91667	1.415460	0.176621
2	1	0.25	-1.5	-0.5	-1.66667	1.015447	0.328263
3	1	0.25	-1.25	-0.25	-1.41667	0.807810	0.464218
4	1	0.0625	-1.18750	-0.31250	-1.47917	0.601112	0.554528
5	2	0.0625	-1.18355	-0.275216	-1.54167	0.581996	0.581512
6	2	0.25	-1.18823	-0.282350	-1.52942	0.581258	0.581241
7	2	0.25	-1.18903	-0.283550	-1.52742	0.581238	0.581238

Table 2

Let us return to the special case of (2.1) given by (1.3). An algorithm for this problem is given by Fan [5] which, if σ_1 is simple, descends using Newton steps (or truncated Newton steps) and so is essentially equivalent in that case to the method suggested here. However, the method deals with multiplicity of σ_1 by taking the steepest descent step, which in contrast to the step defined by the solution of (2.11), can not be calculated in a finite number of steps. Rather remarkably, multiplicity of σ_1 at a solution seems unusual with these structured problems. This observation of Fan and Tits

[6] is supported by experiments involving algorithms based on (2.11), when for a range of problems up to dimension $p=q=30$, and with up to $m=15$ blocks, a solution was always found to be at a point with a simple largest singular value. In such cases the minimum value of (1.3) always gives $\mu(M)$. Doyle [4] establishes this result when $m \leq 3$ independent of k^*: however the fact that $k^* = 1$ appears to be the usual situation shows that this method of computing the structured singular value is potentially important for all values of m. The possibility that $k^* > 1$ cannot be excluded, however, as is shown by the following simple example.

Example 3 Let

$$M = \begin{bmatrix} 1 & 2 \\ -1 & 1 \end{bmatrix}.$$

Then (1.3) is minimized by taking $d_1 = 1/\sqrt{2}$, $d_2 = 1$ with $\sigma_1 = \sigma_2 = \sqrt{3}$.

Some numerical results are given in Tables 3-5 to illustrate the performance of an algorithm based on (2.11) on matrices M generated by the Fortran random number generator RAND. In each case the initial approximations were $\mathbf{d} = (1,1,..,1)^T$, and this is clearly an excellent approximation to the solution. The block sizes were as follows:

Table 3 : $\mathbf{k} = (2,3,4,1)^T$

Table 4 : $\mathbf{k} = (3,2,4,5,1,5)^T$

Table 5 : $\mathbf{k} = (3,5,4,5,3,6,2,2)^T$

The number of seconds of CPU time on a SUN (Fortran program with single precision) were, respectively, 3.2 , 21.9 and 59.3. Because the solution always has $k^* = 1$, Newton steps are effectively being taken and only local properties of the algorithm are being tested. This kind of approach appears to be a very effective one for solving (1.3) (and (1.1)).

i	k'	τ	p	σ_1
0				5.063507
1	1	1	-0.043682	5.042154
2	1	4	-0.001209	5.041533
3	1	4	-0.000002	5.041532
4	1	4	-0.0000001	

Table 3 : $n = 10$, $m = 4$

The method described above was developed independently of work by Overton [13] on a class of problems closely related to (2.1): he derives a method for minimizing the maximum eigenvalue of a symmetric matrix which is an affine function of its free parameters. The methods are similar, both being based on the work of Friedland,

i	k'	τ	p	σ_1
0				10.292997
1	1	1	-0.058446	10.267271
2	1	4	-0.001656	10.266423
3	1	4	-0.000002	10.266422
4	1	4	-0.00000005	

Table 4 : $n = 20$, $m = 6$

i	k'	τ	p	σ_1
0				15.443680
1	1	1	-0.057054	15.416492
2	1	4	-0.001000	15.415984
3	1	4	-0.0000008	

Table 5 : $n = 30$, $m = 8$

Nocedal and Overton [8], and involving the solution of a sequence of quadratic pro-gramming problems. However, the algorithm of [13] uses the correct Hessian matrix at all times, and therefore normally converges at a second order rate; the problem of con-vergence is also dealt with rather more comprehensively than is done here. Thus, although it differs from the approach taken here in that it does not deal directly with the singular values, it is likely to lead to a superior algorithm for solving the general problem (2.1). However, because $k^* = 1$ is the usual situation at solutions to (1.3), it will not necessarily give any improvement in practice on a method of the type described above (or indeed just Newton's method) for solving this problem.

3. An equivalent smooth problem and some generalizations

In this section, another approach is taken to the solution of (1.1). It is based on methods which are available for solving a closely related class of problems, where it is required to find the smallest additive perturbation of a given matrix to produce rank deficiency. This is just one example of a class of matrix nearness problems, and the unstructured problem is classical (for some references, see Higham [11]). It has its ori-gins in statistics, where there are applications in multivariate analysis, in so-called total approximation problems arising in the errors-in-variables context (for example Golub and Van Loan [9], Osborne and Watson [12], Watson [17]), and also in the communality problem of factor analysis (see, for example, Harman [10]). Recent attention has focussed on the introduction of some structure into the perturbation matrix, and also the possibility of using matrix norms other that the spectral or Frobenius norms. For example, Watson [18] and Demmel [2],[3] consider the special case when the original matrix is partitioned into 4 blocks, only some of which can be perturbed, and Watson [19] permits arbitrary sparsity of the perturbation matrix. It is possible to use similar techniques for solving (1.1). In fact we will consider a rather more general problem

$$\text{find } \Delta \in S \text{ to minimize } \|\Delta\|_s \text{ subject to } \det(I + M\Delta) = 0, \tag{3.1}$$

where M and S are as before, but the matrix norm need not be restricted to the spectral norm. In fact a wide class of matrix norms may be dealt with by the techniques to be described here. For our purposes, we will not consider the most general class, but will define a matrix norm on elements of S by

$$\|\Delta\|_s = \|\mathbf{c}\|_m,$$

where $\|.\|_m$ is a *monotonic norm* on $I\!R^m$, so that

$$|u_i| \leq |v_i|, i=1,..,m \Rightarrow \|\mathbf{u}\|_m \leq \|\mathbf{v}\|_m.$$

and where

$$c_i = \|\Delta_i\|_i, \ i=1,..,m,$$

with $\|.\|_i$ any *separable* norm on $k_i \times k_i$ matrices: this class of matrix norms, which includes most norms which occur in practice, was introduced in [12]. The norm $\|.\|_i$ on $k_i \times n_i$ matrices is a separable norm if there exist vector norms $\|.\|_{A_i}$ and $\|.\|_{B_i}$ on vectors of dimension k_i and n_i respectively such that

$$\|\mathbf{u}\mathbf{w}^T\|_i = \|\mathbf{u}\|_{A_i} \|\mathbf{w}\|_{B_i}^*,$$

$$\|\mathbf{u}\mathbf{w}^T\|_i^* = \|\mathbf{u}\|_{A_i}^* \|\mathbf{w}\|_{B_i},$$

where $\mathbf{u} \in I\!R^{k_i}$, $\mathbf{w} \in I\!R^{n_i}$, and the asterisk, as before, denotes the dual norm. For example, if $\|.\|_i$ is the l_2 norm or the Frobenius norm, then both $\|.\|_{A_i}$ and $\|.\|_{B_i}$ are l_2 norms, which of course are self-dual.

Consider now the problem:

$$\min_{\mathbf{x} \in I\!R^n,\ \mathbf{x}^T\mathbf{x}=1} \|\mathbf{h}(\mathbf{x})\|_m, \tag{3.2}$$

where

$$h_i(\mathbf{x}) = \frac{\|\mathbf{x}_i\|_{A_i}}{\|M_i\mathbf{x}\|_{B_i}}, \quad i=1,..,m, \tag{3.3}$$

where $\mathbf{x}_i \in I\!R^{k_i}$, with $\mathbf{x}^T = (\mathbf{x}_1^T,..,\mathbf{x}_m^T)$, and $M_i \in I\!R^{k_i \times n}$ is used to denote the *ith* block row of M, $i=1,..,m$. Then the connection between (3.2) and (3.1) is given in the following theorem.

Theorem 3 Let $\bar{x}$ solve (3.2). Then (3.1) is solved by taking $\Delta = \bar{\Delta}$, where

$$\bar{\Delta}_i = -\frac{\bar{x}_i w_i^T}{\|M_i \bar{x}\|_{B_i}} \quad \text{if } M_i \bar{x} \neq 0, \quad \text{where } w_i \in \partial \|M_i \bar{x}\|_{B_i},$$

otherwise

$$\bar{\Delta}_i = 0.$$

Proof Let Δ^* solve (3.1). Then there exists x^*, $x^{*T} x^* = 1$, such that

$$\Delta_i^* M_i x^* + x_i^* = 0, \quad i=1,..,m.$$

Now if $u \in \mathbb{R}^{k_i}$ with $\|u\|_{B_i} = 1$, then

$$\|\Delta_i u\|_{A_i} \leq \|\Delta_i\|_i, \quad i=1,..,m,$$

(see, for example, [10]). Thus

$$\|x_i^*\|_{A_i} \leq \|\Delta_i^*\|_i \, \|M_i x^*\|_{B_i}, \quad i=1,..,m. \tag{3.4}$$

If either $\bar{x}$ or x^* gives a zero denominator in (3.3), these are accompanied by zero numerators, and appropriate definitions of values of the quotients may be given. The required analysis is similar to that given in [18]; however, for ease of presentation here, it will be assumed in what follows that this situation does not occur. Now

$$(I + \bar{\Delta} M)\bar{x} = 0,$$

so that

$$\det (I + M\bar{\Delta}) = 0.$$

Also

$$\|\Delta^*\|_s \leq \|\bar{\Delta}\|_s = \|\bar{c}\|_m, \quad \text{say}$$

$$= \|h(\bar{x})\|_m$$

$$\leq \|h(x^*)\|_m, \quad \text{by definition of } \bar{x},$$

$$\leq \|\Delta^*\|_s,$$

using the monotonicity of $\|.\|_m$ and (3.4). The result is proved. $\square$

Clearly, if $\|.\|_s$ is the l_2 norm, then the norms $\|.\|_i$ are also l_2 norms, and the norm $\|.\|_m$ is the Chebyshev norm. If $\|.\|_s$ is the Frobenius norm, then the norms $\|.\|_i$ are also Frobenius norms and the norm $\|.\|_m$ is the l_2 norm. In particular for $\|.\|_s$ the l_2 norm, (3.2) can be stated in the form:

$$\text{minimize } h$$

$$\text{subject to } \|\mathbf{x}_i\| \leq h\|M_i\mathbf{x}\|, \quad i=1,\ldots,m, \tag{3.5}$$

$$\mathbf{x}^T\mathbf{x} = 1.$$

In this case Theorem 3 specialises to

Theorem 4 Let $\bar{h},\bar{\mathbf{x}}$ solve (3.5). Then $\mu(M) = \bar{h}^{-1}$, and (1.1) is solved by taking $\Delta = \bar{\Delta}$, where

$$\bar{\Delta}_i = -\frac{\bar{\mathbf{x}}_i\bar{\mathbf{x}}^T M_i^T}{\|M_i\bar{\mathbf{x}}\|^2} \quad \text{if } M_i\bar{\mathbf{x}}\neq 0,$$

otherwise

$$\bar{\Delta}_i = 0.$$

It follows from these results that $\bar{\Delta}_i$ is always a rank one matrix, so that $\bar{\Delta}$ is rank m. However, the solution to (1.1) is not unique: it may be shown that a solution is also given by setting

$$\bar{\Delta}_i = -\frac{\|\bar{\mathbf{x}}_i\|}{\|M_i\bar{\mathbf{x}}\|}R_i, \quad \text{if } M_i\bar{\mathbf{x}}\neq 0, \tag{3.6}$$

otherwise

$$\bar{\Delta}_i = 0,$$

where R_i is a $k_i\times k_i$ orthogonal matrix satisfying

$$\|\bar{\mathbf{x}}_i\|R_i M_i\bar{\mathbf{x}} = \|M_i\bar{\mathbf{x}}\|\bar{\mathbf{x}}_i, \quad \text{if } M_i\bar{\mathbf{x}}\neq 0.$$

The construction given by (3.6) gives solutions to (3.1) for other matrix norms provided that the vector norms which arise are orthogonally invariant.

It is a consequence of Theorem 4 that the structured singular value can be computed by solving a variant of a minimax problem. In fact it may be shown that the inequalities in (3.5) actually hold with equality at a solution, and this is very useful from the point of view of calculation. It follows, therefore, that the problem (3.5) may be posed in the form

$$\text{minimize } \gamma$$

$$\text{subject to } \|\mathbf{x}_i\|^2 = \gamma\|M_i\mathbf{x}\|^2, \quad i=1,\ldots,m, \tag{3.7}$$

$$\|\mathbf{x}\|^2 = 1.$$

This problem may be solved in a standard way through the solution of a sequence of equality constrained quadratic programming problems. There is no need for an artificial penalty function, because descent can be obtained with respect to the natural merit function

$$\max_{i} \frac{\|x_i\|^2}{\|M_i x\|^2}.$$

A second order rate of convergence may be obtained by choosing the Hessian matrix of the quadratic objective function at each iteration to be the Hessian matrix of the Lagrangian function for (3.7). The linear equality constraints may be eliminated in a standard manner by a QR factorization technique, so that it is necessary only to work with the reduced Hessian matrix, and an efficient method can be developed.

The problem (3.7) is very close to another problem also equivalent to (1.1) and derived by Fan and Tits [6]. That problem can be stated as:

$$\text{maximize } \|M x\|^2$$

$$\text{subject to } \|x_i\|^2 \|M x\|^2 = \|M_i x\|^2, \quad i=1,..,m, \tag{3.8}$$

$$\|x\|^2 = 1.$$

It is shown in [6] that if $\hat{x}$ solves (3.8), then

$$\mu(M) = \|M \hat{x}\|.$$

Results given in [6] show that the solution of (3.8) can be considerably more efficient than that of (1.3) in computing the structured singular value. However, because (3.8) does not have a minimax formulation, its solution is not so attractive as that of (3.7).

An algorithm for (3.7) based on the above ideas has been implemented as a Fortran program to run on a SUN system. The intention is to compare this technique with other methods for computing the structured singular value. Full details will appear elsewhere, but some preliminary results are given here. In particular numerical results are given in Tables 6-8 for the same problems solved previously. The tables show the step length (α) and the Chebyshev norm of the step s (that comes from the solution of the quadratic programming problem) at each iteration taking the initial approximation $x = (1,1,..,1)^T$, correctly normalized. The computing times were, respectively, 1.5, 5.3 and 15.3 seconds, and it is clear that this choice of starting point is an excellent first approximation. For the purposes of comparison, notice that the optimal value of γ will be the square of the inverse of the structured singular value as defined before, provided that the global (rather that just a local) minimum has been obtained. In this case values of $1/\sqrt{(\gamma^*)}$ are, respectively, 5.041511, 10.266270 and 15.415660 , showing that the global solution has been correctly obtained. There is of course no guarantee of termination at a global minimum, because (3.7) is not a convex problem, and in this respect the method is not as good as that of the previous section. Nevertheless, it would appear from numerical experiments that it is usual for the global minimum to be attained. Indeed results obtained show that methods based on (3.7) appear to be very competitive as ways of computing the structured singular value.

There are other important norms in (3.1) for which the present analysis leads to much simpler problems. For example it is straightforward to see that if $\|.\|_s$ is the Frobenius norm, then the problem analogous to (3.7) is

$$\text{minimize} \sum_{i=1}^{m} \frac{\|x_i\|^2}{\|M_i x\|^2},$$

again subject to a normalization condition. The analysis given here also leads in a natural way to generalizations in other directions, for example to the treatment of rectangular matrices M, and with randomly structured matrices Δ.

i	$\|s\|_\infty$	α	γ
0			0.045218
1	0.03	1	0.039959
2	0.02	1	0.039686
3	0.05	0.5	0.039574
4	0.02	1	0.039473
5	0.0005	1	0.039344
6	0.0000001		

Table 6 : $n = 10$, $m = 4$

i	$\|s\|_\infty$	α	γ
0			0.011306
1	0.03	1	0.009575
2	0.003	1	0.009526
3	0.03	1	0.009518
4	0.0006	1	0.009488
5	0.0000007		

Table 7 : $n = 20$, $m = 6$

273

i	$\|s\|_\infty$	α	γ
0			0.004889
1	0.01	1	0.004250
2	0.001	1	0.004233
3	0.04	0.5	0.004224
4	0.02	1	0.004218
5	0.0002	1	0.004208
6	0.0000001		

Table 8 : $n = 30$, $m = 8$

References

1.   Berens, H and Finzel, M (1987) A continuous selection of the metric projection in matrix spaces, in Numerical Methods of Approximation Theory, Vol 8, (L Collatz, G Meinardus and G Nurnberger, eds) ISNM 81, Birkhauser Verlag, 21-29.

2   Demmel, J W (1987) The smallest perturbation of a submatrix which lowers the rank and constrained least squares problems, SIAM J Num Anal 24, 199-206.

3.   Demmel, J W (1988) On structured singular values, preprint, Computer Science Dept., Courant Institute of Mathematical Sciences, New York.

4.   Doyle, J C (1982) Analysis of feedback systems with structured uncertainties, Proc IEE-D 129, 242-250.

5.   Fan, M K-H (1988) An algorithm to compute the structured singular value, preprint, Electrical Engineering Department and Systems Research Center, University of Maryland.

6.   Fan, M K-H and Tits, A (1986) Characterization and efficient computation of the structured singular value, IEEE Trans Aut Control AC-31, 734-743.

7.   Friedland, S, Nocedal, J and Overton, M L (1986) Four quadratically convergent methods for solving inverse eigenvalue problems, in Numerical Analysis, Proceedings of the 11th Dundee Biennial Conference (D F Griffiths and G A Watson, eds), Pitman Research Notes in Mathematics No 140, 47-65.

8.   Friedland, S, Nocedal, J and Overton, M L (1987) The formulation and analysis of numerical methods for inverse eigenvalue problems, SIAM J Num Anal 24, 634-667.

9.   Golub, G H and Van Loan, C F (1980) An analysis of the total least squares problem, SIAM J Num Anal 17, 883-893.

10. Harman, H (1967) Modern Factor Analysis, University of Chicago Press, Chicago.

11. Higham, N J (1989) Matrix nearness problems and applications, in Proc IMA Conf on Applications of Matrix Theory (S Barnett and M J C Gover, eds), Oxford University Press (to appear).

12. Osborne, M R and Watson, G A (1985) An analysis of the total approximation problem in separable norms, and an algorithm for the total l_1 problem, SIAMJ Sci Stat Comp 6, 410-424.

13. Overton, M L (1988) On minimizing the maximum eigenvalue of a symmetric matrix, SIAM J Matrix Anal Appl 9, 256-268.

14. Polak, E and Wardi, Y (1982) Nondifferentiable optimization algorithm for designing control systems having singular value inequalities, Automatica 18, 267-283.

15 Rockafellar, R T (1970) Convex Analysis, Princeton University Press, Princeton.

16. Safonov, M G and Doyle, J C (1984) Minimizing conservativeness of robustness singular values, in Multivariable Control (S G Tzafestas, ed), Reidel Publishing Co, 197-207.

17. Watson, G A (1985) On a class of algorithms for total approximation, J Approx Theory 45, 219-231.

18 Watson, G A (1988) The smallest perturbation of a submatrix which lowers the rank of the matrix, IMA J Num Anal 8, 295-303.

19. Watson, G A (1989) On a general class of matrix nearness problems, Numerical Analysis Report NA/121, Department of Mathematics and Computer Science, University of Dundee.

20 Ziętak, K (1988) On the characterization of the extremal points of the unit sphere of matrices, Linear Alg and its Appl 106, 57-75.

G Alistair Watson,
Department of Mathematics and Computer Science,
University of Dundee,
Dundee DD14HN,
Scotland.

Contributed Papers

Towards a Parallel, Commercial Oil Reservoir Simulator.
C. A. Addison, University of Liverpool, Centre for Mathematical Software Research, Brownlow Hill, Victoria Building, Liverpool L69 3BX.

A Criterion Test for Self-scaling Variable Metric Algorithms.
M. Al-Baali, University of Damascus, Department of Mathematics, Damascus, Syria.

A note on the Construction of Continuous Quadric Patches.
M. L. Baart, Potchefstroom University, Department of Mathematics and Applied Mathematics, P.O. Box 1174, Vanderbijlpark 1911, South Africa.

Computing Accurate Eigensystems of Scaled Diagonally Dominant Matrices.
Jesse Barlow Pennsylvania State University, Computer Science Department, University Park, PA 16802, USA, and James Demmel, New York University, Courant Institute, Courant Institute of Mathematical Sciences, 251 Mercer Street, New York, NY 10012, USA.

Finite Element Approximation of a Free Boundary Problem Arising in the Theory of Liquid Drops and Plasma Physics.
J. W. Barrett, Imperial College, Mathematics Department, Queens Gate, London SW7 2BX and C. M. Elliott, University of Sussex, School of Mathematics and Physical Sciences, Falmer, Brighton BN1 9QH.

Conjugate Basis Matrices and Local Convergence for the Algorithm REQP for Constrained Minimisation.
M. C. Bartholomew-Biggs and T. T. Nguyen, Hatfield Polytechnic, Numerical Optimisation Centre, School of Information Science, College Lane, Hatfield AL10 9AB.

Multistep Methods for Ordinary Differential Equations based on Algebraic and First Order Trigonometric Polynomials.
G. Vanden Berghe, Rijksuniversiteit Gent, Lab voor Numerieke Wiskunde, Krijgslaan 281-S9, B-9000 Gent, Belgium.

Error Estimation in Automatic Quadrature Routines.
J. Berntsen and T. O. Espelid, University of Bergen, EDB-senteret, Thormohlens gate 55, N-5006 Bergen, Norway.

Barycentric Formulae for some Optimal Rational Approximants Involving Blaschke Products.
Jean-Paul Berrut, Université de Fribourg, Department de Mathematiques, Perolles, CH-1700 Fribourg, Switzerland.

A Cahn-Hilliard Type Model for Phase Separation.
J. F. Blowey, University of Sussex, School of Mathematics and Physical Sciences, Falmer, Brighton BN1 9QH.

Optimisation of Heavy Oil Recovery.
R. L. Brown, University of Reading, Department of Mathematics, Whiteknights, P.O. Box 220, Reading, RG6 2AX.

Galerkin Methods for Nonlinear Elliptic PDES.
C. Budd and T. Murdoch, Oxford University Computing Laboratory, 8-11 Keeble Road, Oxford OX1 3QD.

Stability of Approximations from Radial Basis Function Spaces.
M. D. Buhmann and M. J. D. Powell, University of Cambridge, DAMPT, Silver Street, Cambridge, CB3 9EW.

An Option for Implicit Runge-Kutta Methods.
J. C. Butcher, University of Auckland, Department of Mathematics & Statistics, Auckland, New Zealand.

The Numerical Solution of Large Scale Systems of Differential Equations.
G. D. Byrne, Exxon Research and Engineering Company, Clinton Township, Route 22 East, Annandale, NJ 08801, USA.

Generalised Padé Approximations to the Exponential Function.
F. H. Chipman Acadia University, Department of Mathematics, Wolfville, N.S., Canada BOP 1X0 and J. C. Butcher, University of Auckland, Department of Mathematics & Statistics, New Zealand.

Estimation of Convergence Orders in Repeated Richardson Extrapolation.
Edmund Christiansen, Odense University, Matematisk Institut, Campusvej 55, DK-5230 Odense M, Denmark.

Approximation of Initial and Boundary Value Problems by Algebraic Functions.
L. Collatz, University of Hamburg, Inst fur Angewandte Math 2 Hamburg 13, Bundestr 55 F.R. Germany.

A Tool for Analysing Ill-conditioned Equations.
W. M. Connelley and J. S. Rollett, Oxford University Computing Laboratory, Numerical Analysis Group, 8-11 Keble Road, Oxford OX1 3QD.

The Construction of Cubature Formulae using Software from Bifurcation Theory.
R. Cools, Katholieke Universiteit Leuven, Dept Computerwetenschappen, Celestijnenlaan 200 A, B3030 Heverlee, Belgium

Numerical Simulation of the Cahn-Hilliard Equation.
M. I. M. Copetti and C. M. Elliott. University of Sussex, School of Mathematics and Physical Sciences, Falmer, Brighton BN1 9QH.

Validated Anti-Derivatives.
G. F. Corliss, Marquette University, Department of Math, Stat & Computer Science, Milwaukee, WI 53233, USA.

Stability Analysis of the Discrete Collocation Methods for Second Kind Volterra Integral Equations and for Integro-differential Equations.
M. R. Crisci, Universita di Napoli, Dipartmento Di Matematica e Applicazioni, Via Mezzocannone 8, 80134 Napoli, Italy, E. Russo Universita di Salerno, Italy and A. Vecchio, National Research Council - Naples, Italy

Multi-Grid Techniques with Non-Standard Coarsening and Group Relaxation Methods.
A. Danaee, International Centre for Theoretical Physics, P.O. Box 586, 34100 Trieste, Italy.

The Relationship Between Theorems of the Alternative, Least Square Problems, and Steepest Descent Directions.
A. Dax, Hydrological Service, P.O. Box 6381, Jerusalem 91060, Israel.

Least Squares Data Smoothing by Non-Negative Second Divided Differences.
I. C. Demetriou, Meg. Alexander 53, GR-45 333 Ionnina, Greece.

Some Preconditioners for Multi-Vector Processing of Nonsymmetric Systems.
S. Doi, INRIA, Rocquencourt BP105, 78153 Le Chesnay Cedex, France.

Parallel Hopscotch Solutions of the Ginzberg-Landau Equation.
D.B. Duncan, Heriot-Watt University, Riccarton, Mathematics Department, Edinburgh, EH14 4AS.

Optimal Distribution of Knots in Tensor-Product Splines Approximation.
N. Dyn and I. Yad-Shalom, Tel-Aviv University, School of Mathematical Sciences, Tel-Aviv, Israel 69978.

On Positive Shock Capturing Schemes.
B. Einfeldt, Cranfield Institute of Technology, College of Aeronautics, Cranfield, Beds. MK43 0AL.

Discretisation and Asymptotic Behaviour of Dissipative Dynamical Systems.
C. M. Elliott, University of Sussex, School of Mathematics and Physical Sciences, Falmer, Brighton BN1 9QH.

Comparison of Roe's Scheme and a Directional Difference Scheme for Gas Network Flows.
S. Emmerson, University of Reading, Department of Mathematics, Whiteknights, P.O. Box 220, Reading, RG6 2AX.

Software for Adaptive Multidimensional Integration.
T. O. Espelid, University of Bergen, Department of Informatics, Allegaten 55, N-5007 Bergen, Norway.

Multistep Methods based on Continued Fractions for solving Initial Value Problems.
D. Eyre, Potchefstroom University, Department of Mathematics and Applied Mathematics, P.O. Box 1174, Vanderbijlpark 1911, South Africa and S. Abelman, University of the Witwatersrand, South Africa.

Thin Plate Spline Approximants,
P. Gonzales-Casanova, University of Oxford, Computing Laboratory, 8-11 Keeble Road, Oxford OX1 3QD.

Stable Solvers for Singular Linear Systems and Application to Block Linear Systems.
W. Govaerts, Rijksuniversiteit Gent, Seminarie voor Hogere Analyse, Galglaan 2, B-9000 Gent, Belgium.

Solution of Elliptic Systems of P.D.E.'s by Cell Discretization.
J. Greenstadt, University of Cambridge, DAMPT, Silver Street, Cambridge, CB3 9EW.

On Approximate Solutions of Integral Equations of the Mixed Type.
L. Hacia, Technical University of Poznan, Institute of Mathematics, Piotrowo 3a, 60-965 Poznan, Poland.

Highly Continuous Runge-Kutta Interpolants.
D. J. Higham, University of Toronto, Department of Computer Science, Toronto, Canada M53 1A4.

Trajectory Optimisation Applications using the Program STOMP for Generating Cost Function and Gradient Evaluations.
K. Horn, Messerschmitt-Bolkow-Blohm, LKE 126, 8012 Ottobrunn, F.R. Germany.

Embedded Runge-Kutta Methods on Parallel Computers.
P. J. van der Houwen, Centre for Mathematics & Computer Science, P.O. Box 4079, 1009 AB Amsterdam, The Netherlands.

Some New Ideas on Conjugate Gradient Methods.
Y. F. Hu and C. Storey, Loughborough University of Technology, Department of Mathematical Studies, Loughborough LE11 3TU.

Unconditional Convergence of some Crank-Nicholson LOD Schemes.
W. Hundsdorfer, Centre for Mathematics & Computer Science, P.O. Box 4079, 1009 AB Amsterdam, The Netherlands.

The Numerical Solution of Parabolic Free Boundary Problems Arising From Thin Film Flows.
R. Hunt, University of Strathclyde, Department of Mathematics, Livingstone Tower, 26 Richmond Street, Glasgow G1.

Equilibria of Runge-Kutta Methods.
A. Iserles University of Cambridge, DAMTP, Silver Street, Cambridge CB3 9EW and J. M. Sanz-Serna, Universidad de Valladolid, Departmento Matematica Aplicada y Computacion, Facultad de Ciencias, Valladolid, Spain.

Constrained Moving Finite Elements using Piecewise Cubics in 2-d.
P. Jimack, University of Bristol, Department of Mathematics, Bristol BS8 1TW.

L_∞ Errors in Finite Volume Approximations.
R. Jones, University of Oxford, Numerical Analysis Group, 8-11 Keble Road, Oxford OX1 3QD.

On Optimally High Order in Time Approximations for the Korteweg-de Vries Equation.
O. Karakashian and W. McKinney, University of Tennessee, Department of Mathematics, Ayres Hall, Knoxville, Ten 37996-1300, USA.

An Interval Step Control for Continuation Methods.
R. B. Kearfott, University of Southwestern Louisiana, Department of Mathematics, Box 4-1010, Lafayette, LA 70504-1010, USA.

Sub-lattices of Integration Lattices.

P Keast, Dalhousie University, Department of Mathematics, Statistics & CS, Halifax, Nova Scotia, Canada B3H 4HN, J. N. Lyness Argonne National Laboratory, USA and T. Sorevik, University of Bergen, Norway.

Parallel Evaluation of Some Recurrence Relations by Recursive Doubling.

Ayse Kiper Middle East Technical University, Department of Computer Science, Inonu Bulvari, Ankara, Turkey and D. J. Evans, Loughborough University of Technology, Department of Computer Studies, Loughborough LE11 3TU.

Solution of Exterior Helmholtz Problems by Improved Boundary Element Methods.

S. M. Kirkup, Brighton Polytechnic, Department of Mathematics, Moulsecoomb, Brighton BN2 4GJ.

A Regularization-Technique for the Numerical Solution of Index-2 - Differential-Algebraic Equations.

Michael Knorrenschild, Simon Fraser University, Department of Mathematics, Burnaby, B.C. V5A 1S6 Canada

Chebyshev Polynomial Solutions of First Kind Integral Equations for Numerical Conformal Mapping.

J. Levesley, D. M. Hough and S. N. Chandler-Wilde, Coventry Polytechnic, Department of Mathematics, Coventry CV1 5FB.

New Stopping Criteria for Some Iterative Methods for a Class of Unsymmetric Linear Systems.

C. Li and D. J. Evans, Loughborough University of Technology, Department of Computer Studies, Loughborough, LE11 3TU.

On a Certain Approximation Problem for Matrices.

W. Light, University of Lancaster, Mathematics Department, Lancaster LA1 4YL.

The Whittaker-Henderson Graduation Method.

A. J. Macleod, Paisley College of Technology, High Street, Paisley PA1 2BE, Glasgow.

Differential and Integro-differential Equations as Population Growth Models with Applications in Agriculture.

A. Makroglou, Agricultural University of Athens, Department of Mathematics, 75 Iera Odos St., Athens, 118 55, Greece.

Fast, Reliable Numerical Algorithms for Sturm-Liouville Computations.

M. Marletta and J. D. Pryce, RMCS Shrivenham, Comp Math Group, Shrivenham, Swindon SN6 8LA, Wilts.

The Effect of Filtering on the Pseudospectral Solution of Evolutionary Partial Differential Equations.

L. S. Mulholland, University of Strathclyde, Department of Mathematics, 26 Richmond Street, Glasgow, G1 1XH.

Galerkin and Petrov-Galerkin Methods for the Semiconductor Problems.
T. Murdoch, University of Oxford, Computing Laboratory, 8-11 Keeble Road, Oxford OX1 3QD.

Preconditioned Conjugate Gradient Methods for Equality Constrained Least Squares Problems.
N. K. Nichols, University of Reading, Department of Mathematics, Whiteknights, P.O. Box 220, Reading, RG6 2AX.

Faber Polynomials on Circular Sectors.
G. Opfer, University of Hamburg, Inst fur Angewandte Math, Bundesstr. 55, D 2000 Hamburg 13, F. R. Germany.

New Families of Convergence Acceleration Methods: The P-algorithms.
K. J. Overholt, University of Bergen, Department of Informatics, Allegaten 57A, N-5007 Bergen, Norway.

A Quadratically Convergent Parallel Jacobi-like Eigenvalue Algorithm.
M. H. C. Paardekooper, Tilburg University, Department of Econometrics, P.O. Box 90153, 5000 LE Tilberg, The Netherlands.

Using Divergent Iterations in High Speed, High Accuracy Computations.
P. Pedersen, Technical University of Denmark, Department of Mathematics, Bg 303, DK 2800 Lyngby, Denmark.

FFT Solution of the Robbins Problem.
W. M. Pickering and P. J. Harley, University of Sheffield, Department of Applied & Computational Mathematics, Sheffield S10 2TN.

Reliable Computation of Weyl's $m(\lambda)$ Function.
J. D. Pryce, M. Brown, RMCS, Shrivenham, Computational Mathematics Group, Shrivenham, Swindon SN6 8LA, and V. Kirby, Cardiff.

The Solution of Semi-infinite Programming Problems by Discretization Methods.
R. Reemtsen, Technical University of Berlin, F.B. Mathematik/(MA 6-3), Str. des 17 Juni 136, D-1000 Berlin, F.R. Germany.

Stability of Wide-angle Absorbing Boundary Conditions for the Wave Equation.
R. A. Renaut, Arizona State University, Department of Mathematics, Tempe, AZ 85287 1804, USA.

Order Results for Runge-Kutta Methods Applied to Differential Algebraic Equations.
M. Roche, Universite de Genéve, 2-4 rue du Lievre, Case Postale 240, CH-1211 Geneve 24, Switzerland.

Starting Methods for the Iteration of IRK's.
J. Sand, University of Copenhagen, Inst. of Datalogy, Universitetsparken 1, DK-2100 Copenhagen 0, Denmark.

Existence and Linear Stability of Period 2 Solutions of Discrete Reaction-Diffusion Models with Various Types of Nonlinear Diffusion.

S. W. Schoombie, University of the Orange Free State, Department of Applied Mathematics, P.O. Box 339, Bloemfontein 9300, South Africa.

Solving Large ODE Problems using the Iterative Matrix Solver WATSIT.

W. L. Seward, University of Waterloo, Department of Computer Science, Waterloo, Ontario N2L 3G1, Canada.

Finite Element Approximation of a Model Reaction-Diffusion Problem with a Non-Lipschitz Nonlinearity.

R. M. Shanahan and J. W. Barrett, Imperial College, Mathematics Department, Queens Gate, London SW7 2BX.

High Order Explicit Runge-Kutta pairs.

P. W. Sharp and E. Smart, University of Toronto, Department of Computer Science, Toronto, Canada M5S 1A4.

Implementation of a Multigrid Method on a Transputer Network.

G. J. Shaw Computing Laboratory, 8-11 Keeble Road, Oxford OX1 3QD and A. Stewart, University of Oxford and Queen's University, Belfast, Ireland.

The Global Error Estimate of Exponential Splitting.

Q. Sheng, University of Cambridge, DAMPT, Silver Street, Cambridge, CB3 9EW.

Some Observations on the Streamline Upwind Petrov Galerkin Method for Convection-Diffusion Problems.

S. Sigurdsson, University of Iceland, Science Institute, Dunhaga 3, IS-107 Reykjavik, Iceland.

Conjugate Gradient Algorithms for Stabilised Mixed Finite Element Methods.

D. J. Silvester and N. Kechkar, UMIST, Department of Mathematics, P.O. Box 88, Manchester M60 1QD.

Accurate Fourier Pseudospectral Ssolution of the RLW Equation.

D. M. Sloan, University of Strathclyde, Department of Mathematics, 26 Richmond Street, Glasgow, G1 1XH.

Block Runge-Kutta Methods on Parallel Computers.

B. P. Sommeijer, Centre for Mathematics & Computer Science, P.O. Box 4079, 1009 AB Amsterdam, The Netherlands.

Optimal Lattice Rules in Dimension 2 and 3.

T. Sorevik University of Bergen, Department of Informatics, N-5014 Bergen, Norway and J. N. Lyness, Argonne National Laboratory, MCSD, Argonne, IL 60439, USA.

Ill - posedness and Singular Integrals in the Boundary Integral Method.

Jacob Steinberg, Technion - Israel Institute of Technology, Department of Mathematics, Haifa 32000, Israel.

Spurious Dynamics in Numerical Simulations of Differential Equations.

A. Stuart, University of Bath, School of Mathematics, Claverton Down, Bath BA2 7AY.

Spectral Transport-diffusion Algorithm for Convection-diffusion Problems.
E. Süli and A. Ware, University of Oxford, Computing Laboratory, 8-11 Keeble Road, Oxford OX1 3QD.

Numerical Solution of Retarded ODE-systems.
P. G. Thomsen, Technical University of Denmark, Institute for Numerical Analysis, DK-2800 Lyngby, Denmark.

The Finite-element Method with Nodes Moving along the Characteristics for Convection-Diffusion Equations.
Y. Tourigny, and E. Süli, University of Oxford, Computing Laboratory, 8-11 Keble Road, Oxford OX1 3QD.

Polynomial Higher Order Predictors in Numerical Path Following Schemes.
Klaus Ulrich, University of Hannover, Institut Angewandte Math, Welfengarten 1, D-3000 Hannover 1, F.R. Germany.

A Shape Preserving Curve Interpolation Scheme using Parametric Rational Cubic Splines.
K. Unsworth, and T.N.T. Goodman, University of Dundee, Department of Mathematics and Computer Science, Dundee DD1 4HN, Scotland.

The Difference Interpolation Method for Splines.
W. Volk, Nymphenburger Str. 11, 1000 Berlin 62, F.R. Germany.

Finite Element Computation on a Multiprocessor Computer.
R. Wait and C. J. Willis, University of Liverpool, Department of Comp. and Stat. Science, P.O. Box 147, Liverpool L69 3BX.

The Use of the Basic Linear Algebra Subprograms (BLAS) in Improving Supercomputer Speeds.
Jerzy Wasniewski, Multiflow Computer, Inc., 31 Business Park Drive, Branford, CT 06405, USA.

Optimal Grids and Solutions for Compressible Flows in Nozzles.
J. Wixcey, University of Reading, Department of Mathematics, Whiteknights, P.O. Box 220, Reading, RG6 2AX.

Hopscotch for Parabolic PDEs.
A. S. Wood, University of Bradford, Maths Department, Bradford BD7 1DP.

A Parallel Algorithm for Banded Linear Systems.
S. J. Wright, Argonne National Laboratory, Mathematics & Computer Science Division, 9700 South Cass Ave., Argonne, IL 60439 USA,

A Moving-Grid Interface for Systems of One-Dimensional Time-Dependent Partial Differential Equations.
P. A. Zegeling, and J. G. Blom, Centre for Mathematics & Computer Science, Kruislaan 413, P.O. Box 4079, 1009 AB Amsterdam, The Netherlands.

Properties of the Approximations of a Matrix which Lower its Rank.
K. Zietak, University of Wroclaw, Institute of Computer Science, Poland.